Ergebnisse der Mathematik und ihrer Grenzgebiete

Band 78

Herausgegeben von P. R. Halmos · P. J. Hilton
R. Remmert · B. Szőkefalvi-Nagy

Unter Mitwirkung von L. V. Ahlfors · R. Baer
F. L. Bauer · A. Dold · J. L. Doob · S. Eilenberg
K. W. Gruenberg · M. Kneser · G. H. Müller
M. M. Postnikov · B. Segre · E. Sperner

Geschäftsführender Herausgeber: P. J. Hilton

János Bognár

Indefinite Inner Product Spaces

Springer-Verlag
Berlin Heidelberg New York 1974

János Bognár

Mathematical Institute of the Hungarian Academy of Sciences,
1053 Budapest, Hungary

AMS Subject Classification (1970)
Primary 46-02, 46 D 05, 47-02, 47 B 50
Secondary 46 C xx, 47 A xx

ISBN-13:978-3-642-65569-2 e-ISBN-13:978-3-642-65567-8
DOI: 10.1007/978-3-642-65567-8

Preface

By definition, an indefinite inner product space is a real or complex vector space together with a symmetric (in the complex case: hermitian) bilinear form prescribed on it so that the corresponding quadratic form assumes both positive and negative values. The most important special case arises when a Hilbert space is considered as an orthogonal direct sum of two subspaces, one equipped with the original inner product, and the other with -1 times the original inner product.

The subject first appeared thirty years ago in a paper of Dirac [1] on quantum field theory (cf. also Pauli [1]). Soon afterwards, Pontrjagin [1] gave the first mathematical treatment of an indefinite inner product space. Pontrjagin was unaware of the investigations of Dirac and Pauli; on the other hand, he was inspired by a work of Sobolev [1], unpublished up to 1960, concerning a problem of mechanics.

The attempts of Dirac and Pauli to apply the concept and elementary properties of indefinite inner product spaces to field theory have been renewed by several authors. At present it is not easy to judge which of their results will contribute to the final form of this part of physics. The following list of references should serve as a guide to the extensive literature: Bleuler [1], Gupta [1], Källén and Pauli [1], Heisenberg [1]—[4], Bogoljubov, Medvedev and Polivanov [1], K.L.Nagy [1]—[3], Berezin [1], Arons, Han and Sudarshan [1], Lee and Wick [1]. In particular, the book of K. L. Nagy [2] contains a critical survey of the literature up to 1966.

The work of Pontrjagin, on the geometry of certain indefinite inner product spaces and on the spectral theory of their linear operators, has also been carried on. This line of research led to a number of deep results connected with the names of M. G. Kreĭn, I. S. Iohvidov, Ju. P. Ginzburg, R. S. Phillips, H. Langer, M. A. Naĭmark, Ju. L. Šmul'jan, G. Wittstock, and others. The present book is an account of this second line of investigations.

The main body of the text offers an introduction, with full proofs, to the subject. Prerequisites do not extend much beyond standard facts of linear algebra, topology, Hilbert and Banach space theory.

The treatment is, whenever possible, invariant, i.e. independent of the choice of bases, decompositions and other additional structures. The presentation proceeds from the general to more and more special cases. Terminology and notation are partially new; in favour of the new terms and because of the large variety of the old ones no glossary is included.

The notes at the end of each chapter contain, besides historical and other comments relating to the main text, a survey of the literature on more advanced topics such as the spectral function, groups and algebras of operators, applications to differential equations and to quasi-definite functions, etc. For applications in stability theory, however, the reader is referred to the monograph of Daleckiĭ and Kreĭn [1]. Applications to characteristic functions of Hilbert space operators are only touched upon, since they do not seem to have reached a final form.

The author expresses his sincere gratitude to B. Sz.-Nagy for proposing to write this book, for constant encouragement and helpful suggestions.

Thanks are also due to the following persons: G. Borg (Royal Institute of Technology, Sweden) and I. S. Louhivaara (University of Jyväskylä, Finland) for invitations to lecture on the subject; H. Langer and G. Wittstock for certain counter-examples that have been included in the text; G. Adler, K. Mályusz and A. Szép for useful discussions.

The author is greatly indebted to his superiors at the Mathematical Institute of the Hungarian Academy of Sciences for affording time to accomplish this work.

Finally, the author is deeply obliged to all members of his family for their patience and understanding with which he could not dispense during the last several years.

Budapest, October 1973 János Bognár

Table of Contents

Chapter IV. Majorant Topologies on Inner Product Spaces

Chapter V. The Geometry of Krein Spaces

Chapter VI. Unitary and Selfadjoint Operators in Krein Spaces

Chapter VII. Positive Operators and Plus-operators in Krein Spaces

Chapter VIII. Invariant Semi-definite Subspaces of Linear Operators in Krein Spaces

Chapter IX. Pontrjagin Spaces and Their Linear Operators

Chapter I. Inner Product Spaces without Topology

The basic notions connected with general inner product spaces are introduced and analysed. In particular, non-degenerate spaces (Sections 4—5), maximal semi-definite subspaces (Sections 6—7), ortho-complemented subspaces (Section 9), and decomposable spaces (Section 11) are investigated. Because of their simplicity, the proofs can alternatively be worked out by the reader. Lemmas 4.4, 5.1, 6.3, 10.1, 10.7 and Corollaries 9.5, 11.9 deserve special attention.

1. Vector Spaces

In this section we fix some conventions (terms and symbols) concerning vector spaces with no additional structure to be used throughout the book.

By a *vector space* we shall always mean a vector space over the field $\boldsymbol{C}$ of complex numbers.

Two vector spaces $\mathfrak{E}$, $\mathfrak{E}'$ are said to be *isomorphic*, if there is a one-to-one mapping T of $\mathfrak{E}$ onto $\mathfrak{E}'$ such that $T(\alpha_1 x_1 + \alpha_2 x_2) = \alpha_1 T x_1 + \alpha_2 T x_2$ for every $x_1, x_2 \in \mathfrak{E}$ and $\alpha_1, \alpha_2 \in \boldsymbol{C}$. The mapping T is called an *isomorphism* between $\mathfrak{E}$ and $\mathfrak{E}'$.

A *subspace* of a vector space is a subset which is a vector space itself. In topological vector spaces eventual closedness requirements will be clearly pointed out.

The whole space is a subspace of itself. The set consisting of the zero vector alone is also a subspace, called the *zero subspace*.

The symbol 0 stands for any of the following objects: the number zero, the zero vector, and the zero subspace. The empty set is denoted by $\emptyset$.

If $\mathfrak{A}$ is a subset of a vector space, then its *span*, denoted by $\langle\mathfrak{A}\rangle$, is the least subspace containing $\mathfrak{A}$. The span $\langle\mathfrak{A}_1, \ldots, \mathfrak{A}_n\rangle$ of several subsets $\mathfrak{A}_j$ $(j = 1, \ldots, n)$ is defined by the equation

$$\langle\mathfrak{A}_1, \ldots, \mathfrak{A}_n\rangle = \langle\mathfrak{A}_1 \cup \cdots \cup \mathfrak{A}_n\rangle .$$

It will frequently occur that some of the sets $\mathfrak{A}_j$ consist of a single element x_j. In this case we denote the subset $\mathfrak{A}_j$ by x_j instead of $\{x_j\}$. We write, for example, $\langle x_1, \mathfrak{A}_2\rangle$ rather than $\langle\{x_1\}, \mathfrak{A}_2\rangle$.

If $\mathfrak{L}_j$ are subspaces $(j = 1, \ldots, n)$, then obviously

$$(1.1)\qquad \langle \mathfrak{L}_1, \ldots, \mathfrak{L}_n \rangle = \left\{ \sum_{j=1}^{n} x_j : \quad x_j \in \mathfrak{L}_j, \quad j = 1, \ldots, n \right\}.$$

Therefore $\langle \mathfrak{L}_1, \ldots, \mathfrak{L}_n \rangle$ can alternatively be denoted by $\mathfrak{L}_1 + \cdots + \mathfrak{L}_n$, and called the *vector sum* of the subspaces $\mathfrak{L}_j$. It is easy to see that for arbitrary sets $\mathfrak{A}_j$ $(j = 1, \ldots, n)$ we have

$$(1.2)\qquad \langle \mathfrak{A}_1, \ldots, \mathfrak{A}_n \rangle = \langle \mathfrak{A}_1 \rangle + \cdots + \langle \mathfrak{A}_n \rangle .$$

The subspaces $\mathfrak{L}_j$ $(j = 1, \ldots, n)$ are said to be *linearly independent*, if the representations $\sum x_j$ in (1.1) are unique or, what is the same, the relations

$$\sum_{j=1}^{n} x_j = 0 , \qquad x_j \in \mathfrak{L}_j \quad (j = 1, \ldots, n)$$

imply $x_1 = x_2 = \cdots = x_n = 0$. Another equivalent condition is the following:

$$\mathfrak{L}_j \cap (\mathfrak{L}_1 + \cdots + \mathfrak{L}_{j-1}) = 0 \qquad (j = 2, \ldots, n) .$$

The vectors v_j $(j = 1, \ldots, n)$ are said to be *linearly independent*, if none of them is zero and the subspaces $\langle v_j \rangle$ $(j = 1, \ldots, n)$ are linearly independent. Equivalently, the vectors $v_1, \ldots, v_n$ are linearly independent if the relation $\sum_{j=1}^{n} \alpha_j v_j = 0$ implies $\alpha_1 = \alpha_2 = \cdots = \alpha_n = 0$.

An infinite system of elements is said to be linearly independent if each of its finite subsystems is so.

The maximal number of linearly independent elements of a vector space $\mathfrak{E}$ is the *dimension* of $\mathfrak{E}$ (in symbols: dim $\mathfrak{E}$). Unless otherwise stated, we allow it to be either 0, or a positive integer, or ∞. Distinction between infinite dimensionalities will only be made in topological vector spaces.

In the case dim $\mathfrak{E} < \infty$ any maximal set of linearly independent elements is called a *basis* of $\mathfrak{E}$.

The vector sum (span) of linearly independent subspaces $\mathfrak{L}_1, \ldots, \mathfrak{L}_n$ is termed also a *direct sum* and denoted alternatively by $\mathfrak{L}_1 \dotplus \cdots \dotplus \mathfrak{L}_n$.

If $\mathfrak{L}$ is a subspace of the vector space $\mathfrak{E}$ then, by virtue of Zorn's lemma, there exists a subspace $\mathfrak{M} \subset \mathfrak{E}$ satisfying the relations

$$(1.3)\qquad \mathfrak{L} \cap \mathfrak{M} = 0 , \qquad \mathfrak{L} \dotplus \mathfrak{M} = \mathfrak{E} .$$

Moreover, given a subspace $\mathfrak{M}_1 \subset \mathfrak{E}$ such that $\mathfrak{L} \cap \mathfrak{M}_1 = 0$, a subspace $\mathfrak{M} \supset \mathfrak{M}_1$ with the properties (1.3) can be found. We say that $\mathfrak{M}$ is a *complementary subspace* for $\mathfrak{L}$ (with respect to $\mathfrak{E}$).

Consider, once again, a subspace $\mathfrak{L}$ of the vector space $\mathfrak{E}$. In $\mathfrak{E}$ an equivalence relation $\sim$ can be introduced by letting $x \sim y$ if

$x - y \in \mathfrak{L}$. The set of equivalence classes becomes a vector space $\mathfrak{E}/\mathfrak{L}$ (the *quotient space* of $\mathfrak{E}$ with respect to $\mathfrak{L}$) once addition and scalar multiplication of classes are defined by the corresponding operations on representatives. The zero vector of $\mathfrak{E}/\mathfrak{L}$ is evidently the class consisting of all elements of $\mathfrak{L}$.

The dimension of the quotient space $\mathfrak{E}/\mathfrak{L}$ will be called the *codimension* of $\mathfrak{L}$ with respect to $\mathfrak{E}$, and denoted by $\operatorname{codim}_{\mathfrak{E}} \mathfrak{L}$ or, when $\mathfrak{E}$ is the whole space under consideration, simply by $\operatorname{codim} \mathfrak{L}$.

The following well-known statement, cited here with a view to later application, shows the close connection between complementary subspaces and quotient spaces.

Lemma 1.1. *If $\mathfrak{L}$ is a subspace of the vector space $\mathfrak{E}$, then every complementary subspace $\mathfrak{M}$ for $\mathfrak{L}$ with respect to $\mathfrak{E}$ is isomorphic to $\mathfrak{E}/\mathfrak{L}$. A suitable isomorphism is given by $Tx = \hat{x}$ $(x \in \mathfrak{M})$, where $\hat{x} \in \mathfrak{E}/\mathfrak{L}$ is the equivalence class containing x.* □

Corollary 1.2. *The dimension of any complementary subspace for a subspace $\mathfrak{L}$ is equal to* $\operatorname{codim} \mathfrak{L}$. □

It is also possible to define the quotient space $\mathfrak{M}/\mathfrak{L}$ of an arbitrary subspace $\mathfrak{M} \subset \mathfrak{E}$ with respect to the subspace $\mathfrak{L} \subset \mathfrak{E}$. In this case the equivalence classes are formed as follows: the elements $x, y \in \mathfrak{M}$ belong to the same class if $x - y \in \mathfrak{L}$. The zero vector of $\mathfrak{M}/\mathfrak{L}$ turns out to be the class consisting of all elements of $\mathfrak{M} \cap \mathfrak{L}$. It is clear that $\mathfrak{M}/\mathfrak{L}$ is isomorphic to a subspace of $\mathfrak{E}/\mathfrak{L}$.

2. Inner Products

In our terminology, an *inner product* on a vector space $\mathfrak{E}$ is a complex-valued function $(\cdot\,,\cdot)$ defined for all pairs $x, y \in \mathfrak{E}$ so that the conditions

$$(\alpha_1 x_1 + \alpha_2 x_2\,,\ y) = \alpha_1(x_1, y) + \alpha_2(x_2, y)\,, \tag{2.1}$$

$$(y, x) = \overline{(x, y)} \tag{2.2}$$

are fulfilled for every $\alpha_1, \alpha_2 \in \boldsymbol{C}$ and $x_1, x_2, x, y \in \mathfrak{E}$. Relations (2.1) and (2.2) obviously imply

$$(y, \alpha_1 x_1 + \alpha_2 x_2) = \bar{\alpha}_1(y, x_1) + \bar{\alpha}_2(y, x_2)\,.$$

The number (x, y) is termed the *inner product* of x and y.

A vector space equipped with an inner product will be called an *inner product space.*

Two inner product spaces $\mathfrak{E}$ and $\mathfrak{E}'$ with inner products $(.\,,.)$ and $(.\,,.)'$, respectively, are said to be *isometrically isomorphic*, if a one-to-one mapping T of $\mathfrak{E}$ onto $\mathfrak{E}'$ can be found so that

$$T(\alpha_1 x_1 + \alpha_2 x_2) = \alpha_1 T x_1 + \alpha_2 T x_2 ,$$
$$(Tx_1, Tx_2)' = (x_1, x_2)$$

for every $x_1, x_2 \in \mathfrak{E}$ and $\alpha_1, \alpha_2 \in \boldsymbol{C}$. The mapping T is called an *isometrical isomorphism* between $\mathfrak{E}$ and $\mathfrak{E}'$.

In an inner product space $\mathfrak{E}$ the number (x, x) may be termed the *inner square* of the element $x \in \mathfrak{E}$. One verifies easily that

$$(2.3)\qquad (x, y) = \frac{1}{4}(x + y, x + y) - \frac{1}{4}(x - y, x - y) + \\ + \frac{i}{4}(x + iy, x + iy) - \frac{i}{4}(x - iy, x - iy)$$

for every $x, y \in \mathfrak{E}$. Thus the inner product is uniquely determined by the values of the inner square. Relation (2.3) is called the *polarization formula*.

Consider an element x of the inner product space $\mathfrak{E}$. As $\overline{(x, x)} = (x, x)$, there are three possibilities: either $(x, x) > 0$, or $(x, x) < 0$, or $(x, x) = 0$. Correspondingly x is said to be *positive, negative,* or *neutral*. It is clear that the zero vector is neutral.

The set of all neutral vectors in $\mathfrak{E}$ is called the *neutral part* of $\mathfrak{E}$, or the *neutral set* of the inner product $(.\,,.)$, and will be denoted by $\mathfrak{P}^0$. The symbol $\mathfrak{P}^{00}$ will stand for the set obtained from $\mathfrak{P}^0$ by omitting the element 0:

$$\mathfrak{P}^{00} = \{x \in \mathfrak{E} : (x, x) = 0,\ x \neq 0\}.$$

We denote by $\mathfrak{P}^{++}$ (resp. $\mathfrak{P}^{--}$) the set consisting of the zero element together with all positive (negative) elements of $\mathfrak{E}$, and by $\mathfrak{P}^{+}$ (resp. $\mathfrak{P}^{-}$) the set of all non-negative (non-positive) elements of $\mathfrak{E}$. Thus e.g.

$$\mathfrak{P}^{++} = \{x \in \mathfrak{E}:\ (x, x) > 0 \quad \text{or} \quad x = 0\},$$
$$\mathfrak{P}^{+} = \{x \in \mathfrak{E}:\ (x, x) \geqq 0\}.$$

If $\mathfrak{E}$ contains positive as well as negative elements, we say that the inner product is *indefinite* on $\mathfrak{E}$, or that $\mathfrak{E}$ is an *indefinite inner product space.*

Lemma 2.1. *Every indefinite inner product space contains non-zero neutral vectors.*

Proof. Let $(x, x) > 0$, $(y, y) < 0$. We set $z = x + \lambda y$, where λ is a (real) solution of the equation

$$(x, x) + 2\lambda\, \mathrm{Re}(x, y) + \lambda^2 (y, y) = 0.$$

Then $(z, z) = 0$. On the other hand, $z = 0$ would imply $(x, x) = |\lambda|^2 (y, y)$, which contradicts our starting point. □

The inner product is said to be *semi-definite* on $\mathfrak{E}$, if it is not indefinite. In this case we speak of a *semi-definite inner product space.*

Lemma 2.2. *In a semi-definite inner product space* $\mathfrak{E}$ *the Schwarz inequality*

$$|(x, y)|^2 \leqq (x, x)\,(y, y) \qquad (x, y \in \mathfrak{E})$$

holds.

The *proof* is the same as in Hilbert space. □

A semi-definite inner product may either be a *positive inner product* ($(x, x) \geqq 0$ for every $x \in \mathfrak{E}$), or a *negative inner product* ($(x, x) \leqq 0$ for $x \in \mathfrak{E}$). A *neutral inner product* ($(x, x) = 0$ for every $x \in \mathfrak{E}$) is positive and negative at the same time.

An inner product on $\mathfrak{E}$ is said to be *definite,* if $(x, x) = 0$ implies $x = 0$. In view of Lemma 2.1 a definite inner product is semi-definite. Hence we have either $(x, x) > 0$ for $x \neq 0$ (*positive definite inner product*), or $(x, x) < 0$ for $x \neq 0$ (*negative definite inner product*).

A *positive* (*negative, neutral, definite, positive definite, negative definite*) *inner product space* is a vector space with an inner product of the respective kind.

Example 2.3. Let $\varepsilon_1\ \varepsilon_2, \ldots$ be real numbers. Denote by $\mathfrak{E}$ the vector space of sequences $\{\xi_1, \xi_2, \ldots\}$ of complex numbers satisfying $\sum_{j=1}^{\infty}{}' |\varepsilon_j|\, |\xi_j|^2 < \infty$. The formula

$$(x, y) = \sum_{j=1}^{\infty}{}' \varepsilon_j \xi_j \overline{\eta_j} \qquad (x = \{\xi_1, \xi_2, \ldots\} \in \mathfrak{E}, \qquad y = \{\eta_1, \eta_2, \ldots\} \in \mathfrak{E})$$

defines an inner product on $\mathfrak{E}$, which may be indefinite, semi-definite etc. depending on the numbers ε_j.

Example 2.4. We consider a real-valued function σ of bounded variation on the real line $\boldsymbol{R}$, and the vector space $\mathfrak{E}$ of all complex-valued functions on $\boldsymbol{R}$ that are measurable and square-summable with respect to the total variation of σ. The relation

$$(x, y) = \int_{\boldsymbol{R}} x(\xi)\, \overline{y(\xi)}\, d\sigma(\xi) \qquad (x, y \in \mathfrak{E})$$

defines an inner product on $\mathfrak{E}$, which may be indefinite, semi-definite etc. depending on the function σ.

A subspace $\mathfrak{L}$ of an inner product space $\mathfrak{E}$ is an inner product space with respect to the restriction of $(.,.)$ to $\mathfrak{L}$. If the inner product is indefinite on $\mathfrak{L}$, we say that $\mathfrak{L}$ is an *indefinite subspace* of $\mathfrak{E}$. Similarly we can speak of *semi-definite, positive, negative, neutral, definite, positive definite*, and *negative definite subspaces*. In particular, the zero subspace is positive definite as well as negative definite.

Positive definite inner product spaces are well-known objects ("prehilbert spaces", "inner product spaces" in the classical sense). Negative definite inner product spaces do not possess any new properties, and semi-definite inner product spaces can be reduced to definite ones (see Lemma 4.4 and Corollary 5.3 below). For this reason indefinite inner product spaces will be our main concern. Nevertheless, in order that the results would be applicable to any subspace, semi-definite as well as indefinite, we shall formulate them for arbitrary inner product spaces whenever possible. In such cases it will be understood that the space under consideration may be a subspace of another inner product space.

Throughout this chapter, when not specified, $\mathfrak{E}$ denotes an arbitrary inner product space.

Lemma 2.5. *The span of a positive (negative, neutral) vector is a positive definite (negative definite, neutral) subspace.*

This statement is trivial. □

Most of the results concerning inner product spaces have two variants obtained from each other by interchanging the words "positive" and "negative". Making use of the next definition it is always sufficient to treat only one variant.

Let $\mathfrak{E}$ be an inner product space. Replacing each value (x, y) of the inner product by $(x, y)' = -(x, y)$ we obtain an inner product space $\mathfrak{E}'$. We shall say that $\mathfrak{E}'$ is the *anti-space* of $\mathfrak{E}$.

Lemma 2.6. *If the inner product space $\mathfrak{E}$ contains at least one positive (negative) vector, then every element of $\mathfrak{E}$ is the sum of two positive (negative) vectors.*

Proof. If $x, x_0 \in \mathfrak{E}$, $(x_0, x_0) > 0$, then for sufficiently large positive numbers α the quantity

$$(x + \alpha x_0,\ x + \alpha x_0) = (x, x) + 2\alpha\ \mathrm{Re}(x, x_0) + \alpha^2 (x_0, x_0)$$

will be positive. Hence the vectors $x_1 = x + \alpha x_0$, $x_2 = -\alpha x_0$ are positive and we have $x = x_1 + x_2$. The other half of the lemma follows by taking the anti-space of $\mathfrak{E}$. □

Corollary 2.7. *If the inner product is indefinite on $\mathfrak{E}$, then none of the sets $\mathfrak{P}^+$, $\mathfrak{P}^-$, $\mathfrak{P}^{++}$, $\mathfrak{P}^{--}$ is a subspace.* □

3. Orthogonality

If $(x,y) = 0$ or, equivalently, $(y, x) = 0$ for a pair of vectors x,y in the inner product space $\mathfrak{E}$, we say that x and y are *orthogonal* to each other, and we write $x \perp y$.

Two sets $\mathfrak{A}, \mathfrak{B} \subset \mathfrak{E}$ are said to be *orthogonal* (in symbols: $\mathfrak{A} \perp \mathfrak{B}$), if $x \perp y$ for every $x \in \mathfrak{A}$, $y \in \mathfrak{B}$.

The relation $x \perp y$ implies $(x + y, x + y) = (x, x) + (y, y)$. Consequently, we have the following result.

Lemma 3.1. *If each of the pairwise orthogonal subspaces $\mathfrak{L}_1, \ldots, \mathfrak{L}_n$ is positive (negative, neutral, positive definite, negative definite), then so is the span $\langle \mathfrak{L}_1, \ldots, \mathfrak{L}_n \rangle$.* □

Remark 3.2. Making use of Lemma 2.5 and equation (1.2) Lemma 3.1 can be extended to orthogonal families of sets $\mathfrak{A}_1, \ldots, \mathfrak{A}_n$, where each $\mathfrak{A}_j$ is either a subspace or a single vector.

In connection with Lemma 3.1 we introduce the following definitions and notations.

If $\mathfrak{L}$ is the vector sum (resp. direct sum) of pairwise orthogonal subspaces $\mathfrak{L}_j$ $(j = 1, \ldots, n)$, we say that $\mathfrak{L}$ is the *orthogonal sum* (resp. *orthogonal direct sum*) of the subspaces $\mathfrak{L}_j$ and we write $\mathfrak{L} = \mathfrak{L}_1 (+) \cdots (+) \mathfrak{L}_n$ (resp. $\mathfrak{L} = \mathfrak{L}_1 (\dot{+}) \cdots (\dot{+}) \mathfrak{L}_n$).

We now turn our attention to a concept which is the natural generalization of orthogonal complement in Hilbert space, but has, as a rule, less favourable properties.

For any set $\mathfrak{A} \subset \mathfrak{E}$ we put

$$\mathfrak{A}^\perp = \{x \in \mathfrak{E} \colon x \perp \mathfrak{A}\},$$

and say that $\mathfrak{A}^\perp$ is the *orthogonal companion* of $\mathfrak{A}$.

It is clear that a) the orthogonal companion of any set is a subspace; b) $\mathfrak{A} \subset \mathfrak{B}$ implies $\mathfrak{B}^\perp \subset \mathfrak{A}^\perp$; c) if $\mathfrak{L}_1$, $\mathfrak{L}_2$ are subspaces of $\mathfrak{E}$, then $(\mathfrak{L}_1 + \mathfrak{L}_2)^\perp = \mathfrak{L}_1^\perp \cap \mathfrak{L}_2^\perp$.

Denoting the *second orthogonal companion* $(\mathfrak{A}^\perp)^\perp$ of the set $\mathfrak{A} \subset \mathfrak{E}$ by $\mathfrak{A}^{\perp\perp}$ we have

$$\mathfrak{A} \subset \mathfrak{A}^{\perp\perp}. \tag{3.1}$$

The case $\mathfrak{A} = \mathfrak{A}^{\perp\perp}$ will be characterized in Chapter III.

The third orthogonal companion $\mathfrak{A}^{\perp\perp\perp}$ of $\mathfrak{A}$ always coincides with $\mathfrak{A}^\perp$:

$$\mathfrak{A}^{\perp\perp\perp} = \mathfrak{A}^\perp. \tag{3.2}$$

Really, taking the orthogonal companions of both sides of (3.1) we obtain $\mathfrak{A}^\perp \supset \mathfrak{A}^{\perp\perp\perp}$. On the other hand, replacing of $\mathfrak{A}$ in (3.1) by $\mathfrak{A}^\perp$ yields $\mathfrak{A}^\perp \subset \mathfrak{A}^{\perp\perp\perp}$.

Lemma 3.3. *If* $\mathfrak{L}$, $\mathfrak{M}$ *are subspaces of* $\mathfrak{E}$ *and* $\dim \mathfrak{M} = m < \infty$, *then*

$$\dim (\mathfrak{L} \cap \mathfrak{M}^\perp) \geqq \dim \mathfrak{L} - \dim \mathfrak{M} .$$

Proof. We may assume that $\dim \mathfrak{L} = l < \infty$, since in the opposite case the result follows by considering subspaces of $\mathfrak{L}$ with increasing finite dimensions. We may also assume that $l > m \geqq 1$.

Let $e_1, \ldots, e_l$ and $f_1, \ldots, f_m$ be bases of $\mathfrak{L}$ and $\mathfrak{M}$, respectively. The elements of $\mathfrak{L} \cap \mathfrak{M}^\perp$ are of the form $\alpha_1 e_1 + \cdots + \alpha_l e_l$, where the coefficients α_j satisfy the equations

$$\sum_{j=1}^{l} (e_j, f_k)\alpha_j = 0 \qquad (k = 1, \ldots, m) .$$

These are m homogeneous linear equations for l unknowns. Consequently, the number of linearly independent solutions is at least $l - m$. □

Corollary 3.4. *If* $\mathfrak{L}$, $\mathfrak{M}$ *are subspaces of* $\mathfrak{E}$ *satisfying* $\mathfrak{L} \cap \mathfrak{M}^\perp = 0$, *then* $\dim \mathfrak{L} \leqq \dim \mathfrak{M}$. □

Lemma 3.5. *For every subspace* $\mathfrak{M} \subset \mathfrak{E}$ *we have* $\operatorname{codim} \mathfrak{M}^\perp \leqq \dim \mathfrak{M}$.

Proof. If $\operatorname{codim} \mathfrak{M}^\perp \geqq n$ for some non-negative integer n, then by Corollary 1.2 a subspace $\mathfrak{L} \subset \mathfrak{E}$ with $\dim \mathfrak{L} = n$, $\mathfrak{L} \cap \mathfrak{M}^\perp = 0$ can be found. Applying Corollary 3.4 we obtain the relation $\dim \mathfrak{M} \geqq n$. □

We end this section by introducing one more notion related to orthogonality.

Let $\mathfrak{E}_1, \ldots, \mathfrak{E}_n$ be inner product spaces. Consider the set $\mathfrak{E}_1 \times \cdots \times \mathfrak{E}_n$ of all n-tuples $\{x_1, \ldots, x_n\}$, where $x_j \in \mathfrak{E}_j$ $(j = 1, \ldots, n)$. The conventions

$$\{x_j\}_1^n + \{y_j\}_1^n = \{x_j + y_j\}_1^n \qquad (x_j, y_j \in \mathfrak{E}_j) ,$$

$$\alpha \{x_j\}_1^n = \{\alpha x_j\}_1^n \qquad (\alpha \in \boldsymbol{C},\ x_j \in \mathfrak{E}_j) ,$$

$$(\{x_j\}_1^n, \{y_j\}_1^n) = \sum_{j=1}^{n} (x_j, y_j) \qquad (x_j, y_j \in \mathfrak{E}_j)$$

turn $\mathfrak{E}_1 \times \cdots \times \mathfrak{E}_n$ into an inner product space, called the *cartesian product* of the inner product spaces $\mathfrak{E}_1, \ldots, \mathfrak{E}_n$.

It is easy to see that the elements $\{x_j\}_1^n \in \mathfrak{E}_1 \times \cdots \times \mathfrak{E}_n$ with $x_j = 0$ for $j \neq k$ form a subspace $\mathfrak{E}_{(k)}$, which is isometrically isomorphic to $\mathfrak{E}_k$ $(k = 1, \ldots, n)$, and $\mathfrak{E}_1 \times \cdots \times \mathfrak{E}_n$ is the orthogonal direct sum of these subspaces:

$$\mathfrak{E}_1 \times \cdots \times \mathfrak{E}_n = \mathfrak{E}_{(1)} \dotplus \cdots \dotplus \mathfrak{E}_{(n)} .$$

4. Isotropic Vectors

Example 4.1. Let $\mathfrak{E}$ be a two-dimensional vector space with basis e_1, e_2. Define an inner product on $\mathfrak{E}$ by the relations $(e_1, e_2) = 0$, $(e_1, e_1) = 1$, $(e_2, e_2) = -1$. It is easy to see that for the one-dimensional subspace $\mathfrak{L} = \langle e_1 + e_2 \rangle$ we have $\mathfrak{L}^\perp = \mathfrak{L}$. This shows that a) a subspace may have a non-zero intersection with its orthogonal companion; b) even for a subspace of finite dimension the orthogonal companion need not contain a complementary subspace.

Example 4.1 motivates the following definitions.

Let $\mathfrak{L}$ be a subspace of an inner product space $\mathfrak{E}$. The subspace $\mathfrak{L} \cap \mathfrak{L}^\perp$, to be denoted by $\mathfrak{L}^0$, is called the *isotropic part*, and its elements the *isotropic vectors* of $\mathfrak{L}$. If $\mathfrak{L}^0 \neq 0$, we say that $\mathfrak{L}$ is *degenerate*, or that the inner product is degenerate on $\mathfrak{L}$.

In particular, the whole space $\mathfrak{E}$ is degenerate if the isotropic part $\mathfrak{E}^0 = \mathfrak{E}^\perp$ is not equal to 0.

The subspace $\mathfrak{L}$ of Example 4.1 is a degenerate subspace in a non-degenerate space. The space $\mathfrak{E}$ of Example 2.3 is degenerate if and only if $\varepsilon_j = 0$ for at least one j.

The next lemma is a simple consequence of the definition of isotropic part.

Lemma 4.2. *If* $\mathfrak{L} = \mathfrak{L}_1 (\dot{+}) \cdots (\dot{+}) \mathfrak{L}_n$, *then for the isotropic parts we have* $\mathfrak{L}^0 = \mathfrak{L}_1^0 (\dot{+}) \cdots (\dot{+}) \mathfrak{L}_n^0$. □

Corollary 4.3. *The orthogonal direct sum of non-degenerate subspaces is non-degenerate.* □

As isotropic vectors are orthogonal to themselves, the isotropic part of any subspace is neutral. For this reason every definite subspace is non-degenerate.

In Example 4.1 the vector $e_1 + e_2$ is neutral, but it is not an isotropic vector of $\mathfrak{E}$ (however, it is isotropic for $\mathfrak{L}$). The following lemma asserts that vectors of this kind may only occur in indefinite spaces.

Lemma 4.4. *The isotropic part* $\mathfrak{E}^0$ *of a semi-definite inner product space* $\mathfrak{E}$ *consists of all neutral elements of* $\mathfrak{E}$.

Proof. We have already mentioned that an isotropic vector is always neutral. On the other hand, according to Lemma 2.2 the relations $x \in \mathfrak{E}$, $(x, x) = 0$ imply $(x, y) = 0$ for every $y \in \mathfrak{E}$. □

As a consequence, a neutral inner product on a vector space is necessarily the trivial one:

Corollary 4.5. *If* $\mathfrak{E}$ *is a neutral inner product space, then* $(x, y) = 0$ *for every pair* $x, y \in \mathfrak{E}$. □

Corollary 4.6. *If two vectors x, y of an inner product space satisfy the relations $(x, x) = 0$, $(x, y) \neq 0$, then the subspace $\langle x, y \rangle$ is indefinite.* $\square$

Lemma 4.7. *If $\mathfrak{L}$ is a neutral subspace, so is $\mathfrak{L}^{\perp\perp}$.*

Proof. On account of Corollary 4.5 our statement is equivalent to the following: $\mathfrak{L} \subset \mathfrak{L}^{\perp}$ implies $\mathfrak{L}^{\perp\perp} \subset (\mathfrak{L}^{\perp\perp})^{\perp}$. But this implication follows by a repeated application of property b) of the orthogonal companion. $\square$

The next example shows that if $\mathfrak{L}$ is positive, $\mathfrak{L}^{\perp\perp}$ need not be positive.

Example 4.8. Consider the vector space $\mathfrak{E}$ of complex sequences $x = \{\xi_0, \xi_1, \ldots\}$ with $\sum_{j=0}^{\infty} |\xi_j|^2 < \infty$. For $y = \{\eta_0, \eta_1, \ldots\} \in \mathfrak{E}$ put

$$(x, y) = - \xi_0 \overline{\eta_0} + \sum_{j=1}^{\infty} \frac{1}{2^j} \xi_j \overline{\eta_j}.$$

One verifies easily that this inner product is non-degenerate and indefinite on $\mathfrak{E}$. The subspace $\mathfrak{L} \subset \mathfrak{E}$ characterized by the relation $\xi_0 = \sum_{j=1}^{\infty} \frac{1}{2^j} \xi_j$ is positive definite. Let $y = \{\eta_j\}_0^{\infty} \in \mathfrak{L}^{\perp}$. Since the element $x^{(n)}$ with 0-th coordinate equal to 1, n-th coordinate equal to 2^n, and all other coordinates equal to 0 belongs to $\mathfrak{L}$ $(n = 1, 2, \ldots)$, we have

$$(y, x^{(n)}) = - \eta_0 + \eta_n = 0 \qquad (n = 1, 2, \ldots).$$

In view of the relation $\sum_{j=0}^{\infty} |\eta_j|^2 < \infty$ we obtain $y = 0$. Thus $\mathfrak{L}^{\perp} = 0$, $\mathfrak{L}^{\perp\perp} = \mathfrak{E}$.

It is also untrue that $\mathfrak{L}^{\perp\perp}$ would be non-degenerate whenever $\mathfrak{L}$ is so.

Example 4.9. Let, again, $\mathfrak{E}$ be the vector space of square-summable numerical sequences $x = \{\xi_j\}_0^{\infty}$. For $y = \{\eta_j\}_0^{\infty} \in \mathfrak{E}$ put

$$(x, y) = - \xi_0 \overline{\eta_0} + \sum_{j=1}^{\infty} \xi_j \overline{\eta_j}.$$

Denote by $\mathfrak{L}$ the subspace of $\mathfrak{E}$ consisting of all finite sequences $\{\xi_j\}_0^{\infty}$ with $\xi_0 = \xi_1 = \sum_{j=2}^{\infty} \xi_j$. Then $\mathfrak{L}^{\perp}$ is the 1-dimensional subspace spanned by the neutral vector $y_0 = \{1, 1, 0, 0, 0, \ldots\}$, and

$$\mathfrak{L}^{\perp\perp} = \{\{\xi_j\}_0^{\infty} \in \mathfrak{E} : \ \xi_0 = \xi_1\}.$$

It is easy to see that y_0 is an isotropic vector of $\mathfrak{L}^{\perp\perp}$ though $\mathfrak{L}$ is positive definite, hence non-degenerate.

A possible extension of Lemma 4.7 reads as follows.

Lemma 4.10. *If $\mathfrak{L}$ is a degenerate subspace, so is $\mathfrak{L}^{\perp\perp}$.*

Proof. Owing to the relations (3.1), (3.2) we have:

$$\mathfrak{L} \cap \mathfrak{L}^{\perp} \subset \mathfrak{L}^{\perp\perp} \cap \mathfrak{L}^{\perp} = \mathfrak{L}^{\perp\perp} \cap \mathfrak{L}^{\perp\perp\perp}. \quad \square$$

5. Maximal Non-degenerate Subspaces

Non-degenerate spaces (subspaces) have several convenient properties not shared by degenerate ones. With the aid of the following lemmas many problems can be reduced to problems concerning non-degenerate spaces (or subspaces).

Lemma 5.1. *Let $\mathfrak{E}^0$ be the isotropic part of the inner product space $\mathfrak{E}$, and let $\mathfrak{E}^1$ be a complementary subspace for $\mathfrak{E}^0$. Then $\mathfrak{E}^1$ is non-degenerate, and we have $\mathfrak{E} = \mathfrak{E}^0 \dotplus \mathfrak{E}^1$.*

Proof. The relation $\mathfrak{E}^0 \perp \mathfrak{E}^1$ follows from $\mathfrak{E}^0 \perp \mathfrak{E}$. Since $\mathfrak{E}^0$ is the isotropic part of itself, $\mathfrak{E}^1$ is non-degenerate by Lemma 4.2. $\square$

The contents of Lemma 5.1 can be expressed in another way if we consider the *quotient space* $\mathfrak{E}/\mathfrak{E}^0$. The latter has up to now been merely a vector space, but it can be equipped with an inner product by setting $(\hat{x}, \hat{y})^\wedge = (x, y)$, where $\hat{x}, \hat{y} \in \mathfrak{E}/\mathfrak{E}^0$, and x, y are representatives of $\hat{x}$ and $\hat{y}$, respectively: $x, y \in \mathfrak{E}$, $x \in \hat{x}$, $y \in \hat{y}$.

Let $x_1, x_2 \in \hat{x}$; $y_1, y_2 \in \hat{y}$. Then $x_1 - x_2, y_1 - y_2 \in \mathfrak{E}^0$. It follows that $(x_1, y_1) - (x_2, y_2) = (x_1 - x_2, y_1) + (x_2, y_1 - y_2) = 0$. Thus the definition of the function $(.,.)^\wedge$ is correct. It is easy to see that $(.,.)^\wedge$ is really an inner product on the vector space $\mathfrak{E}/\mathfrak{E}^0$. For an inner product space $\mathfrak{E}$ the quotient space $\mathfrak{E}/\mathfrak{E}^0$ will always be understood to carry this inner product.

The inner product space $\mathfrak{E}/\mathfrak{L}$ can also be defined in the more general case $\mathfrak{L} \subset \mathfrak{E}^0$, but not for an arbitrary subspace $\mathfrak{L} \subset \mathfrak{E}$.

Lemma 5.2. *If $\mathfrak{E}^1$ is a complementary subspace for the isotropic part $\mathfrak{E}^0$ of $\mathfrak{E}$, then $\mathfrak{E}^1$ is isometrically isomorphic to the quotient space $\mathfrak{E}/\mathfrak{E}^0$. The isometrical isomorphism is given by the formula $Tx = \hat{x}$ $(x \in \mathfrak{E}^1)$, where $\hat{x} \in \mathfrak{E}/\mathfrak{E}^0$ is the equivalence class containing x.*

Proof. According to Lemma 1.1 the mapping T serves as an isomorphism between $\mathfrak{E}^1$ and $\mathfrak{E}/\mathfrak{E}^0$. It remains to note that, by definition, $(Tx, Ty)^\wedge = (\hat{x}, \hat{y})^\wedge = (x, y)$. $\square$

From Lemmas 5.1—5.2 we obtain:

Corollary 5.3. *$\mathfrak{E}/\mathfrak{E}^0$ is non-degenerate.* □

The complementary subspaces to $\mathfrak{E}^0$ can also be characterized as *maximal non-degenerate subspaces* of $\mathfrak{E}$, i.e. non-degenerate subspaces which are not contained in any other non-degenerate subspace of $\mathfrak{E}$.

Theorem 5.4. *The set $\mathfrak{E}^1 \subset \mathfrak{E}$ is a complementary subspace for $\mathfrak{E}^0$ if and only if $\mathfrak{E}^1$ is a maximal non-degenerate subspace of $\mathfrak{E}$.*

Proof. Let $\mathfrak{E} = \mathfrak{E}^0 \dotplus \mathfrak{E}^1$. By Lemma 5.1 $\mathfrak{E}^1$ is non-degenerate. If $\mathfrak{L} \supset \mathfrak{E}^1$, $\mathfrak{L} \neq \mathfrak{E}^1$, then the subspace $\mathfrak{L} \cap \mathfrak{E}^0$ is non-zero, and its elements being orthogonal to the whole space $\mathfrak{E}$, they are orthogonal to $\mathfrak{L}$; hence $\mathfrak{L}$ is degenerate.

Conversely, assume that $\mathfrak{E}^1$ is a maximal non-degenerate subspace of $\mathfrak{E}$. Then $\mathfrak{E}^1 \cap \mathfrak{E}^0 = 0$. Let $\mathfrak{M}$ be a complementary subspace for $\mathfrak{E}^0 \dotplus \mathfrak{E}^1$, i.e. $\mathfrak{E} = \mathfrak{E}^0 \dotplus \mathfrak{E}^1 \dotplus \mathfrak{M}$. On account of Lemma 5.1 the subspace $\mathfrak{E}^1 \dotplus \mathfrak{M}$ is non-degenerate. The maximality of $\mathfrak{E}^1$ implies $\mathfrak{M} = 0$. As a result, $\mathfrak{E} = \mathfrak{E}^0 \dotplus \mathfrak{E}^1$. □

Remark 5.5. From Theorem 5.4 and the existence of complementary subspaces, or by a direct application of Zorn's lemma, we obtain that every inner product space contains maximal non-degenerate subspaces. Moreover, every non-degenerate subspace admits a maximal non-degenerate extension.

6. Maximal Semi-definite Subspaces

A subspace $\mathfrak{L}$ of the inner product space $\mathfrak{E}$ is said to be *maximal semi-definite,* if $\mathfrak{L}$ is semi-definite and $\mathfrak{L}$ is not contained in any other semi-definite subspace of $\mathfrak{E}$. The definitions of *maximal positive, maximal negative, maximal neutral, maximal definite, maximal positive definite,* and *maximal negative definite* subspaces are similar.

By Zorn's lemma every semi-definite (resp. definite etc.) subspace can be extended to a maximal one. Applying this statement to the zero subspace we obtain that $\mathfrak{E}$ always contains maximal subspaces of each kind introduced above.

Most of the following lemmas will be used later.

Lemma 6.1. *If each of the subspaces $\mathfrak{L}_1, \mathfrak{L}_2 \subset \mathfrak{E}$ is either maximal positive or maximal negative, then $\langle \mathfrak{L}_1, \mathfrak{L}_2 \rangle^0$ consists of the neutral elements of $\mathfrak{L}_1 \cap \mathfrak{L}_2$.*

Proof. Let $x \in \mathfrak{L}_1 \cap \mathfrak{L}_2$, $(x, x) = 0$. The vector x is a neutral element of the semi-definite subspace $\mathfrak{L}_1$. Hence, by virtue of Lemma 4.4,

$x \in \mathfrak{L}_1 \cap \mathfrak{L}_1^\perp$. In the same way we obtain the relation $x \in \mathfrak{L}_2 \cap \mathfrak{L}_2^\perp$. Thus $x \in (\mathfrak{L}_1 + \mathfrak{L}_2) \cap (\mathfrak{L}_1 + \mathfrak{L}_2)^\perp$.

Let, conversely, $x \in \langle \mathfrak{L}_1, \mathfrak{L}_2 \rangle^0$. It is evident that x is a neutral vector orthogonal to $\mathfrak{L}_1$ and $\mathfrak{L}_2$. Let $\mathfrak{L}_1$ be maximal positive (maximal negative). According to Remark 3.2 the span $\langle x, \mathfrak{L}_1 \rangle$ is a positive (resp. negative) extension of $\mathfrak{L}_1$. Therefore $x \in \mathfrak{L}_1$. The relation $x \in \mathfrak{L}_2$ follows in the same manner. □

Lemma 6.2. *If* $\mathfrak{E} = \mathfrak{L}_1 + \mathfrak{L}_2$ *with a positive subspace* $\mathfrak{L}_1$ *and a negative definite subspace* $\mathfrak{L}_2$, *then* $\mathfrak{L}_1$ *is maximal positive and* $\mathfrak{L}_2$ *is maximal negative definite. For negative* $\mathfrak{L}_1$ *and positive definite* $\mathfrak{L}_2$ *a similar statement holds.*

Proof. For any subspace $\mathfrak{L}_1' \supset \mathfrak{L}_1$ $(\mathfrak{L}_1' \neq \mathfrak{L}_1)$ we have $\mathfrak{L}_1' \cap \mathfrak{L}_2 \neq 0$. Thus $\mathfrak{L}_1'$ cannot be positive. Analogously, if $\mathfrak{L}_2'$ is a subspace which is a proper extension of $\mathfrak{L}_2$, then $\mathfrak{L}_2'$ has a non-zero intersection with the positive subspace $\mathfrak{L}_1$, so that it cannot be negative definite.

If $\mathfrak{L}_1$ is negative and $\mathfrak{L}_2$ is positive definite, we consider the anti-space of $\mathfrak{E}$. □

Lemma 6.3. *The orthogonal companion of a maximal positive (maximal negative) subspace is negative (positive).*

Proof. Let $\mathfrak{L}$ be a maximal positive subspace. Suppose that for some $x \in \mathfrak{L}^\perp$ we have $(x, x) > 0$. Then, by Remark 3.2, $\langle x, \mathfrak{L} \rangle$ is a positive extension of $\mathfrak{L}$. Moreover, it is a proper extension, since $x \in \mathfrak{L}$ would imply $x \in \mathfrak{L}^0$, $(x, x) = 0$. This contradicts the maximality of $\mathfrak{L}$.

If $\mathfrak{L}$ is maximal negative, we consider the anti-space. □

As a positive subspace is not necessarily positive definite and a maximal positive definite subspace is not always maximal positive (namely it may happen to have positive extensions without having positive definite extensions), the next lemma neither implies nor is implied by the preceding one.

Lemma 6.4. *The orthogonal companion of a maximal positive definite (maximal negative definite) subspace is negative (positive).*

The *proof* is similar to that of Lemma 6.3. □

Remark 6.5. The orthogonal companion of a maximal positive or maximal positive definite subspace need not be maximal negative. For instance, the subspace $\mathfrak{L}$ of Example 4.8 is maximal positive and maximal positive definite on account of Lemma 6.2, but $\mathfrak{L}^\perp = 0$ is not maximal negative.

Finally we mention a result related to Lemma 6.3.

Lemma 6.6. *The orthogonal companion of a positive definite, maximal positive (negative definite, maximal negative) subspace is negative definite (positive definite).*

Proof. Let $\mathfrak{L}$ be a positive definite, maximal positive subspace. By Lemma 6.3 $\mathfrak{L}^\perp$ is negative. If $x \in \mathfrak{L}^\perp$, $(x,x) = 0$, $x \neq 0$ then, in view of Remark 3.2, $\langle x, \mathfrak{L}\rangle$ is a proper positive extension of $\mathfrak{L}$ contrary to the assumption. □

7. Maximal Neutral Subspaces

Maximal neutral subspaces do not play such an important role in further developments as, say, maximal positive ones. Nevertheless, they have some interesting properties which deserve to be mentioned.

Lemma 7.1. *If $\mathfrak{L}$ is a maximal neutral subspace, then $\mathfrak{L}^\perp$ is semi-definite and its isotropic part coincides with $\mathfrak{L}$.*

Proof. Suppose that $x_1, x_2 \in \mathfrak{L}^\perp$, $(x_1, x_1) > 0$, $(x_2, x_2) < 0$. According to Lemma 2.1 the two-dimensional subspace $\langle x_1, x_2\rangle$ contains a non-zero neutral vector x_0. We have $x_0 \notin \mathfrak{L}$, since otherwise x_2 would be an element of the subspace $\langle x_1, \mathfrak{L}\rangle$, which is positive by Remark 3.2. Therefore $\langle x_0, \mathfrak{L}\rangle \neq \mathfrak{L}$. Furthermore, owing to the same Remark 3.2, $\langle x_0, \mathfrak{L}\rangle$ is a neutral subspace. This contradicts the maximality of $\mathfrak{L}$. Consequently, $\mathfrak{L}^\perp$ is semi-definite. Replacing x_0 by any neutral element of $\mathfrak{L}^\perp$ in our reasoning we find that x_0 must belong to $\mathfrak{L}$. Now the assertion $\mathfrak{L}^\perp \cap \mathfrak{L}^{\perp\perp} = \mathfrak{L}$ follows from Lemma 4.4. □

Lemmas 7.1 and 4.7 yield necessary conditions in order that a neutral subspace be maximal neutral. It turns out that both conditions together are sufficient too.

Theorem 7.2. *The neutral subspace $\mathfrak{L} \subset \mathfrak{E}$ is maximal neutral if and only if $\mathfrak{L}^\perp$ is semi-definite and $\mathfrak{L}^{\perp\perp} = \mathfrak{L}$.*

Proof. If $\mathfrak{L}$ is maximal neutral, then $\mathfrak{L}^\perp$ is semi-definite by Lemma 7.1, whereas the relation $\mathfrak{L}^{\perp\perp} = \mathfrak{L}$ follows from Lemma 4.7. Now let $\mathfrak{L}$ be a neutral subspace with $\mathfrak{L}^\perp$ semi-definite and $\mathfrak{L}^{\perp\perp} = \mathfrak{L}$. Let $\mathfrak{L}_1$ be a neutral subspace, $\mathfrak{L}_1 \supset \mathfrak{L}$. Corollary 4.5 implies $\mathfrak{L} \perp \mathfrak{L}_1$ i.e. $\mathfrak{L}_1 \subset \mathfrak{L}^\perp$. Applying Lemma 4.4 to the semi-definite subspace $\mathfrak{L}^\perp$ we obtain the relation $\mathfrak{L}_1 \perp \mathfrak{L}^\perp$. Hence $\mathfrak{L}_1 \subset \mathfrak{L}^{\perp\perp} = \mathfrak{L}$ and, consequently, $\mathfrak{L}_1 = \mathfrak{L}$. □

The following theorem asserts that a maximal neutral subspace is either maximal positive or maximal negative (or both maximal positive and maximal negative).

Theorem 7.3. *If a maximal neutral subspace $\mathfrak{L} \subset \mathfrak{E}$ admits a positive proper extension, then $\mathfrak{L}$ does not admit any negative proper extension.*

Proof. Suppose there exist a positive subspace $\mathfrak{L}_1 \subset \mathfrak{E}$ and a negative subspace $\mathfrak{L}_2 \subset \mathfrak{E}$ that contain $\mathfrak{L}$ properly. Let $x_1 \in \mathfrak{L}_1$, $x_1 \notin \mathfrak{L}$. By Lemma 4.4 we have $\mathfrak{L}_1 \perp \mathfrak{L}$. In particular, $x_1 \perp \mathfrak{L}$. Therefore, in view of Remark 3.2 and the maximality of $\mathfrak{L}$, the vector x_1 must be positive. Similarly, if $x_2 \in \mathfrak{L}_2$, $x_2 \notin \mathfrak{L}$, then $x_2 \in \mathfrak{L}^\perp$, $(x_2, x_2) < 0$. This contradicts Lemma 7.1. □

If a maximal neutral subspace $\mathfrak{L}$ has neither positive nor negative proper extensions, then $\mathfrak{L}$ is said to be a *hypermaximal neutral* subspace.

Theorem 7.4. *The subspace $\mathfrak{L} \subset \mathfrak{E}$ is hypermaximal neutral if and only if $\mathfrak{L}^\perp = \mathfrak{L}$.*

Proof. Let $\mathfrak{L}$ be hypermaximal neutral. Corollary 4.5 yields $\mathfrak{L} \subset \mathfrak{L}^\perp$. To prove the inclusion in the opposite direction, let $x \in \mathfrak{L}^\perp$ be arbitrary. x can be positive, negative, or neutral. In each case the span $\langle x, \mathfrak{L}\rangle$ is semi-definite (see Remark 3.2). Therefore the hypermaximality of $\mathfrak{L}$ implies $x \in \mathfrak{L}$.

Conversely, assume that $\mathfrak{L}^\perp = \mathfrak{L}$. Then $\mathfrak{L} \perp \mathfrak{L}$, hence $\mathfrak{L}$ is neutral. Denote a semi-definite extension of $\mathfrak{L}$ by $\mathfrak{L}_1$. On account of Lemma 4.4 $\mathfrak{L}$ is orthogonal to $\mathfrak{L}_1$. Thus $\mathfrak{L}_1 \subset \mathfrak{L}^\perp = \mathfrak{L}$. □

The subspace $\mathfrak{L}$ appearing in Example 4.1 is hypermaximal neutral. On the other hand, there are inner product spaces that do not contain hypermaximal neutral subspaces. For instance, in a one-dimensional definite inner product space 0 is the only neutral subspace, and it is not hypermaximal neutral.

8. Projections of Vectors on Subspaces

Let $\mathfrak{L}$ be a subspace and x an element of the inner product space $\mathfrak{E}$. If $x = y + z$, where $y \in \mathfrak{L}$ and $z \in \mathfrak{L}^\perp$, we say that y is a *projection* of x on $\mathfrak{L}$.

Neither the existence, nor the uniqueness of the projection of a vector on a subspace is guaranteed. For instance, in Example 4.1 the elements of $\mathfrak{L}$ have infinitely many projections on $\mathfrak{L}$ (namely every $y \in \mathfrak{L}$ is a projection), while the vectors outside $\mathfrak{L}$ have none.

The problem of uniqueness can easily be settled.

Lemma 8.1. *Two projections of the vector $x \in \mathfrak{E}$ on the subspace $\mathfrak{L} \subset \mathfrak{E}$ differ in an arbitrary isotropic vector of $\mathfrak{L}$.*

Proof. Let $x = y_j + z_j$, where $y_j \in \mathfrak{L}$, $z_j \in \mathfrak{L}^\perp$ $(j = 1, 2)$. Then $y_1 - y_2 \in \mathfrak{L}$ and $y_1 - y_2 = z_2 - z_1 \in \mathfrak{L}^\perp$. Thus $y_1 - y_2 \in \mathfrak{L}^0$.

On the other hand, if $x = y_1 + z_1$, where $y_1 \in \mathfrak{L}$, $z_1 \in \mathfrak{L}^\perp$, then for every $u \in \mathfrak{L}^0$ we have $x = (y_1 + u) + (z_1 - u)$, where $y_1 + u \in \mathfrak{L}$, $z_1 - u \in \mathfrak{L}^\perp$. □

Corollary 8.2. *If there is an element in $\mathfrak{E}$ having exactly one projection on $\mathfrak{L}$, then $\mathfrak{L}$ is non-degenerate. If $\mathfrak{L}$ is non-degenerate, every element of $\mathfrak{E}$ has at most one projection on $\mathfrak{L}$.* □

The question concerning the existence of projections is more difficult. We are going to treat it in a few special cases.

Theorem 8.3. *Let $\mathfrak{L}$ be a neutral subspace. The vector x admits a projection on $\mathfrak{L}$ if and only if $x \perp \mathfrak{L}$. If this condition is satisfied, every element of $\mathfrak{L}$ is a projection of x on $\mathfrak{L}$.*

Proof. By Lemma 8.1 and Corollary 4.5 x has a projection on $\mathfrak{L}$ if and only if every element of $\mathfrak{L}$, in particular the zero vector, is a projection of x on $\mathfrak{L}$; i.e., $x = 0 + z$, where $0 \in \mathfrak{L}$, $z \in \mathfrak{L}^\perp$. The proof is complete. □

Theorem 8.4. *Let $\mathfrak{L}$ be a positive definite subspace of the inner product space $\mathfrak{E}$. The element $x \in \mathfrak{E}$ admits a projection on $\mathfrak{L}$ if and only if the function $\varphi(y) = (x-y, x-y)$, considered for $y \in \mathfrak{L}$, attains its minimum at some $y_0 \in \mathfrak{L}$:*

$$(x-y_0, x-y_0) = \min_{y \in \mathfrak{L}} (x-y, x-y) . \tag{8.1}$$

The element y_0 defined by (8.1) is unique and it is the projection of x on $\mathfrak{L}$. For a negative definite $\mathfrak{L}$ a similar condition with max *in place of* min *holds.*

Proof. Let $\mathfrak{L}$ be positive definite. If $x = y_0 + z$, where $y_0 \in \mathfrak{L}$, $z \in \mathfrak{L}^\perp$, then for every $y \in \mathfrak{L}$, $y \neq y_0$, we find:

$$(x-y, x-y) = (z+y_0-y, z+y_0-y) =$$
$$= (z, z) + (y_0-y, y_0-y) > (z, z) = (x-y_0, x-y_0) .$$

If, conversely, for some $y_0 \in \mathfrak{L}$ the condition (8.1) is fulfilled, then for every $y \in \mathfrak{L}$, $\lambda \in \boldsymbol{C}$ it follows that

$$(x-y_0+\lambda y, x-y_0+\lambda y) \geqq (x-y_0, x-y_0) .$$

Hence

$$\lambda (y, x-y_0) + \bar{\lambda} (x-y_0, y) + |\lambda|^2 (y, y) \geqq 0 .$$

Setting $\lambda = \frac{1}{\mu}(x-y_0, y)$, where μ runs through all non-zero real numbers, we obtain:

$$2\mu\,|(x-y_0, y)|^2 + |(x-y_0, y)|^2\,(y, y) \geqq 0 \qquad (\mu \in \boldsymbol{R};\ \mu \neq 0).$$

Thus $(x-y_0, y) = 0$, $x-y_0 \perp \mathfrak{L}$.

The case of a negative definite $\mathfrak{L}$ can be reduced to the preceding one by passing to the anti-space of $\mathfrak{E}$. □

Theorem 8.5. *Let the subspace $\mathfrak{L} \subset \mathfrak{E}$ be the orthogonal direct sum of n subspaces:*

$$\mathfrak{L} = \mathfrak{L}_1\,(\dot{+})\cdots(\dot{+})\,\mathfrak{L}_n\,. \tag{8.2}$$

The vector $y \in \mathfrak{L}$ is a projection of the vector $x \in \mathfrak{E}$ on the subspace $\mathfrak{L}$ if and only if

$$y = y_1 + \cdots + y_n\,, \tag{8.3}$$

where y_j is a projection of x on $\mathfrak{L}_j$ $(j = 1, \ldots, n)$. In particular, x has a projection on $\mathfrak{L}$ if and only if it has a projection on each $\mathfrak{L}_j$ $(j = 1, \ldots, n)$.

Proof. Suppose that y is a projection of x on $\mathfrak{L}$. Then $x = y + z$, where $y \in \mathfrak{L}$, $z \in \mathfrak{L}^\perp$. Let (8.3) be the decomposition of y corresponding to (8.2). Using the notation

$$z_j' = \sum_{k \neq j} y_k \qquad (j = 1, \ldots, n) \tag{8.4}$$

we have $x = y_j + (z + z_j')$, where $y_j \in \mathfrak{L}_j$, $z + z_j' \in \mathfrak{L}_j^\perp$ for every j. Thus y_j is a projection of x on $\mathfrak{L}_j$.

Let, conversely, $x = y_j + z_j$, where $y_j \in \mathfrak{L}_j$, $z_j \in \mathfrak{L}_j^\perp$ for every j. Then with the notations (8.3), (8.4) we obtain $x = y + (z_j - z_j')$, where $y \in \mathfrak{L}$, $z_j - z_j' \in \mathfrak{L}_j^\perp$ for every j. Since $z_j - z_j' = x - y$ does not depend on j, it is orthogonal to every $\mathfrak{L}_j$ and, in view of (8.2), it is orthogonal to $\mathfrak{L}$. □

Theorems 8.3 — 8.5 enable us to find the projections of any vector x on a subspace $\mathfrak{L}$ that can be decomposed into the orthogonal direct sum of a neutral, a positive definite, and a negative definite subspace. As we shall see later (Example 11.3), not every $\mathfrak{L}$ belongs to this class. Nevertheless, it is always possible to represent $\mathfrak{L}$ as the orthogonal direct sum of a neutral and a non-degenerate subspace (cf. Lemma 5.1). Therefore the problem of finding projections on $\mathfrak{L}$ can always be reduced to the case where $\mathfrak{L}$ is non-degenerate.

9. Ortho-complemented Subspaces

In the last section we were looking for projections of different vectors x on a fixed subspace $\mathfrak{L}$. One may also raise the following question: what are the conditions on the subspace $\mathfrak{L}$ in order that every x in the space $\mathfrak{E}$ would have at least one projection on $\mathfrak{L}$? Clearly, a necessary and sufficient condition is $\langle\mathfrak{L}, \mathfrak{L}^\perp\rangle = \mathfrak{E}$.

If a subspace $\mathfrak{L}$ of the inner product space $\mathfrak{E}$ satisfies the condition $\langle\mathfrak{L}, \mathfrak{L}^\perp\rangle = \mathfrak{E}$, we shall say that $\mathfrak{L}$ is *ortho-complemented*. In other words, the subspace $\mathfrak{L} \subset \mathfrak{E}$ is ortho-complemented, if it admits an orthogonal complementary subspace, i.e. a complementary subspace contained in $\mathfrak{L}^\perp$. For a non-degenerate $\mathfrak{L}$ this happens if and only if $\mathfrak{L} \dotplus \mathfrak{L}^\perp = \mathfrak{E}$.

Lemma 9.1. *If $\mathfrak{L}$ is ortho-complemented, so is $\mathfrak{L}^\perp$.*

Proof. Since $\mathfrak{L} \subset \mathfrak{L}^{\perp\perp}$, the relation $\langle\mathfrak{L}, \mathfrak{L}^\perp\rangle = \mathfrak{E}$ implies $\langle\mathfrak{L}^\perp, \mathfrak{L}^{\perp\perp}\rangle = \mathfrak{E}$. □

We remark that $\mathfrak{L}^\perp$ may be ortho-complemented without $\mathfrak{L}$ being so. An example is a non-closed subspace in Hilbert space.

The following result is a special case of Theorem 8.5.

Lemma 9.2. *Let $\mathfrak{L}$ be the orthogonal direct sum of n subspaces: $\mathfrak{L} = \mathfrak{L}_1 \dotplus \cdots \dotplus \mathfrak{L}_n$. Then $\mathfrak{L}$ is ortho-complemented if and only if each $\mathfrak{L}_j$ $(j = 1, \ldots, n)$ is so.* □

Lemma 9.3. *A neutral subspace $\mathfrak{L} \subset \mathfrak{E}$ is ortho-complemented if and only if $\mathfrak{L} \subset \mathfrak{E}^0$.*

Proof. According to Corollary 4.5 we have $\mathfrak{L} \subset \mathfrak{L}^\perp$. Therefore the property $\langle\mathfrak{L}, \mathfrak{L}^\perp\rangle = \mathfrak{E}$ is now equivalent to $\mathfrak{L}^\perp = \mathfrak{E}$ or, what is the same, to $\mathfrak{L} \perp \mathfrak{E}$. □

Theorem 9.4. *The following two conditions are necessary and sufficient in order that the subspace $\mathfrak{L} \subset \mathfrak{E}$ would be ortho-complemented:*

a) $\mathfrak{L}^0 \subset \mathfrak{E}^0$;

b) $\mathfrak{L}/\mathfrak{L}^0$ *is ortho-complemented in* $\mathfrak{E}/\mathfrak{E}^0$.

Proof. Let $\mathfrak{L}$ be ortho-complemented in $\mathfrak{E}$. Then, in view of Lemmas 5.1 and 9.2, also $\mathfrak{L}^0$ is ortho-complemented. Thus the relation a) is a consequence of Lemma 9.3.

To prove b) we first remark that $\mathfrak{L}/\mathfrak{L}^0$ is isometrically isomorphic to the subspace $\hat{\mathfrak{L}} \subset \mathfrak{E}/\mathfrak{E}^0$ consisting of those classes $\hat{x} \in \mathfrak{E}/\mathfrak{E}^0$ which contain at least one element of $\mathfrak{L}$. This can be seen as follows. Let $\hat{x}_{\mathfrak{L}^0} \in \mathfrak{L}/\mathfrak{L}^0$. Let $x \in \mathfrak{L}$ be a representative of $\hat{x}_{\mathfrak{L}^0}$, and let $\hat{x} \in \mathfrak{E}/\mathfrak{E}^0$ be the class containing x. On account of a) the mapping $\hat{x}_{\mathfrak{L}^0} \to \hat{x}$ is single-valued. Moreover, it is one-to-one, since $\mathfrak{L} \cap \mathfrak{E}^0 = \mathfrak{L}^0$. Finally, it preserves sums, scalar multiples, and inner products.

Next we show that $\hat{\mathfrak{L}}$ is ortho-complemented in $\mathfrak{E}/\mathfrak{E}^0$. Let $\hat{x} \in \mathfrak{E}/\mathfrak{E}^0$. For $x \in \hat{x}$, $x \in \mathfrak{E}$ we have a decomposition

$$x = y + z; \quad y \in \mathfrak{L}, \quad z \perp \mathfrak{L}. \tag{9.1}$$

Therefore

$$\hat{x} = \hat{y} + \hat{z}; \quad \hat{y} \in \hat{\mathfrak{L}}, \quad \hat{z} \perp \hat{\mathfrak{L}}, \tag{9.2}$$

where $y \in \hat{y}$, $z \in \hat{z}$.

Let, conversely, a) and b) be satisfied (the latter in the sense that $\hat{\mathfrak{L}}$ is ortho-complemented in $\mathfrak{E}/\mathfrak{E}^0$). Then for any $x \in \mathfrak{E}$ the respective class $\hat{x} \in \mathfrak{E}/\mathfrak{E}^0$ admits a decomposition of the form (9.2). Choosing a representative $y \in \mathfrak{L}$ of $\hat{y}$ and an arbitrary representative z of $\hat{z}$ we obtain relation (9.1). □

Corollary 9.5. *In a non-degenerate space $\mathfrak{E}$ every ortho-complemented subspace is non-degenerate.* □

According to Theorem 9.4 and Corollary 5.3, in the study of ortho-complemented subspaces it is sufficient to concentrate upon non-degenerate subspaces of non-degenerate inner product spaces. In many cases (namely when the subspace under consideration is the orthogonal direct sum of a positive definite and a negative definite subspace) Lemma 9.2 makes possible a further reduction: we may restrict our attention to definite subspaces of non-degenerate spaces. Since most of the results known for this situation are of a topological nature, they will be given in subsequent chapters.

Here we mention only some simple facts which can be obtained without topological tools.

Lemma 9.6. *If $\mathfrak{L}$ is an ortho-complemented subspace of the non-degenerate space $\mathfrak{E}$, then $\mathfrak{L}^{\perp\perp} = \mathfrak{L}$.*

Proof. Our assumptions imply the relation $\mathfrak{L} \,(\dot{+})\, \mathfrak{L}^{\perp} = \mathfrak{E}$. Consequently, if $\mathfrak{L}^{\perp\perp}$ is a proper extension of $\mathfrak{L}$, then $\mathfrak{L}^{\perp\perp} \cap \mathfrak{L}^{\perp} \neq 0$. On the other hand, by Lemma 9.1 and Corollary 9.5 the subspace $\mathfrak{L}^{\perp}$ cannot be degenerate. □

It is well known that in Hilbert space the condition $\mathfrak{L}^{\perp\perp} = \mathfrak{L}$ is not only necessary, but also sufficient for a subspace $\mathfrak{L}$ to be ortho-complemented. In an arbitrary non-degenerate inner product space both of the conditions $\mathfrak{L}^{\perp\perp} = \mathfrak{L}$, $\mathfrak{L} \cap \mathfrak{L}^{\perp} = 0$ are necessary (cf. Lemma 9.6 and Corollary 9.5), but, as the following example shows, even together they are not sufficient for the ortho-complementedness of $\mathfrak{L}$.

Example 9.7. Consider the space $\mathfrak{E}$ of Example 2.3 with $\varepsilon_j = (-1)^j$ ($j = 1, 2, \ldots$). Thus $\mathfrak{E}$ is the vector space of complex numerical sequences $x = \{\xi_1, \xi_2, \ldots\}$ satisfying $\sum_{j=1}^{\infty} |\xi_j|^2 < \infty$, and the inner product is defined by the relation

$$(x, y) = \sum_{j=1}^{\infty} (-1)^j \xi_j \bar{\eta}_j ,$$

where $y = \{\eta_1, \eta_2, \ldots\}$, $\sum_{j=1}^{\infty} |\eta_j|^2 < \infty$. Set

$$\mathfrak{L} = \left\{ \{\xi_j\}_1^\infty \in \mathfrak{F}: \xi_{2j} = \frac{2j}{2j-1} \xi_{2j-1} \ (j = 1, 2, \ldots) \right\}.$$

The subspace $\mathfrak{L}$ is positive definite. In particular, $\mathfrak{L}^0 = 0$. Furthermore,

$$\mathfrak{L}^\perp = \left\{ \{\eta_j\}_1^\infty \in \mathfrak{F}: \eta_{2j} = \frac{2j-1}{2j} \eta_{2j-1} \ (j = 1, 2, \ldots) \right\}, \quad \mathfrak{L}^{\perp\perp} = \mathfrak{L} .$$

But $\mathfrak{L}$ is not ortho-complemented. Really, suppose that the element

$$z = \{\zeta_1, \zeta_2, \ldots\}; \ \zeta_{2j-1} = 0 , \ \zeta_{2j} = \frac{1}{2j} \quad (j = 1, 2, \ldots)$$

can be written in the form $z = x + y$, where $x = \{\xi_j\}_1^\infty \in \mathfrak{L}$, $y = \{\eta_j\}_1^\infty \in \mathfrak{L}^\perp$. Then an easy calculation yields $\xi_{2j-1} = \frac{2j-1}{4j-1}$ $(j = 1, 2, \ldots)$, so that $\sum_{j=1}^{\infty} |\xi_j|^2 = \infty$, contradicting the definition of $\mathfrak{F}$.

Lemma 9.8. *Every definite subspace of finite dimension is ortho-complemented.*

Proof. If $\mathfrak{L}$ is an n-dimensional positive definite subspace of the inner product space $\mathfrak{F}$, then we can find a basis $e_1, \ldots, e_n$ for $\mathfrak{L}$ which is orthonormal: $(e_j, e_k) = \delta_{jk}$ $(j, k = 1, \ldots, n)$. Given a vector $x \in \mathfrak{F}$ it is easy to verify that $\sum_{j=1}^{n} (x, e_j) e_j$ is the projection of x on $\mathfrak{L}$.

The case of a negative definite $\mathfrak{L}$ can be reduced to the above one by considering the anti-space of $\mathfrak{F}$. □

10. Dual Pairs of Subspaces

In Section 5 it was shown that every subspace $\mathfrak{L}$ of an inner product space $\mathfrak{F}$ is the orthogonal direct sum of the isotropic part $\mathfrak{L}^0$ and a non-degenerate subspace $\mathfrak{L}^1$. In certain problems, e.g. those concerning intrinsic properties of $\mathfrak{L}$ or the restriction of an operator to $\mathfrak{L}$, the isotropic part can be neglected or easily handled, and it remains to consider the non-degenerate component. In other problems, e.g. when a decomposition of $\mathfrak{E}$ corresponding to $\mathfrak{L}$ is required, such a neglection is unjustified, and one would like to proceed by extending $\mathfrak{L}^0$ to a non-degenerate subspace. Sometimes, mainly if $\dim \mathfrak{L}^0 < \infty$, this can be done with the aid of so-called dual companions of $\mathfrak{L}^0$.

We will say that the subspaces $\mathfrak{L}, \mathfrak{M} \subset \mathfrak{E}$ form a *dual pair*, or that $\mathfrak{L}$ and $\mathfrak{M}$ are *dual companions* for each other (in symbols: $\mathfrak{L} \# \mathfrak{M}$), if $\mathfrak{L} \cap \mathfrak{M}^\perp = \mathfrak{L}^\perp \cap \mathfrak{M} = 0$. In other words, two subspaces of the inner product space $\mathfrak{E}$ form a dual pair if there is no non-zero vector in any of them that would be orthogonal to the whole of the other.

Every non-degenerate subspace is a dual companion for itself. In Example 4.1 the neutral subspace $\mathfrak{M} = \langle e_1 - e_2 \rangle$ is a dual companion for the neutral subspace $\mathfrak{L} = \langle e_1 + e_2 \rangle$.

Lemma 10.1. *If* $\mathfrak{L} \# \mathfrak{M}$, *where* $\mathfrak{L}$ *or* $\mathfrak{M}$ *is neutral, then* $\mathfrak{L} \cap \mathfrak{M} = 0$, *and* $\mathfrak{L} \dot{+} \mathfrak{M}$ *is non-degenerate.*

Proof. Let $\mathfrak{L} \# \mathfrak{M}$, where e.g. $\mathfrak{M} \subset \mathfrak{M}^\perp$ (cf. Corollary 4.5). Then $\mathfrak{L} \cap \mathfrak{M} \subset \mathfrak{L} \cap \mathfrak{M}^\perp = 0$. Further let $x + y$ ($x \in \mathfrak{L}$, $y \in \mathfrak{M}$) denote an isotropic vector of $\mathfrak{L} \dot{+} \mathfrak{M}$. From the relations $x+y \in \mathfrak{M}^\perp$, $y \in \mathfrak{M}^\perp$ we deduce $x \in \mathfrak{M}^\perp$. Therefore $x \in \mathfrak{L} \cap \mathfrak{M}^\perp$, that is, $x = 0$. Thus $x+y \in \mathfrak{L}^\perp$ yields $y \in \mathfrak{L}^\perp$, $y \in \mathfrak{L}^\perp \cap \mathfrak{M}$, $y = 0$. □

In order to get a converse of this statement we must require that both of the subspaces $\mathfrak{L}, \mathfrak{M}$ be neutral.

Lemma 10.2. *If* $\mathfrak{L} \cap \mathfrak{M} = 0$ *and* $(\mathfrak{L} \dot{+} \mathfrak{M})^0 = 0$, *where* $\mathfrak{L}, \mathfrak{M}$ *are neutral subspaces of* $\mathfrak{E}$, *then* $\mathfrak{L} \# \mathfrak{M}$.

Proof. On account of Corollary 4.5 we have $\mathfrak{L} \subset \mathfrak{L}^\perp$, $\mathfrak{M} \subset \mathfrak{M}^\perp$. Therefore the elements of $\mathfrak{L} \cap \mathfrak{M}^\perp$ as well as the elements of $\mathfrak{L}^\perp \cap \mathfrak{M}$ are isotropic vectors of $\mathfrak{L} \dot{+} \mathfrak{M}$. □

Lemma 10.3. *If* $\mathfrak{L} \# \mathfrak{M}$, *then* $\dim \mathfrak{L} = \dim \mathfrak{M}$.

Proof. The conclusion follows from Corollary 3.4. □

When studying dual pairs of finite-dimensional subspaces it is often convenient to introduce bases. In this connection we need the following definition.

Two systems $\{e_j\}_1^n \subset \mathfrak{E}$, $\{f_j\}_1^n \subset \mathfrak{E}$ will be said to form a *dual pair*, or to be *dual companions* of each other, if $(e_j, f_k) = \delta_{jk}$ $(j,k = 1, \ldots, n)$.

Let the systems $\{e_j\}_1^n$, $\{f_j\}_1^n$ form a dual pair. The following assertions can immediately be proved: a) the vectors $e_1, \ldots, e_n$ are linearly independent and the vectors $f_1, \ldots, f_n$ are also linearly independent, b) the subspaces $\langle e_1, \ldots, e_n\rangle$, $\langle f_1, \ldots, f_n\rangle$ form a dual pair. The next lemma enables us to show that a converse of b) holds too.

Lemma 10.4. *Let $\mathfrak{L}$ be an n-dimensional* ($1 \leqq n < \infty$) *subspace of $\mathfrak{E}$, and let $\mathfrak{M} \subset \mathfrak{E}$ be a subspace satisfying $\mathfrak{L} \cap \mathfrak{M}^\perp = 0$. There exist two systems $\{e_j\}_1^n$, $\{f_j\}_1^n$ forming a dual pair such that $\langle e_1, \ldots, e_n\rangle = \mathfrak{L}$, $\langle f_1, \ldots, f_n\rangle \subset \mathfrak{M}$.*

Proof. Choose a basis $\{g_j\}_1^n$ of $\mathfrak{L}$, and put $e_1 = g_1$. As $e_1 \notin \mathfrak{M}^\perp$, one can find an f_1 in $\mathfrak{M}$ with $(e_1, f_1) = 1$.

Suppose that two dual companions $\{e_j\}_1^k$, $\{f_j\}_1^k$ ($k < n$) have already been constructed, where $\langle e_1, \ldots, e_k\rangle = \langle g_1, \ldots, g_k\rangle$ and $f_j \in \mathfrak{M}$ ($j = 1, \ldots, k$). Put

$$e_{k+1} = g_{k+1} - \sum_{j=1}^{k} (g_{k+1}, f_j)\, e_j\,.$$

Then $\langle e_1, \ldots, e_{k+1}\rangle = \langle g_1, \ldots, g_{k+1}\rangle$, $(e_{k+1}, f_j) = 0$ ($j = 1, \ldots, k$). Furthermore, since $e_{k+1} \notin \mathfrak{M}^\perp$, there is a vector $h_{k+1} \in \mathfrak{M}$ satisfying the relation $(e_{k+1}, h_{k+1}) = 1$. We set

$$f_{k+1} = h_{k+1} - \sum_{j=1}^{k} (h_{k+1}, e_j)\, f_j\,,$$

and verify easily that $f_{k+1} \in \mathfrak{M}$, $(f_{k+1}, e_j) = \delta_{j,k+1}$ ($j = 1, \ldots, k+1$).

In the n-th step we obtain the desired dual pair. □

Corollary 10.5. *If $\mathfrak{L}$, $\mathfrak{M}$ are subspaces of $\mathfrak{E}$ such that $1 \leqq \dim \mathfrak{L} = \dim \mathfrak{M} < \infty$ and $\mathfrak{L} \cap \mathfrak{M}^\perp = 0$, then some basis of $\mathfrak{L}$ forms a dual pair with some basis of $\mathfrak{M}$.* □

From property b) of dual pairs of systems it follows that under the assumptions of Corollary 10.5 we have $\mathfrak{L} \# \mathfrak{M}$. Therefore in verifying that two finite-dimensional subspaces $\mathfrak{L}$, $\mathfrak{M}$ are dual companions one of the requirements $\mathfrak{L} \cap \mathfrak{M}^\perp = 0$, $\mathfrak{L}^\perp \cap \mathfrak{M} = 0$ may be replaced by the equation $\dim \mathfrak{L} = \dim \mathfrak{M}$.

Combining Corollary 10.5 with Lemma 10.3 we obtain a result on bases of dual pairs of finite-dimensional subspaces. However, this result can be improved.

Lemma 10.6. *Let $\mathfrak{L} \# \mathfrak{M}$, where $\mathfrak{L}, \mathfrak{M}$ are non-zero subspaces of finite dimension. Then for any basis $\{e_j\}_1^n$ of $\mathfrak{L}$ some basis $\{f_j\}_1^n$ of $\mathfrak{M}$ serves as a dual companion.*

Proof. Denote by $\mathfrak{L}_k$ ($k = 1, \ldots, n$) the span of all vectors e_j with $j \neq k$. On account of Lemmas 3.3 and 10.3 we have

$$\dim (\mathfrak{M} \cap \mathfrak{L}_k^\perp) \geqq 1 \qquad (k = 1, \ldots, n)\,.$$

Let $f_k \in \mathfrak{M} \cap \mathfrak{L}_k^\perp$, $f_k \neq 0$. As f_k cannot be orthogonal to the whole of $\mathfrak{L}$, it is not orthogonal to e_k. Hence we may assume that $(e_k, f_k) = 1$. Consequently, $(e_j, f_k) = \delta_{jk}$ $(j = 1, \ldots, n)$, and this is true for every k. It follows that f_k $(k = 1, \ldots, n)$ are linearly independent; thus they form a basis of $\mathfrak{M}$. □

Lemma 10.7. *In a non-degenerate inner product space* $\mathfrak{E}$ *every finite-dimensional subspace has a dual companion.*

Proof. Let $\mathfrak{L}$ be a subspace of $\mathfrak{E}$ with $\dim \mathfrak{L} = n < \infty$. If $n = 0$, then $\mathfrak{M} = 0$ is a dual companion for $\mathfrak{L}$. If $n > 0$, we may apply Lemma 10.4 to $\mathfrak{L}$ and $\mathfrak{E}$. Thus we obtain a basis $\{e_j\}_1^n$ of $\mathfrak{L}$ and a system $\{f_j\}_1^n \subset \mathfrak{E}$ which form a dual pair. The span $\langle f_1, \ldots, f_n \rangle$ will be a dual companion for $\mathfrak{L}$. □

The following lemma can be applied in some cases as a substitute for Lemma 10.1.

Lemma 10.8. *If* $\mathfrak{L} \# \mathfrak{M}$, *where* $\mathfrak{L}$ *and* $\mathfrak{M}$ *are finite-dimensional subspaces of the inner product space* $\mathfrak{E}$, *then* $\mathfrak{L} \dotplus \mathfrak{M}^\perp = \mathfrak{E}$.

Proof. If $\mathfrak{L} = 0$, then necessarily $\mathfrak{M} = 0$, hence the assertion is true. Otherwise we make use of Corollary 10.5 in order to obtain a basis $\{e_j\}_1^n$ of $\mathfrak{L}$ and a basis $\{f_j\}_1^n$ of $\mathfrak{M}$ such that $(e_j, f_k) = \delta_{jk}$ $(j,k = 1, \ldots, n)$. For every $x \in \mathfrak{E}$ we set

$$x_{\mathfrak{L}} = \sum_{j=1}^{n} (x, f_j)\, e_j\,,$$

and verify immediately that $x_{\mathfrak{L}} \in \mathfrak{L}$, $x - x_{\mathfrak{L}} \in \mathfrak{M}^\perp$. □

By way of application we consider a partial improvement of Lemma 3.5.

Theorem 10.9. *For every subspace* $\mathfrak{M}$ *of a non-degenerate inner product space* $\mathfrak{E}$ *we have* $\operatorname{codim} \mathfrak{M}^\perp = \dim \mathfrak{M}$.

Proof. If $\dim \mathfrak{M} < \infty$, then the statement follows from Lemmas 10.7, 10.8, 10.3 and Corollary 1.2. If $\dim \mathfrak{M} = \infty$, let $\mathfrak{M}_1$ denote an n-dimensional $(n < \infty)$ subspace of $\mathfrak{M}$. Applying the part just proved of our theorem we obtain the relation $\operatorname{codim} \mathfrak{M}_1^\perp = n$. Therefore $\operatorname{codim} \mathfrak{M}^\perp \geqq n$. This being true for every n, it follows that $\operatorname{codim} \mathfrak{M}^\perp = \infty$. □

Theorem 10.10. *If* $\mathfrak{L}$ *is a finite-dimensional subspace of a non-degenerate inner product space, then* $\mathfrak{L}^{\perp\perp} = \mathfrak{L}$.

Proof. According to Theorem 10.9 and relation (3.2) we have: $\dim \mathfrak{L}^{\perp\perp} = \operatorname{codim} \mathfrak{L}^{\perp\perp\perp} = \operatorname{codim} \mathfrak{L}^\perp = \dim \mathfrak{L}$. It remains to observe that $\mathfrak{L} \subset \mathfrak{L}^{\perp\perp}$ (see (3.1)). □

11. Fundamental Decompositions

The inner product space $\mathfrak{E}$ is said to be *decomposable*, if it can be represented as the orthogonal direct sum of a neutral subspace $\mathfrak{E}^0$, a positive definite subspace $\mathfrak{E}^+$, and a negative definite subspace $\mathfrak{E}^-$:

$$(11.1) \qquad \mathfrak{E} = \mathfrak{E}^0 (\dot{+}) \mathfrak{E}^+ (\dot{+}) \mathfrak{E}^-; \quad \mathfrak{E}^0 \subset \mathfrak{P}^0, \quad \mathfrak{E}^+ \subset \mathfrak{P}^{++}, \quad \mathfrak{E}^- \subset \mathfrak{P}^{--}.$$

Every decomposition of the type (11.1) is called a *fundamental decomposition* of $\mathfrak{E}$.

We first justify the use of the symbol $\mathfrak{E}^0$ for the neutral component.

Lemma 11.1. *If* $\mathfrak{E}^0$, $\mathfrak{E}^+$, $\mathfrak{E}^-$ *are subspaces satisfying* (11.1), *then* $\mathfrak{E}^0$ *coincides with the isotropic part of* $\mathfrak{E}$.

Proof. This is a consequence of Lemmas 4.2 and 4.4. □

Corollary 11.2. *Every fundamental decomposition of a non-degenerate inner product space* $\mathfrak{E}$ *is of the form*

$$(11.2) \qquad \mathfrak{E} = \mathfrak{E}^+ (\dot{+}) \mathfrak{E}^-; \quad \mathfrak{E}^+ \subset \mathfrak{P}^{++}, \quad \mathfrak{E}^- \subset \mathfrak{P}^{--}. \quad \square$$

Not every inner product space is decomposable.

Example 11.3. Denote by $\mathfrak{E}$ the vector space of complex numerical sequences $x = \{\xi_j\}_{j=-\infty}^{\infty}$ that are finite from the left: $\xi_j = 0$ for $j \leqq j_0(x)$. Define an inner product on $\mathfrak{E}$ by the formula

$$(11.3) \qquad (x, y) = \sum_{j=-\infty}^{\infty}{}' \xi_j \bar{\eta}_{-j-1},$$

where $y = \{\eta_j\}_{-\infty}^{\infty} \in \mathfrak{E}$.

It is easy to see that this inner product is non-degenerate. Therefore, according to Corollary 11.2, every fundamental decomposition of $\mathfrak{E}$ must be of the form (11.2).

Consider the linear mapping $\{\xi_j\} \to \{\xi'_j\}$ of $\mathfrak{E}$ to $\mathfrak{E}$, where $\xi'_j = \xi_j$ for $j < 0$, and $\xi'_j = 0$ for $j \geqq 0$. The restriction of this mapping to any definite subspace is one-to-one. Really, if $x = \{\xi_j\} \to 0$, then $\xi_{-1} = \xi_{-2} = \cdots = 0$ so that, in view of (11.3), $(x, x) = 0$. Thus every definite subspace of $\mathfrak{E}$ is isomorphic to some part of the subspace of $\mathfrak{E}$ characterized by the relations $\xi_j = 0$ $(j \geqq 0)$. In particular, if (11.2) holds, then $\mathfrak{E}^+$ and $\mathfrak{E}^-$ are isomorphic to certain subspaces of the vector space of finite numerical sequences. On the other hand, $\mathfrak{E}$ is by definition isomorphic to the vector space of all infinite sequences. Therefore (11.2) leads to a contradiction.

Lemma 11.4. *Let* $\mathfrak{L}$ *be a positive definite (negative definite) subspace of the inner product space* $\mathfrak{E}$. *In order that* $\mathfrak{E}$ *have a fundamental decomposition of the form* (11.1) *with positive definite component* $\mathfrak{E}^+ = \mathfrak{L}$

(negative definite component $\mathfrak{E}^- = \mathfrak{L}$) *the following two conditions are necessary and sufficient:*

a) $\mathfrak{L}$ *is maximal positive definite (maximal negative definite);*

b) $\mathfrak{L}$ *is ortho-complemented.*

Proof. Suppose that $\mathfrak{E}$ admits a fundamental decomposition (11.1) with positive definite component $\mathfrak{E}^+ = \mathfrak{L}$. Since, owing to Lemma 3.1, the subspace $\mathfrak{E}^0(\dot{+})\mathfrak{E}^-$ is negative, Lemma 6.2 guarantees that $\mathfrak{L}$ is maximal positive definite. Condition b) is a consequence of the relation $\mathfrak{L}^\perp = \mathfrak{E}^0(\dot{+})\mathfrak{E}^-$.

Let, conversely, $\mathfrak{L}$ be an ortho-complemented maximal positive definite subspace of $\mathfrak{E}$. Then $\mathfrak{E} = \mathfrak{L}(\dot{+})\mathfrak{L}^\perp$, since $\mathfrak{L}$ is non-degenerate. By Lemma 6.4 $\mathfrak{L}^\perp$ is negative. Applying Lemma 5.1 to $\mathfrak{L}^\perp$ and taking Lemma 4.4 into account we obtain an orthogonal decomposition $\mathfrak{L}^\perp = \mathfrak{M}^0(\dot{+})\mathfrak{M}^1$, where $\mathfrak{M}^0$ is neutral and $\mathfrak{M}^1$ is negative definite. Thus $\mathfrak{E}^+ = \mathfrak{L}$, $\mathfrak{E}^0 = \mathfrak{M}^0$, $\mathfrak{E}^- = \mathfrak{M}^1$ make up a fundamental decomposition of $\mathfrak{E}$.

If $\mathfrak{L}$ is negative definite, we consider the anti-space of $\mathfrak{E}$. □

Corollary 11.5. *An inner product space is decomposable if and only if it contains ortho-complemented maximal definite subspaces.* □

Remark 11.6. If $\mathfrak{E}$ is non-degenerate and admits a fundamental decomposition (11.2), then Lemma 6.2 yields not only that $\mathfrak{E}^+$ is maximal positive definite, but also that it is maximal positive. Similarly, $\mathfrak{E}^-$ is maximal negative in this case.

Lemma 11.4 indicates that the fundamental decomposition need not be unique. In Example 4.1, for instance, any one-dimensional positive definite subspace can play the role of $\mathfrak{E}^+$. Making use of this example one verifies easily that, in fact, the fundamental decomposition of a decomposable indefinite inner product space is never unique.

Invariant properties of fundamental decompositions will be discussed in Chapter IV.

The next theorem singles out an important class of decomposable spaces.

We say that the inner product $(.\,,.)$ is *quasi-positive* (*quasi-negative*) on $\mathfrak{E}$, or that $\mathfrak{E}$ is a quasi-positive (quasi-negative) inner product space, if $\mathfrak{E}$ does not contain any negative definite (positive definite) subspace of infinite dimension.

Theorem 11.7. *Every quasi-positive (quasi-negative) inner product space is decomposable.*

Proof. Let $\mathfrak{E}$ be quasi-positive (in the opposite case we consider the anti-space of $\mathfrak{E}$). Let $\mathfrak{L}$ be a maximal negative definite subspace of $\mathfrak{E}$.

Then $\dim \mathfrak{L} < \infty$. It remains to apply Lemma 9.8 and Corollary 11.5. □

Corollary 11.8. *Every inner product space of finite dimension is decomposable.* □

From Corollaries 11.8, 11.2 and Lemmas 9.2, 9.8 we obtain the following strengthening of Lemma 9.8.

Corollary 11.9. *Every finite-dimensional non-degenerate subspace of an inner product space is ortho-complemented.* □

Notes to Chapter I

Most of the results collected in this chapter are natural extensions of classical theorems concerning quadratic forms in finite-dimensional spaces. During the past decades some of the latter have repeatedly been reformulated within the framework of the axiomatic theory of vector spaces, and included in textbooks of linear algebra (see e.g. Greub [1], Mal'cev [1]).

The systematic study of the geometry of special indefinite inner product spaces of infinite dimension began with the works of Pontrjagin [1], Hestenes [1], Nevanlinna [1] — [4], and those of Iohvidov and Kreĭn [1], [2]. Greub and Hestenes consider inner products on real vector spaces, but this difference is irrelevant for the questions treated by them.

The first comprehensive treatments of general inner product spaces are due to Scheibe [1] and to Ginzburg and Iohvidov [1]. Our exposition follows mainly these two papers. The article of Ginzburg and Iohvidov contains a survey, also some criticism, of earlier results.

The space of Example 2.3 with $\varepsilon_j = -1$ $(j \leqq \varkappa)$, $\varepsilon_j = 1$ $(j > \varkappa)$ has been considered by Pontrjagin [1], while a special case of Example 2.4 appears in a note by Iohvidov [2]. The facts contained in Lemma 2.6 and Corollary 2.7 have been observed by Kreĭn and Šmul'jan [1].

Lemma 4.7 was proved by Scheibe [1]. Example 4.8 belongs to H. Langer (personal communication).

The contents of Section 5 are taken from the paper of Scheibe [1].

A special case of Lemma 6.1 was obtained by Bognár [6].

Lemma 7.1 and Theorem 7.2 are due to Scheibe [1]. Theorem 7.3 was proved by Phillips [3], and independently by Ginzburg and Iohvidov [1]. Theorem 7.4 belongs to Iohvidov (see Ginzburg and Iohvidov [1]).

The results of Section 8 were found by Nevanlinna [4].

Theorem 9.4, though evident, seems to be formulated for the first time. Example 9.7 is, in fact, merely an illustration to a theorem characterizing ortho-complemented definite subspaces in a particular type of inner product spaces (see Theorem V.5.2 below).

Lemma 10.1 (for two neutral subspaces) and Lemma 10.2 were obtained by Iohvidov and Kreĭn [1], and independently by Scheibe [1]. For a special case of Lemma 10.8 see Iohvidov and Kreĭn [2].

Example 11.3 (with real scalars) is due to G. W. Mackey, and cited by Savage [1]. Lemma 11.4 appears in the paper of Ginzburg and Iohvidov [1].

We mention that Mal'cev [1], Aronszajn [1], Ginzburg and Iohvidov [1] permit the inner product to be non-hermitian. Fischer and Gross [1] consider bilinear forms on vector spaces over fields different from $\boldsymbol{R}$ and $\boldsymbol{C}$ and give references to earlier work.

An extension of the Gram-Schmidt orthogonalization process to indefinite inner product spaces will be given in Section IV.3.

Chapter II. Linear Operators in Inner Product Spaces without Topology

In Sections 2–3 algebraic properties of isometric and symmetric operators of general inner product spaces are examined. Sections 6–7 constitute the common foundation for the study of plus-operators (an extension of the class of isometric operators; Section 8) and Pesonen operators (a subclass of symmetric operators; Section 9). Sections 4–5 on Cayley transformations and Sections 10–11 on fundamental decompositions serve applications in subsequent chapters. Special attention should be payed to Theorems 2.5, 3.3, 6.2, 9.5, 10.1, and 11.7.

1. Linear Operators in Vector Spaces

Before considering linear operators in inner product spaces, we collect some notations and definitions concerning linear operators in vector spaces.

A *linear operator* T from a vector space $\mathfrak{E}_1$ to a vector space $\mathfrak{E}_2$ is a mapping of a subspace $\mathfrak{D}(T) \subset \mathfrak{E}_1$ into $\mathfrak{E}_2$ such that

$$T(\alpha_1 x_1 + \alpha_2 x_2) = \alpha_1 Tx_1 + \alpha_2 Tx_2 \quad (\alpha_1, \alpha_2 \in \boldsymbol{C};\ x_1, x_2 \in \mathfrak{D}(T)).$$

The subspace $\mathfrak{D}(T)$ is called the *domain* of T. We say also that T is *defined on* $\mathfrak{D}(T)$. The set $\mathfrak{R}(T) = T\,\mathfrak{D}(T)$, evidently a subspace of $\mathfrak{E}_2$, is the *range* of T. If $\mathfrak{E}_1 = \mathfrak{E}_2 = \mathfrak{E}$, we say that T is a linear operator *in* $\mathfrak{E}$.

If T is a linear operator, then $\dim \mathfrak{R}(T)$ is called the *rank* of T.

A *linear form* is a linear operator T with $\mathfrak{R}(T) \subset \boldsymbol{C}$.

The *identity operator* I of the vector space $\mathfrak{E}$ is defined by the relation $Ix = x$ $(x \in \mathfrak{E})$. If confusion can arise, the underlying space will be indicated by writing $I_{\mathfrak{E}}$ instead of I.

The *zero operator* 0 of $\mathfrak{E}$ is defined by $0x = 0$ $(x \in \mathfrak{E})$, where the symbol 0 stands for an operator on the left-hand, but for a vector on the right-hand side. (This will be the fourth meaning of the character 0.)

Let T be a linear operator from $\mathfrak{E}_1$ to $\mathfrak{E}_2$. The set

$$\mathfrak{N}(T) = \{x \in \mathfrak{D}(T):\ Tx = 0\}$$

is a subspace of $\mathfrak{E}_1$, termed the *null space* of T.

If $\mathfrak{N}(T) = 0$, then T is said to be *invertible*. In this case one can define a linear operator T^{-1} with $\mathfrak{D}(T^{-1}) = \mathfrak{R}(T)$, setting $T^{-1}y = x$ if and only if $Tx = y$. The operator T^{-1} is called the *inverse* of T.

If T is an invertible operator in $\mathfrak{E}$, whereas $\mathfrak{D}(T) = \mathfrak{R}(T) = \mathfrak{E}$, we shall say that T is *completely invertible*.

Let T be a linear operator in $\mathfrak{E}$. If for a $\lambda \in \boldsymbol{C}$ the operator $T - \lambda I$ is not invertible, we say that λ is an *eigenvalue* of T. The subspace $\mathfrak{N}(T - \lambda I)$ is called the *eigenspace*, its non-zero elements the *eigenvectors* of T belonging to λ.

The vector $x \neq 0$ is a *principal vector* of T belonging to the eigenvalue λ, if there is a positive integer r such that $x \in \mathfrak{D}(T^r)$ and $(T - \lambda I)^r x = 0$.

The span of all principal vectors of T belonging to λ is the *principal subspace* $\mathfrak{S}_\lambda(T)$. Equivalently,

$$\mathfrak{S}_\lambda(T) = \bigcup_{j=0}^{\infty} \mathfrak{N}((T - \lambda I)^j) .$$

In particular, $\mathfrak{N}(T - \lambda I) \subset \mathfrak{S}_\lambda(T)$. If λ is not an eigenvalue of T, we set $\mathfrak{S}_\lambda(T) = 0$.

The dimension of $\mathfrak{S}_\lambda(T)$ will be denoted by $m_\lambda(T)$ and called the *algebraic multiplicity* of λ as an eigenvalue of T.

An eigenvalue λ of T is said to be *semi-simple*, if $\mathfrak{S}_\lambda(T) = \mathfrak{N}(T - \lambda I)$, i.e., if every principal vector of T belonging to λ is an eigenvector.

If $x \in \mathfrak{S}_\lambda(T)$, and $p \geqq 1$ is the least exponent such that $(T - \lambda I)^p x = 0$, then the vectors

$$x_j = (T - \lambda I)^{p-1-j}\, x \qquad (j = 0, 1, \ldots, p-1)$$

are linearly independent and belong to $\mathfrak{S}_\lambda(T)$; in particular, $Tx_0 = \lambda x_0$. We say that $x_0, x_1, \ldots, x_{p-1}$ form a *Jordan chain* of T to the eigenvalue λ. Equivalently, the system $\{x_j\}_0^{p-1} \subset \mathfrak{E}$ is a Jordan chain of the operator T corresponding to the eigenvalue λ, if $x_0 \neq 0$ and

$$Tx_j = \lambda x_j + x_{j-1} \quad (j = 0, 1, \ldots, p-1;\ x_{-1} = 0) .$$

The positive integer p is called the *length* of the Jordan chain.

It is well known that the principal subspaces of a linear operator T are linearly independent. Moreover, if $\mathfrak{D}(T) = \mathfrak{E}$, $\mathfrak{R}(T) \subset \mathfrak{E}$, and $0 < \dim \mathfrak{E} < \infty$, then a) $\mathfrak{E}$ is the direct sum of the principal subspaces of T; b) each principal subspace $\mathfrak{S}_\lambda(T)$ has a basis consisting of Jordan chains of T to the eigenvalue λ; c) the set of the lengths of these Jordan chains is uniquely determined by the operator T.

A subspace $\mathfrak{L} \subset \mathfrak{E}$ is said to be *invariant* for the linear operator T, if $T(\mathfrak{L} \cap \mathfrak{D}(T)) \subset \mathfrak{L}$. We say also that $\mathfrak{L}$ is an invariant subspace of T. Principal subspaces are a particular type of invariant subspaces.

The *restriction* $T|\mathfrak{L}$ of the operator T to the subspace $\mathfrak{L}$ has, by definition, domain $\mathfrak{L} \cap \mathfrak{D}(T)$, its values there coinciding with those of T. An eigenvalue of $T|\mathfrak{L}$ is also called an *eigenvalue of T in $\mathfrak{L}$*.

Two operators T_1, T_2 defined on the vector space $\mathfrak{E}$ and having range in $\mathfrak{E}$ are said to *commute*, if $T_1 T_2 = T_2 T_1$. The members of an operator family are said to commute, if they commute in pairs.

The following two results are well known and will be referred to later.

Lemma 1.1. *If two linear operators commute, then every eigenspace of the one is an invariant subspace of the other.*

Proof. Denoting the two operators by T_1 and T_2 we find that the relation $(T_1 - \lambda I)x = 0$ implies $(T_1 - \lambda I) T_2 x = T_2 (T_1 - \lambda I) x = 0$. □

Lemma 1.2. *Let $\mathfrak{E}$ be a vector space with $0 < \dim \mathfrak{E} < \infty$. If $T_1, \ldots, T_n$ are commuting linear operators on $\mathfrak{E}$, then they have a common eigenvector.*

Proof. Restricting $T_1, \ldots, T_{n-1}$ to an eigenspace of T_n and making use of Lemma 1.1 the problem can be reduced to the case of $n - 1$ operators. Thus the conclusion follows by induction. □

Let T be a linear operator in $\mathfrak{E}$. Suppose that the subspace $\mathfrak{L} \subset \mathfrak{E}$ is invariant for T. On the quotient space $\mathfrak{D}(T)/\mathfrak{L}$ one can define a linear operator $\hat{T}$ in the following way: for $\hat{x} \in \mathfrak{D}(T)/\mathfrak{L}$ let $\hat{T}\hat{x} = (Tx)\hat{}$, where x is a representative of the class $\hat{x}$. We say, $\hat{T}$ is the operator *induced* by T in $\mathfrak{E}/\mathfrak{L}$.

A direct decomposition $\mathfrak{E} = \mathfrak{L}_1 \dotplus \cdots \dotplus \mathfrak{L}_n$ is said to *reduce* the linear operator T, if 1) the components $\mathfrak{L}_1, \ldots, \mathfrak{L}_n$ are invariant subspaces of T, and 2) $\mathfrak{D}(T) = (\mathfrak{D}(T) \cap \mathfrak{L}_1) \dotplus \cdots \dotplus (\mathfrak{D}(T) \cap \mathfrak{L}_n)$. In this case T is called the *direct sum* of the linear operators $T|\mathfrak{L}_1, \ldots, T|\mathfrak{L}_n$.

Let the vector space $\mathfrak{E}$ be the direct sum of n subspaces:

$$\mathfrak{E} = \mathfrak{L}_1 \dotplus \cdots \dotplus \mathfrak{L}_n . \tag{1.1}$$

If T is a linear operator in $\mathfrak{E}$ such that

$$\mathfrak{D}(T) = (\mathfrak{D}(T) \cap \mathfrak{L}_1) \dotplus \cdots \dotplus (\mathfrak{D}(T) \cap \mathfrak{L}_n)$$

then, with respect to (1.1), T can be represented by an *operator matrix* $(T_{jk})_{j,k=1}^{n}$. The entries T_{jk} are linear operators from $\mathfrak{L}_k$ to $\mathfrak{L}_j$, respectively, defined by the relations

$$y_j = \sum_{k=1}^{n} T_{jk} x_k \qquad (j = 1, \ldots, n) , \tag{1.2}$$

where $x = x_1 + \cdots + x_n$ and $Tx = y_1 + \cdots + y_n$ are the decompositions of the vectors $x \in \mathfrak{D}(T)$ and Tx corresponding to (1.1). In the case $x = x_k$ (1.2) reduces to $y_j = T_{jk} x_k$, giving a more explicit definition of T_{jk}.

2. Isometric Operators

Let $\mathfrak{E}$ be an inner product space with inner product $(.\,,.)$. A linear operator U in $\mathfrak{E}$ is said to be *isometric,* if $(Ux, Uy) = (x, y)$ for every pair $x, y \in \mathfrak{D}(U)$.

If $\mathfrak{E}$ is decomposable, and for some fundamental decomposition

$$\mathfrak{E} = \mathfrak{E}^0 (\dot{+}) \mathfrak{E}^+ (\dot{+}) \mathfrak{E}^-; \qquad \mathfrak{E}^0 \subset \mathfrak{P}^0, \quad \mathfrak{E}^+ \subset \mathfrak{P}^{++}, \quad \mathfrak{E}^- \subset \mathfrak{P}^{--}$$

the linear operator U is the direct sum of an operator in $\mathfrak{E}^0$, an isometric operator in $\mathfrak{E}^+$, and an isometric operator in $\mathfrak{E}^-$, then, as it is easy to see, U is isometric. From Example 2.3 below it turns out, however, that even in a decomposable space not every isometric operator can be obtained in this way.

In contrast to the situation in Hilbert space, an isometric operator in an inner product space need not be invertible; a trivial example is the zero operator on a neutral inner product space $\mathfrak{E} \neq 0$. For an invertible isometric operator U the inverse U^{-1} is obviously isometric.

Lemma 2.1. *An isometric operator U transforms isotropic and non-isotropic elements of $\mathfrak{D}(U)$ into isotropic and non-isotropic elements of $\mathfrak{R}(U)$, respectively.*

Proof. Let $x \in \mathfrak{D}(U)$. If $(x, y) = 0$ for every $y \in \mathfrak{D}(U)$, then $(Ux, Uy) = 0$ for every $Uy \in \mathfrak{R}(U)$. If $(x, y) \neq 0$ for some $y \in \mathfrak{D}(U)$, then $(Ux, Uy) \neq 0$ for the corresponding $Uy \in \mathfrak{R}(U)$. □

Corollary 2.2. *If the domain of an isometric operator U is non-degenerate, then U is invertible.* □

The condition appearing in Corollary 2.2 is not necessary for invertibility; e.g. the restriction of an invertible isometric operator U to a degenerate subspace $\mathfrak{L} \subset \mathfrak{D}(U)$ remains to be invertible.

We are going to study the eigenvalues and principal subspaces of isometric operators.

In Hilbert space, as it is well known, an isometric operator has only unimodular, semi-simple eigenvalues. (We say that the number $\alpha \in \boldsymbol{C}$ is *unimodular,* if $|\alpha| = 1$.) In an indefinite inner product space this is not necessarily true, not even in the most "regular" case of an isometric operator that maps a finite-dimensional non-degenerate space onto itself.

Example 2.3. Consider the two-dimensional vector space $\mathfrak{E}=\langle e, f\rangle$. Define an inner product $(\,.\,,\,.\,)$ on $\mathfrak{E}$ by the relations

$$(e, e) = (f, f) = 0\,, \quad (e, f) = i\,,$$

and a linear operator U in $\mathfrak{E}$ by

$$Ue = e\,, \quad Uf = e + f\,.$$

We have $(U-I)f \neq 0$, $(U-I)^2 f = 0$; so the eigenvalue $\nu = 1$ of U is not semi-simple. Nevertheless, U is easily verified to be isometric.

Example 2.4. In the inner product space of Example 2.3 we define a linear operator U by the equations

$$Ue = \frac{1}{2}e\,, \qquad Uf = 2f\,.$$

Then U is an isometric operator with eigenvalues $\nu_1 = 1/2$, $\nu_2 = 2$.

The Hilbert space theorem on the orthogonality of eigenvectors belonging to different eigenvalues μ, ν of an isometric operator remains valid for the principal vectors of an isometric operator in an arbitrary inner product space under the assumption that μ and ν do not lie symmetrically with respect to the unit circle.

Theorem 2.5. *Let U be an isometric operator in the inner product space $\mathfrak{E}$. If μ, ν are eigenvalues of U such that $\mu\bar{\nu} \neq 1$, then the principal subspace $\mathfrak{S}_\mu(U)$ is orthogonal to $\mathfrak{S}_\nu(U)$.*

Proof. Let

$$(U-\mu I)^r\, x = 0\,, \quad (U-\nu I)^s\, y = 0\,.$$

We must show that $(x, y) = 0$. For $r + s = 0$ this is evident. Assuming the validity of the same conclusion for $r + s < n$, we prove it for $r + s = n$.

Introduce the notations

$$(U-\mu I)\, x = x_1\,, \quad (U-\nu I)\, y = y_1\,.$$

As the cases where $r = 0$ or $s = 0$ can be neglected, we have

$$(U-\mu I)^{r-1}\, x_1 = 0\,, \quad (U-\nu I)^{s-1}\, y_1 = 0\,.$$

Hence, by the assumption,

$$(x, y_1) = (x_1, y) = (x_1, y_1) = 0\,.$$

Making use of these relations we obtain

$$(x, y) = (Ux, Uy) = (\mu x + x_1,\ \nu y + y_1) = \mu\bar{\nu}\,(x, y)\,.$$

The theorem is proved. □

Corollary 2.6. *The principal subspace belonging to a non-unimodular eigenvalue of an isometric operator is neutral.* □

For certain classes of isometric operators it is known that the set of eigenvalues is symmetric with respect to the unit circle. Here we consider only the simplest case.

Theorem 2.7. *Let $\mathfrak{E}$ be a finite-dimensional, non-degenerate inner product space. Let U be an isometric operator in $\mathfrak{E}$ with $\mathfrak{D}(U) = \mathfrak{E}$. If ν is an eigenvalue of U, then the number $\nu^* = 1/\bar{\nu}$ is also an eigenvalue of U. Moreover, $\mathfrak{S}_\nu(U) \# \mathfrak{S}_{\nu^*}(U)$.*

Proof. Owing to Corollary 2.2, ν cannot be 0; thus ν^* is defined. Denoting the direct sum of all principal subspaces $\mathfrak{S}_\mu(U)$ with $\mu \neq \nu$, $\mu \neq \nu^*$ by $\mathfrak{L}_\nu(U)$, we have

$$\mathfrak{E} = \big(\mathfrak{S}_\nu(U) + \mathfrak{S}_{\nu^*}(U)\big) \dotplus \mathfrak{L}_\nu(U) .$$

Here, according to Theorem 2.5, the last component is orthogonal to the first and second ones. (Note that the case $|\nu| = 1$, $\mathfrak{S}_\nu(U) = \mathfrak{S}_{\nu^*}(U)$ is not excluded.) Applying Lemma I.4.2 we obtain that the subspace $\mathfrak{S}_\nu(U) + \mathfrak{S}_{\nu^*}(U)$ is non-degenerate. This proves the assertion for $|\nu| = 1$, while for $|\nu| \neq 1$ the desired conclusion follows from Corollary 2.6, Lemma I.10.2, and Lemma I.10.3. □

The next theorem complements the above one by showing that the structure of $U|\mathfrak{S}_\nu(U)$ and $U|\mathfrak{S}_{\nu^*}(U)$ is the same.

Theorem 2.8. *Let $\mathfrak{E}$ be a finite-dimensional, non-degenerate inner product space. Let U be an isometric operator in $\mathfrak{E}$ with $\mathfrak{D}(U) = \mathfrak{E}$, and let ν be an eigenvalue of U. If the basis $\{f_j\}_1^n$ of $\mathfrak{S}_{\nu^*}(U)$ is a dual companion for the basis $\{e_j\}_1^n$ of $\mathfrak{S}_\nu(U)$, then the matrices of $U|\mathfrak{S}_\nu(U)$ and $U|\mathfrak{S}_{\nu^*}(U)$ relative to $\{e_j\}_1^n$ and $\{f_j\}_1^n$, respectively, are the inverses of the transposed conjugates of each other.*

Proof. The existence of the dual pair $\{e_j\}_1^n$, $\{f_j\}_1^n$ follows from Theorem 2.7 and Lemma I.10.6. Let

$$Ue_j = \sum_{r=1}^{n} \varepsilon_{rj}\, e_r\,, \quad Uf_j = \sum_{r=1}^{n} \varphi_{rj} f_r \quad (j = 1, \ldots, n)\,.$$

Making use of the relations

$$(Ue_j\,,\ Uf_k) = (e_j\,, f_k) = \delta_{jk} \quad (j, k = 1, \ldots, n)$$

we obtain

$$\sum_{r=1}^{n} \varepsilon_{rj}\, \bar{\varphi}_{rk} = \delta_{jk} \quad (j, k = 1, \ldots, n)\,. \quad □$$

The following result is a simple extension of a property of isometric operators in Hilbert space.

Lemma 2.9. *Let U be an isometric operator in the inner product space $\mathfrak{E}$. If $\mathfrak{L}$ is a subspace of $\mathfrak{E}$ such that $U(\mathfrak{L} \cap \mathfrak{D}(U)) \supset \mathfrak{L}$, then $\mathfrak{L}^\perp$ is invariant for U.*

Proof. Let $x \in \mathfrak{L}^{\perp} \cap \mathfrak{D}(U)$, $y \in \mathfrak{L}$. The assumption guarantees the existence of a $y_1 \in \mathfrak{L} \cap \mathfrak{D}(U)$ with $Uy_1 = y$. Hence $(Ux, y) = (Ux, Uy_1) = (x, y_1) = 0$. □

Finally, we mention a sufficient condition for an eigenvalue of an isometric operator to be semi-simple.

Lemma 2.10. *Let ν be an eigenvalue of the isometric operator U. If the eigenspace $\mathfrak{N}(U - \nu I)$ is definite, then the eigenvalue ν is unimodular and semi-simple.*

Proof. If ν is non-unimodular then, by Corollary 2.6, $\mathfrak{N}(U - \nu I)$ is neutral. Next take $|\nu| = 1$. Suppose that $(U - \nu I)^r x = 0$, $(U - \nu I)^{r-1} x \neq 0$ for some vector x and integer $r \geqq 2$. Then $x_1 = (U - \nu I)^{r-1} x$ is a non-zero element of $\mathfrak{N}(U - \nu I)$, and we have

$$(x_1, x_1) = \left((U - \nu I)^{r-1} x,\ (U - \nu I)^{r-1} x\right) = = -\left((U - \nu I)^r x,\ \nu U (U - \nu I)^{r-2} x\right) = 0\,. \quad \square$$

3. Symmetric Operators

A linear operator A in the inner product space $\mathfrak{G}$ is said to be *symmetric* if $(Ax, y) = (x, Ay)$ for every pair $x, y \in \mathfrak{D}(A)$.

If

$$\mathfrak{G} = \mathfrak{G}^0 \dotplus \mathfrak{G}^+ \dotplus \mathfrak{G}^-; \qquad \mathfrak{G}^0 \subset \mathfrak{P}^0,\ \mathfrak{G}^+ \subset \mathfrak{P}^{++},\ \mathfrak{G}^- \subset \mathfrak{P}^{--}$$

is a fundamental decomposition of $\mathfrak{G}$, and the linear operator A is the direct sum of an operator in $\mathfrak{G}^0$, a symmetric operator in $\mathfrak{G}^+$, and a symmetric operator in $\mathfrak{G}^-$, then A is a symmetric operator in the space $\mathfrak{G}$. Nevertheless, there are symmetric operators in decomposable spaces which cannot be obtained in this way for any fundamental decomposition.

Example 3.1. Let $\{e, f\}$ be a basis of the inner product space $\mathfrak{G}$, where

$$(e, e) = (f, f) = 0\,, \quad (e, f) = 1\,.$$

The linear operator A defined by the relations

$$Ae = e\,, \quad Af = e + f$$

is symmetric, but no fundamental decomposition of $\mathfrak{G}$ reduces A.

Example 3.1 shows also that the eigenvalues of a symmetric operator in an inner product space need not be semi-simple. The next example exhibits a symmetric operator with non-real eigenvalues.

Example 3.2. We define the space and the inner product as in Example 3.1 and observe that the linear operator specified by the relations

$$Ae = ie\,, \quad Af = -if$$

is also symmetric.

The analogy between isometric and symmetric operators known from Hilbert space theory remains valid in inner product spaces. For the sake of completeness, we are going to formulate the analoga of the assertions 2.5—2.8 of the preceding section. The proofs will be omitted, since the only new point is the replacement of the characteristic property of isometric operators by that of symmetric ones.

Theorem 3.3. *Let A be a symmetric operator in the inner product space $\mathfrak{E}$. If λ, μ are eigenvalues of A such that $\lambda \neq \bar{\mu}$, then $\mathfrak{S}_\lambda(A)$ is orthogonal to $\mathfrak{S}_\mu(A)$.* □

Corollary 3.4. *The principal subspace belonging to a non-real eigenvalue of a symmetric operator is neutral.* □

Theorem 3.5. *Let $\mathfrak{E}$ be a finite-dimensional, non-degenerate inner product space. Let A be a symmetric operator in $\mathfrak{E}$ with $\mathfrak{D}(A)=\mathfrak{E}$. If λ is an eigenvalue of A, then $\bar{\lambda}$ is also an eigenvalue of A. Moreover, $\mathfrak{S}_\lambda(A) \# \mathfrak{S}_{\bar{\lambda}}(A)$.* □

Theorem 3.6. *Let $\mathfrak{E}$ be a finite-dimensional, non-degenerate inner product space. Let A be a symmetric operator in $\mathfrak{E}$ with $\mathfrak{D}(A)=\mathfrak{E}$, and let λ be an eigenvalue of A. If the basis $\{f_j\}_1^n$ of $\mathfrak{S}_{\bar{\lambda}}(A)$ is a dual companion for the basis $\{e_j\}_1^n$ of $\mathfrak{S}_\lambda(A)$, then the matrices of $A|\mathfrak{S}_\lambda(A)$ and $A|\mathfrak{S}_{\bar{\lambda}}(A)$ relative to $\{e_j\}_1^n$ and $\{f_j\}_1^n$, respectively, are the transposed conjugates of each other.* □

The next result corresponds to Lemma 2.9.

Theorem 3.7. *If A is a symmetric operator in the inner product space $\mathfrak{E}$, and $\mathfrak{L}$ is an invariant subspace of A such that $\mathfrak{L} \subset \mathfrak{D}(A)$, then $\mathfrak{L}^\perp$ is also an invariant subspace of A.*

Proof. For $x \in \mathfrak{L}^\perp \cap \mathfrak{D}(A)$, $y \in \mathfrak{L}$ we have $(Ax, y) = (x, Ay) = 0$. □

We also state the analogue of Lemma 2.10; in the proof obvious changes are only needed.

Lemma 3.8. *Let λ be an eigenvalue of the symmetric operator A. If the eigenspace $\mathfrak{N}(A - \lambda I)$ is definite, then the eigenvalue λ is real and semi-simple.* □

Let, again, $\mathfrak{E}$ be an inner product space, and let A be a symmetric operator in $\mathfrak{E}$ with $\mathfrak{D}(A) = \mathfrak{E}$. Set

$$(3.1) \qquad (x, y)_A = (Ax, y) \qquad (x, y \in \mathfrak{E}) .$$

Then $(.\,,.)_A$ is also an inner product on $\mathfrak{E}$. We shall refer to it as the *A-inner product* on $\mathfrak{E}$.

When we consider $\mathfrak{E}$ as an inner product space with inner product $(.\,,.)_A$ rather than $(.\,,.)$, we use the symbol $\mathfrak{E}_A$. In $\mathfrak{E}_A$ one can introduce all notions defined for inner product spaces. These notions when considered in the original space $\mathfrak{E}$ will be distinguished by a prefix A, and the corresponding symbols by a subscript A.

For instance, a vector $x \in \mathfrak{E}$ is said to be *A-positive,* if it is a positive vector of $\mathfrak{E}_A$, and we set

$$\mathfrak{P}_A^{++} = \{x \in \mathfrak{E}:\ (Ax, x) > 0 \quad \text{or} \quad x = 0\} .$$

Similarly we can speak of *A-positive subspaces, A-orthogonality* (to be denoted by $\perp_A$), *A-orthogonal companion* of a set $\mathfrak{A}$ (in symbols: $\mathfrak{A}^{\perp}_A$), *A-isotropic part* of a subspace $\mathfrak{L}$ (in symbols: $\mathfrak{L}^0_A$), *A-orthogonal sums* and *A-orthogonal direct sums* (the respective symbols being $(+)_A$ and $(\dot{+})_A$), *A-fundamental decompositions* (with components $\mathfrak{E}^0_A$, $\mathfrak{E}^+_A$, $\mathfrak{E}^-_A$), *quasi-A-positive* spaces, *A-isometric* and *A-symmetric* operators etc.

The following simple fact is often useful.

Lemma 3.9. *Let A be an everywhere defined symmetric operator in the inner product space $\mathfrak{E}$. If the subspaces $\mathfrak{L}, \mathfrak{M} \subset \mathfrak{E}$ are orthogonal to each other, while one of them is invariant under A, then $\mathfrak{L}$ and $\mathfrak{M}$ are A-orthogonal to each other.*

Proof. The relations $\mathfrak{L} \perp \mathfrak{M}$, $A\mathfrak{L} \subset \mathfrak{L}$ imply $A\mathfrak{L} \perp \mathfrak{M}$ i. e. $\mathfrak{L} \perp_A \mathfrak{M}$. □

An important subclass of symmetric operators is that of orthogonal projectors.

An *orthogonal projector* in an inner product space $\mathfrak{E}$ is a symmetric operator P in $\mathfrak{E}$ such that $\mathfrak{D}(P) = \mathfrak{E}$, $P^2 = P$. This definition is justified by the following theorem.

Theorem 3.10. *If P is an orthogonal projector in the inner product space $\mathfrak{E}$, then $\mathfrak{R}(P)$ is ortho-complemented, and for every $x \in \mathfrak{E}$ the vector Px is a projection of x on $\mathfrak{R}(P)$. If, conversely, $\mathfrak{L}$ is an ortho-complemented subspace of the non-degenerate inner product space $\mathfrak{E}$, and $P_{\mathfrak{L}}$ is the mapping that carries each vector of $\mathfrak{E}$ into its projection on $\mathfrak{L}$, then $P_{\mathfrak{L}}$ is an orthogonal projector with $\mathfrak{R}(P_{\mathfrak{L}}) = \mathfrak{L}$.*

Proof. Let P be an orthogonal projector in $\mathfrak{E}$. Then for every $x, y \in \mathfrak{E}$ we have

$$(x - Px, Py) = (Px - P^2x, y) = 0 .$$

Let, conversely, $\mathfrak{E}$ be non-degenerate, and let $\mathfrak{L}$ be an ortho-complemented subspace of $\mathfrak{E}$. Then $P_{\mathfrak{L}} = P$ is everywhere defined, single-valued (Corollary I.8.2), and satisfies the relation $(x - Px, Py) = 0$ $(x, y \in \mathfrak{E})$. Thus

$$(x, Py) = (Px, Py) \quad (x, y \in \mathfrak{E}) . \tag{3.2}$$

Interchanging x and y and taking complex conjugates we obtain:

$$(Px, y) = (Px, Py) \quad (x, y \in \mathfrak{E}) . \tag{3.3}$$

From (3.2) and (3.3) it follows that P is symmetric. Further, the symmetry of P along with equation (3.3) yield:

$$(P^2x, y) = (Px, Py) = (Px, y) \quad (x, y \in \mathfrak{E}) . \tag{3.4}$$

The space $\mathfrak{E}$ being non-degenerate, (3.4) implies $P^2 = P$. □

To finish with, we mention an interesting property of commuting orthogonal projectors.

Theorem 3.11. *Let P_1, P_2 be commuting orthogonal projectors in the inner product space $\mathfrak{E}$. If $\mathfrak{R}(P_1)$ and $\mathfrak{R}(P_2)$ are positive subspaces, then so is their span $\mathfrak{R}(P_1) + \mathfrak{R}(P_2)$.*

Proof. For any $x_1 \in \mathfrak{E}$ the orthogonal decomposition $P_1x_1 = P_2P_1x_1 + (I - P_2)P_1x_1$ (see Theorem 3.10) yields

$$(P_1x_1, P_1x_1) = (P_2P_1x_1, P_2P_1x_1) + ((I - P_2)P_1x_1, (I - P_2)P_1x_1).$$

Hence, making use of the relations $P_1P_2 = P_2P_1$, $\mathfrak{R}(P_1) \subset \mathfrak{P}^+$, we obtain:

$$(P_1x_1, P_1x_1) \geqq (P_1P_2x_1, P_1P_2x_1) . \tag{3.5}$$

Similarly, for every $x_2 \in \mathfrak{E}$ we have

$$(P_2x_2, P_2x_2) \geqq (P_1P_2x_2, P_1P_2x_2) . \tag{3.6}$$

Furthermore, a simple calculation shows that

$$(P_1x_1, P_2x_2) = (P_1P_2x_1, P_1P_2x_2) , \tag{3.7}$$

$$(P_2x_2, P_1x_1) = (P_1P_2x_2, P_1P_2x_1) . \tag{3.8}$$

Adding up the relations (3.5) — (3.8) we find:

$$(P_1x_1 + P_2x_2, P_1x_1 + P_2x_2) \geqq \geqq (P_1P_2(x_1 + x_2), P_1P_2(x_1 + x_2)) . \tag{3.9}$$

Since $\mathfrak{R}(P_1) \subset \mathfrak{P}^+$, the right-hand side of (3.9) is non-negative. □

4. Cayley Transformations

The analogy between isometric and symmetric operators in an inner product space finds its explanation in some kind of correspondence between these two classes of operators. The correspondence in question is closely related to the Cayley transformation known from Hilbert space theory.

Lemma 4.1. *Let A be a symmetric operator in the inner product space $\mathfrak{E}$. Let ε be a unimodular number, and ζ a non-real number which is not an eigenvalue of A:*

$$|\varepsilon| = 1\,, \quad \bar{\zeta} \neq \zeta\,, \tag{4.1}$$

$$\mathfrak{N}(A - \zeta I) = 0\,. \tag{4.2}$$

Then the operator

$$U = \varepsilon(A - \bar{\zeta} I)(A - \zeta I)^{-1} \tag{4.3}$$

is isometric, has domain

$$\mathfrak{D}(U) = \mathfrak{R}(A - \zeta I)\,, \tag{4.4}$$

and ε is not an eigenvalue of U:

$$\mathfrak{N}(U - \varepsilon I) = 0\,. \tag{4.5}$$

Let, conversely, U be an isometric operator in $\mathfrak{E}$. Suppose that the numbers $\varepsilon, \zeta \in \boldsymbol{C}$ satisfy the conditions (4.1), (4.5). *Then the operator*

$$A = (\zeta U - \varepsilon\bar{\zeta} I)(U - \varepsilon I)^{-1} \tag{4.6}$$

is symmetric, satisfies (4.2), *and its domain is*

$$\mathfrak{D}(A) = \mathfrak{R}(U - \varepsilon I)\,. \tag{4.7}$$

The transformations (4.3) *and* (4.6) *are the inverses of each other.*

Proof. Consider a symmetric operator A and two numbers ε, ζ satisfying the conditions (4.1), (4.2). We observe that the definition (4.3) makes sense and can be written in the equivalent form

$$x = Af - \zeta f\,, \qquad Ux = \varepsilon Af - \varepsilon\bar{\zeta} f\,, \tag{4.8}$$

where f is any element of $\mathfrak{D}(A)$, while x is an arbitrary element of $\mathfrak{D}(U)$. Setting

$$y = Ag - \zeta g\,, \qquad Uy = \varepsilon Ag - \varepsilon\bar{\zeta} g$$

we obtain:

$$(Ux, Uy) - (x, y) = (\bar{\zeta} - \zeta)\big((Af, g) - (f, Ag)\big)\,. \tag{4.9}$$

Therefore U is isometric. The relations (4.4), (4.5) also follow directly from (4.8), taking into account (4.1).

In order to find the inverse transformation we solve the equations (4.8) for f and Af:

$$(4.10)\quad f = \frac{1}{\varepsilon(\zeta - \bar{\zeta})}(Ux - \varepsilon x)\,, \qquad Af = \frac{1}{\varepsilon(\zeta - \bar{\zeta})}(\zeta U x - \varepsilon\bar{\zeta}x)\,.$$

Thus (4.6) is the expression for A in terms of U.

Moreover, if U is any isometric operator in $\mathfrak{E}$ with the property (4.5), then (4.6) or (4.10) define a linear operator A in $\mathfrak{E}$ and, (4.8) and (4.10) being equivalent, U can be expressed in terms of A by the formula (4.3). Consequently, (4.9) even now holds for any $x, y \in \mathfrak{D}(U)$ and $f, g \in \mathfrak{D}(A)$. Therefore the isometry of U implies the symmetry of A. Relations (4.2), (4.7) follow easily from (4.10). □

Remark 4.2. If A is a non-symmetric linear operator, then U will be a non-isometric linear operator, and conversely. The other statements of Lemma 4.1 do not involve the inner product and remain valid for linear operators A and U in an arbitrary vector space.

If two linear operators A and U in a vector space are related by the formulas (4.3) and (4.6) (the conditions (4.1), (4.2), (4.5) being satisfied), we say that A and U are *Cayley transforms* of each other. The mutually inverse transformations (4.3), (4.6) are called *Cayley transformations*.

Let $\mathfrak{E}$ be an inner product space, and let ε, ζ be fixed numbers satisfying the conditions (4.1). Consider the class of all symmetric operators A in $\mathfrak{E}$ such that ζ is not an eigenvalue of A, and the class of all isometric operators U in $\mathfrak{E}$ such that ε is not an eigenvalue of U. Lemma 4.1 states that the Cayley transformations (4.3), (4.6) induce a one-to-one correspondence between these classes. However, a symmetric operator A may have several Cayley transforms belonging to different values of ε and ζ. It may also happen that A has no Cayley transforms at all, since every non-real number is an eigenvalue of A. Similarly, an isometric operator U admits an infinite number of Cayley transforms provided there is a unimodular ε which is not an eigenvalue of U; otherwise it admits none.

In certain situations the existence of a Cayley transform can be guaranteed. If this is not the case or is difficult to establish, independent proofs for the isometric and symmetric versions of a theorem are justified.

5. Principal Vectors of Cayley Transforms

In this section we examine the behaviour of eigenvalues and principal subspaces under a Cayley transformation.

Lemma 5.1. *Let A be a linear operator in a vector space. Let λ be an eigenvalue of A, and f a principal vector of A belonging to λ:*

$$(5.1)\qquad f \in \mathfrak{D}(A^r), \quad (A - \lambda I)^r f = 0, \quad (A - \lambda I)^{r-1} f \neq 0 .$$

Let U be a Cayley transform of A, defined by (4.3) with parameters ε, ζ satisfying the conditions (4.1) — (4.2). Set

$$(5.2)\qquad \nu = \varepsilon \frac{\lambda - \bar{\zeta}}{\lambda - \zeta},$$

$$(5.3)\qquad x = \alpha^r (A - \zeta I)^r f,$$

where

$$(5.4)\qquad \alpha = \frac{\lambda - \zeta}{\varepsilon (\bar{\zeta} - \zeta)} .$$

Then ν is an eigenvalue of U, and x is a principal vector of U belonging to ν such that

$$(5.5)\qquad x \in \mathfrak{D}(U^r), \quad (U - \nu I)^r x = 0, \quad (U - \nu I)^{r-1} x \neq 0 .$$

Moreover, we have

$$(5.6)\qquad \langle x, (U - \nu I)x, \ldots, (U - \nu I)^{r-1} x\rangle = \\ = \langle f, (A - \lambda I) f, \ldots, (A - \lambda I)^{r-1} f\rangle .$$

Proof. First observe that the definition of α is correct because of the assumptions (4.1). The definition of x is justified by the inclusion $f \in \mathfrak{D}(A^r)$ (see (5.1)). Also ν is properly defined, since λ is, while ζ is not an eigenvalue of A (cf. (5.1) and (4.2)). For the same reason $\alpha \neq 0$.

From equations (4.3), (5.2), (5.3), (5.4) for $s = 0, 1, 2, \ldots, r$ we obtain:

$$(U - \nu I)^s x = \\ = \alpha^r \left[\varepsilon (A - \bar{\zeta} I)(A - \zeta I)^{-1} - \varepsilon (\lambda - \bar{\zeta})(\lambda - \zeta)^{-1} I\right]^s (A - \zeta I)^r f = \\ = \alpha^r \varepsilon^s (\lambda - \zeta)^{-s} \left[(\lambda - \zeta)(A - \bar{\zeta} I) - (\lambda - \bar{\zeta})(A - \zeta I)\right]^s (A - \zeta I)^{r-s} f = \\ = \alpha^r \varepsilon^s (\lambda - \zeta)^{-s} (\bar{\zeta} - \zeta)^s (A - \lambda I)^s (A - \zeta I)^{r-s} f = \\ = \alpha^{r-s} (A - \zeta I)^{r-s} (A - \lambda I)^s f .$$

Consequently, (5.1) implies (5.5). One also sees that $(U - \nu I)^s x$ is a linear combination of the vectors $f, Af, A^2 f, \ldots, A^{r-1} f$ $(s = 0, 1, \ldots, r - 1)$. Therefore the left-hand member of the relation (5.6) is contained in the right-hand member. But, owing to (5.1) and (5.5), both members have dimension r. Thus (5.6) holds. □

The counterpart of Lemma 5.1 reads as follows.

Lemma 5.2. *Let U be a linear operator in a vector space. Let ν be an eigenvalue of U, and x a principal vector of U belonging to ν such that the relations* (5.5) *are valid. Let A be a Cayley transform of U, defined by* (4.6) *with parameters ε, ζ satisfying the conditions* (4.1), (4.5). *Set*

$$\lambda = \frac{\zeta\nu - \bar{\zeta}\varepsilon}{\nu - \varepsilon}, \tag{5.7}$$

$$f = \beta^r (U - \varepsilon I)^r x, \tag{5.8}$$

where

$$\beta = \frac{\nu - \varepsilon}{\varepsilon(\bar{\zeta} - \zeta)}. \tag{5.9}$$

Then λ is an eigenvalue of A, and f is a principal vector to λ satisfying (5.1). *Moreover, relation* (5.6) *holds.*

The *proof* proceeds along the same lines as that of Lemma 5.1. □

Remark 5.3. The transformations (5.2) and (5.7) are the numerical analogues of the Cayley transformations (4.3) and (4.6), respectively. Therefore, according to Lemma 4.1 and Remark 4.2, they are inverse to each other. Making use of, say, (5.2) and (4.3), one verifies that (5.3) and (5.8) are mutually inverse transformations too.

As a consequence of Lemmas 5.1—5.2, the finite-dimensional invariant subspaces of a linear operator and its Cayley transforms coincide. More precisely, we have:

Lemma 5.4. *Assume that the linear operators A and U in the vector space $\mathfrak{E}$ are related by the Cayley transformations* (4.3), (4.6), *where $|\varepsilon| = 1$, $\bar{\zeta} \neq \zeta$. Let $\mathfrak{L} \subset \mathfrak{E}$ be a subspace of finite dimension. Then the relations $\mathfrak{L} \subset \mathfrak{D}(A)$, $A\mathfrak{L} \subset \mathfrak{L}$ imply the relations $\mathfrak{L} \subset \mathfrak{D}(U)$, $U\mathfrak{L} \subset \mathfrak{L}$, and vice versa.*

Proof. Let $\mathfrak{L} \subset \mathfrak{D}(A)$, $A\mathfrak{L} \subset \mathfrak{L}$. Then $\mathfrak{L}$ can be decomposed into a direct sum

$$\mathfrak{L} = \mathfrak{L}_1 \dotplus \cdots \dotplus \mathfrak{L}_n,$$

where

$$\mathfrak{L}_j = \langle f_j, (A - \lambda_j I) f_j, \ldots, (A - \lambda_j I)^{r_j - 1} f_j \rangle \quad (j = 1, \ldots, n)$$

with some $f_j \in \mathfrak{E}$, $\lambda_j \in \boldsymbol{C}$ and positive integer r_j satisfying the relations

$$(A - \lambda_j I)^{r_j} f_j = 0, \quad (A - \lambda_j I)^{r_j - 1} f_j \neq 0.$$

By Lemma 5.1 we have $\mathfrak{L}_j \subset \mathfrak{D}(U)$, $U\mathfrak{L}_j \subset \mathfrak{L}_j$.

The second half of the lemma can be proved in a similar way, making use of Lemma 5.2. □

Another consequence of Lemmas 5.1—5.2 is that the principal subspaces of an operator coincide with those of its Cayley transforms.

Lemma 5.5. *Let A and U be linear operators in a vector space, related to each other by the Cayley transformations* (4.3) *and* (4.6), *where* $|\varepsilon| = 1$, $\bar{\zeta} \neq \zeta$. *Let λ and ν be eigenvalues of A and U, respectively, related by the formulas* (5.2) *and* (5.7). *Then* $\mathfrak{S}_\lambda(A) = \mathfrak{S}_\nu(U)$.

Proof. Consider a non-zero vector $f \in \mathfrak{S}_\lambda(A)$. For a suitable positive integer r it satisfies the conditions (5.1). With the notations and by the conclusion of Lemma 5.1 we have:

$$f \in \langle x, (U - \nu I)x, \ldots, (U - \nu I)^{r-1} x \rangle \subset \mathfrak{S}_\nu(U).$$

Thus $\mathfrak{S}_\lambda(A) \subset \mathfrak{S}_\nu(U)$. The complementary inclusion can be obtained from Lemma 5.2 in a similar way. □

6. Pairs of Inner Products: Semi-boundedness

In this section and the next one we consider the mutual behaviour of two inner products $(.\,,.)$ and $(.\,,.)_1$ defined on the same vector space $\mathfrak{E}$, when $(.\,,.)_1$ satisfies a sign condition on the neutral set of $(.\,,.)$. The results have applications in the special case $(x, y)_1 = (Tx, Ty)$ $(x, y \in \mathfrak{E})$ as well as in the case $(x,y)_1 = (Ax, y)$ $(x,y \in \mathfrak{E})$, the operator T (resp. A) being linear (resp. symmetric) and defined everywhere in $\mathfrak{E}$.

Throughout the present section we assume that $(.\,,.)$ is an indefinite inner product, while $(.\,,.)_1$ is an arbitrary inner product on $\mathfrak{E}$.

Lemma 6.1. *If* $(x, x) = 0$ *implies* $(x, x)_1 \geqq 0$, *then*

$$\frac{(y, y)_1}{(y, y)} \leqq \frac{(z, z)_1}{(z, z)}$$

for every pair $y, z \in \mathfrak{E}$ *with* $(y, y) < 0$, $(z, z) > 0$.

Proof. If the assertion is false, there exist two elements $y, z \in \mathfrak{E}$ such that

$$(y, y) = -1, \quad (z, z) = 1, \quad -(y, y)_1 > (z, z)_1.$$

Setting $x = \varepsilon y + z$, where $|\varepsilon| = 1$, we obtain:

$$(x, x) = 2 \operatorname{Re} \{\varepsilon (y, z)\},$$

$$(x, x)_1 < 2 \operatorname{Re} \{\varepsilon (y, z)_1\}.$$

Thus for a suitable ε we have $(x,x) = 0$, $(x,x)_1 < 0$ in contradiction with the hypothesis. □

Theorem 6.2. *If* $(x, x) = 0$ *implies* $(x, x)_1 \geqq 0$, *then*

$$(x, x)_1 \geqq \mu_1(x, x) \quad (x \in \mathfrak{G}) , \tag{6.1}$$

where

$$\mu_1 = \inf_{(z, z) = 1} (z, z)_1 > -\infty . \tag{6.2}$$

Proof. Lemma 6.1 and the indefiniteness of $(.,.)$ (required throughout the section) yield $-\infty < \mu_1 < \infty$. As to the relation (6.1), for vectors x with $(x, x) > 0$ it follows from the definition (6.2), for $(x, x) = 0$ it coincides with the assumption, while for $(x, x) < 0$ it is a consequence of Lemma 6.1. □

Theorem 6.2 can be given the following loose formulation. If the inner product $(.,.)_1$ is semi-bounded on the neutral set of the indefinite inner product $(.,.)$, then it is semi-bounded with respect to $(.,.)$ on the whole space.

Theorem 6.3. *If* $(x, x) = 0$ *implies* $(x, x)_1 = 0$, *then for some real number* μ_1 *we have*

$$(x, y)_1 = \mu_1(x, y) \quad (x,y \in \mathfrak{G}) .$$

Proof. According to Theorem 6.2,

$$(x, x)_1 \geqq \mu_1(x, x) , \qquad (x, x)_1' = -(x, x)_1 \geqq \mu_1'(x, x)$$

for suitable real numbers μ_1, μ_1' and every vector x. By addition we obtain $(\mu_1 + \mu_1')(x,x) \leqq 0$ for every x. Since $(.,.)$ is indefinite, $\mu_1' = -\mu_1$. Thus $(x,x)_1 = \mu_1(x,x)$ identically in x. It remains to apply the polarization formula (I.2.3). □

The next result is a modification of Lemma 6.1, and will be needed later.

Lemma 6.4. *If the relations* $(x, x) = 0, x \neq 0$ *imply* $(x, x)_1 > 0$, *then*

$$\frac{(y, y)_1}{(y, y)} < \frac{(z, z)_1}{(z, z)}$$

for every pair $y,z \in \mathfrak{G}$ *with* $(y, y) < 0$, $(z, z) > 0$.

The *proof* can be modelled on that of Lemma 6.1, keeping in mind that a positive and a negative vector are always linearly independent. □

7. Pairs of Inner Products: Sign

The results contained in this section are more or less evident from the intuitive picture of the neutral set of an inner product as a surface separating the positive and negative elements of the space. The symbols $(.\,,.)$ and $(.\,,.)_1$ will denote two arbitrary inner products on the vector space $\mathfrak{E}$.

Lemma 7.1. *Set, as usually,*

$$\mathfrak{P}^{00} = \{x \in \mathfrak{E}\colon\ (x, x) = 0\,,\ x \neq 0\}\,, \tag{7.1}$$

and assume that $x \in \mathfrak{P}^{00}$ *implies* $(x, x)_1 \neq 0$. *Then either* $(x, x)_1 > 0$ *for every* $x \in \mathfrak{P}^{00}$, *or* $(x, x)_1 < 0$ *for every* $x \in \mathfrak{P}^{00}$.

Proof. Let y, z be two vectors satisfying the conditions

$$(y, y) = (z, z) = 0\,, \tag{7.2}$$

$$(y, y)_1 = -1\,, \qquad (z, z)_1 = 1\,. \tag{7.3}$$

For $x = \varepsilon y + \alpha z$, $|\varepsilon| = 1$, $\alpha \in \boldsymbol{R}$, we find

$$(x, x) = 2\alpha \operatorname{Re}\{\varepsilon(y, z)\}\,,$$

$$(x, x)_1 = -1 + 2\alpha \operatorname{Re}\{\varepsilon(y, z)_1\} + \alpha^2\,.$$

Thus for a suitable ε and every α we have $(x, x) = 0$, and by a subsequent choice of α also $(x, x)_1 = 0$ can be achieved. Since (7.3) guarantees that $x \neq 0$, our result contradicts the assumption. Therefore equations (7.2)—(7.3) cannot hold simultaneously. □

Lemma 7.2. *Suppose that* $x \in \mathfrak{P}^{00}$ *implies* $(x, x)_1 > 0$. *Then either* $(y, y) < 0$ *implies* $(y, y)_1 > 0$, *or* $(z, z) > 0$ *implies* $(z, z)_1 > 0$.

Proof. If the conclusion is false, there exist two vectors y, z such that

$$(y, y) = -1\,, \qquad (y, y)_1 \leqq 0\,,$$

$$(z, z) = 1\,, \qquad (z, z)_1 \leqq 0\,.$$

Setting $x = \varepsilon y + z$, where $|\varepsilon| = 1$, we obtain:

$$x \neq 0\,, \quad (x, x) = 2 \operatorname{Re}\{\varepsilon(y, z)\}\,, \quad (x, x)_1 \leqq 2 \operatorname{Re}\{\varepsilon(y, z)_1\}\,.$$

Thus for a suitable ε we have $x \neq 0$, $(x, x) = 0$, $(x, x)_1 \leqq 0$, contrary to the assumption. □

Lemma 7.3. *Suppose that the relation* $(x, x) = 0$ *implies* $(x, x)_1 \geqq 0$. *Then either* $(y,y) < 0$ *implies* $(y,y)_1 \geqq 0$, *or* $(z,z) > 0$ *implies* $(z,z)_1 \geqq 0$.

The *proof* is a slight modification of the preceding one. Another possibility is to notice that the assertion follows immediately from Theorem 6.2 in the case of an indefinite $(.\,,.)$, while for semi-definite $(.\,,.)$ it is trivial. □

We finally mention a result where the conclusion rather than the assumption is concerned with the neutral set of $(.\,,.)$.

Lemma 7.4. *If the inner product* $(.\,,.)$ *is not negative semi-definite and the relation* $(z, z) > 0$ *implies* $(z, z)_1 \geqq 0$, *then* $(x, x) = 0$ *implies* $(x, x)_1 \geqq 0$. *Similarly, if* $(.\,,.)$ *is not positive semi-definite and the relation* $(y,y) < 0$ *implies* $(y,y)_1 \geqq 0$, *then* $(x,x) = 0$ *implies* $(x, x)_1 \geqq 0$.

Proof. Suppose that some vector x satisfies the equations

$$(x, x) = 0\,, \qquad (x, x)_1 = -1\,. \tag{7.4}$$

Then choosing an element z with

$$(z, z) = 1\,, \qquad \mathrm{Re}\,(x, z) \geqq 0$$

and setting $z_\alpha = x + \alpha z \quad (\alpha \in \boldsymbol{R})$, we find:

$$(z_\alpha, z_\alpha) = 2\alpha\,\mathrm{Re}\,(x, z) + \alpha^2\,,$$

$$(z_\alpha, z_\alpha)_1 = -1 + 2\alpha\,\mathrm{Re}\,(x, z)_1 + \alpha^2 (z, z)_1\,.$$

Hence for sufficiently small positive values of α we obtain $(z_\alpha, z_\alpha) > 0$, $(z_\alpha, z_\alpha)_1 < 0$.

From assumption (7.4) one also derives the existence of a vector y_α with the properties $(y_\alpha, y_\alpha) < 0$, $(y_\alpha, y_\alpha)_1 < 0$. □

Lemmas 7.1—7.4 can be formulated in terms of inclusion relations between the sets $\mathfrak{P}^0$, $\mathfrak{P}^+$ etc. on the one hand, and $\mathfrak{P}_1^0$, $\mathfrak{P}_1^+$ etc. on the other, the latter sets being defined by the aid of the inner product $(.\,,.)_1$.

8. Plus-operators

In this section we assume that the inner product $(.\,,.)$ is indefinite on the space $\mathfrak{E}$.

A linear operator T in the indefinite inner product space $\mathfrak{E}$ is said to be a *plus-operator*, if it is defined everywhere in $\mathfrak{E}$ and carries non-negative vectors into non-negative ones: $T\mathfrak{P}^+ \subset \mathfrak{P}^+$.

By Lemma 7.4 applied to the inner products (x, y) and $(x, y)_1 = (Tx, Ty)$ $(x, y \in \mathfrak{E})$ it is sufficient to require that positive vectors would be transformed into non-negative ones: $T\mathfrak{P}^{++} \subset \mathfrak{P}^+$.

Isometric operators U with $\mathfrak{D}(U) = \mathfrak{E}$ are examples of plus-operators.

The most important property of plus-operators reads as follows.

Theorem 8.1. *Let T be a plus-operator in $\mathfrak{E}$. Then*

(8.1) $(Tx, Tx) \geqq \mu(T)\,(x, x)$ $(x \in \mathfrak{E})$,

where

(8.2) $\mu(T) = \inf_{(x,x)=1} (Tx, Tx)$.

Proof. In Theorem 6.2 we put $(x, y)_1 = (Tx, Ty)$ $(x, y \in \mathfrak{E})$.   □

For every plus-operator T the number $\mu(T)$ defined by (8.2) is non-negative. If $\mu(T) > 0$, we say T is a *strict plus-operator*. In the opposite case (i.e., when $\mu(T) = 0$) T is said to be a *non-strict plus-operator*.

Corollary 8.2. *A strict plus-operator carries positive vectors into positive vectors.*   □

Corollary 8.3. *The range of a non-strict plus-operator is a positive subspace.*   □

The properties of strict and non-strict plus-operators contained in Corollaries 8.2 and 8.3 are not characteristic.

Example 8.4. Let $\mathfrak{E}$ be the vector space of complex sequences $x = \{\xi_j\}_{j=1}^{\infty}$ with $\sum_{j=1}^{\infty} |\xi_j|^2 < \infty$. For $y = \{\eta_j\} \in \mathfrak{E}$ we define

$$(x, y) = -\,\xi_1\,\bar{\eta}_1 + \sum_{j=2}^{\infty}{}' \xi_j\,\bar{\eta}_j \,.$$

The linear operator T that maps $\{\xi_j\}_{j=1}^{\infty}$ into $\{\alpha_j\,\xi_{j-1}\}_{j=1}^{\infty}$, where $\alpha_1 = 0$ and $\alpha_j = \dfrac{1}{j-1}$ $(j > 1)$, is an invertible non-strict plus-operator transforming positive vectors into positive ones.

Example 8.5. With the notations of Example 8.4, let T_1 be the operator that carries $\{\xi_j\}_{j=1}^{\infty} \in \mathfrak{E}$ into $\{\beta_j\,\xi_j\}_{j=1}^{\infty}$, where $\beta_1 = 0$ and $\beta_j = 1$ $(j > 1)$. Then T_1 is a strict plus-operator with $\mathfrak{R}(T_1) \subset \mathfrak{P}^+$.

In order to make the survey of different possibilities complete we mention that in an indefinite inner product space the zero operator is a non-strict plus-operator which does not map every positive vector into a positive vector, and the identity operator is a strict plus-operator which does not map every vector into a non-negative vector.

By Corollary 8.3 a sufficient condition for a plus-operator T to be strict is $\mathfrak{R}(T) = \mathfrak{E}$. Hence, making use of Lemma I.2.6, we obtain:

Theorem 8.6. *If T is a plus-operator and $\mathfrak{R}(T) \supset \mathfrak{P}^{++}$, then T is a strict plus-operator.* □

Our next aim is to characterize a subclass of strict plus-operators (see Theorem 8.9 below).

Lemma 8.7. *If T is a plus-operator in $\mathfrak{E}$ as well as in the anti-space of $\mathfrak{E}$ (i.e. $T\mathfrak{P}^+ \subset \mathfrak{P}^+$, $T\mathfrak{P}^- \subset \mathfrak{P}^-$), then with the notation* (8.2) *we have:*

$$(8.3) \qquad (Tx, Ty) = \mu(T)\,(x, y) \qquad (x, y \in \mathfrak{E}) .$$

Proof. By assumption, $(x, x) = 0$ implies $(Tx, Tx) = 0$. Therefore we may apply Theorem 6.3 to the inner products (x, y) and $(x, y)_1 = (Tx, Ty)$ $(x, y \in \mathfrak{E})$. □

Corollary 8.8. *If T is a plus-operator in $\mathfrak{E}$ as well as in the anti-space of $\mathfrak{E}$, then either T is a positive multiple of an isometric operator, or $\mathfrak{R}(T) \subset \mathfrak{P}^0$.* □

Theorem 8.9. *Let T be a linear operator defined on the indefinite inner product space $\mathfrak{E}$. Then the following statements* a)—c) *are equivalent.*

a) *T maps $\mathfrak{P}^+$ onto itself in a one-to-one manner.*

b) *T maps $\mathfrak{P}^{++}$ onto itself in a one-to-one manner.*

c) *$T = \alpha U$, where U is a completely invertible isometric operator and α is a positive number.*

Proof. First suppose that T satisfies a).

Let $Tx = 0$ for some $x \in \mathfrak{E}$. On account of Lemma I.2.6 x is the sum of two positive vectors x_1, x_2. Then $Tx_1 = T(-x_2)$; hence a) yields $x_1 = -x_2$, i.e. $x = 0$. Thus T is invertible. Furthermore, a) implies the relation $\mathfrak{R}(T) \supset \mathfrak{P}^{++}$. Owing to Lemma I.2.6, $\mathfrak{R}(T) = \mathfrak{E}$. In other words, T is completely invertible.

Let $(x, x) < 0$ for some $x \in \mathfrak{E}$. Then $(Tx, Tx) < 0$. Really, according to a) a non-negative Tx must be the image of a non-negative x, and by the invertibility just proved it cannot be the image of any other element of $\mathfrak{E}$. Applying Lemma 7.4 to the inner products (x, y) and $(x, y)_1 = -(Tx, Ty)$ $(x, y \in \mathfrak{E})$ we obtain that $(x, x) \leqq 0$ implies $(Tx, Tx) \leqq 0$.

To sum up, T is completely invertible and satisfies the relations $T\mathfrak{P}^+ \subset \mathfrak{P}^+$, $T\mathfrak{P}^- \subset \mathfrak{P}^-$. Therefore, by Corollary 8.8, T satisfies c).

That b) implies c) can be established quite similarly. The only difference is the application of Lemma 7.4 for verifying the inclusion $T\mathfrak{P}^+ \subset \mathfrak{P}^+$ rather than $T\mathfrak{P}^- \subset \mathfrak{P}^-$.

That c) implies a) and b) is obvious. □

The statements as well as the proofs of Lemma 8.7, Corollary 8.8 and Theorem 8.9 easily generalize to the case where $\mathfrak{R}(T)$ is not contained in $\mathfrak{E}$. In particular, the following fact will be needed later.

Theorem 8.10. *Let $\mathfrak{E}_j$ $(j = 1, 2)$ be indefinite inner product spaces with respective inner products $(.\,,.)_j$. Put*

$$\mathfrak{P}_j^+ = \{x \in \mathfrak{E}_j\colon\ (x, x)_j \geqq 0\}\,, \quad \mathfrak{P}_j^{++} = \{x \in \mathfrak{E}_j\colon\ (x, x)_j > 0 \ \text{ or } \ x = 0\}\,.$$

Let T be a linear operator from $\mathfrak{D}(T) = \mathfrak{E}_1$ to $\mathfrak{E}_2$. Then the assertions a)—c) *below are equivalent.*

a) *T maps $\mathfrak{P}_1^+$ onto $\mathfrak{P}_2^+$ in a one-to-one manner.*

b) *T maps $\mathfrak{P}_1^{++}$ onto $\mathfrak{P}_2^{++}$ in a one-to-one manner.*

c) *$T = \alpha U$, where U is an invertible operator with $\mathfrak{R}(U) = \mathfrak{E}_2$ such that $(Ux, Uy)_2 = (x, y)_1$ $(x, y \in \mathfrak{E}_1)$, and α is a positive number.* □

9. Pesonen Operators

We are going to study a subclass of symmetric operators in $\mathfrak{E}$ whose members behave very much like symmetric operators in Hilbert space.

Let the inner product space $\mathfrak{E}$ be indefinite. Let A be a symmetric operator in $\mathfrak{E}$ with $\mathfrak{D}(A) = \mathfrak{E}$. If the relations $(x, x) = 0$, $(Ax, x) = 0$ do not hold simultaneously for any $x \neq 0$, we say A is a *Pesonen operator*.

Applying Lemmas 7.1—7.2 to the inner products $(.\,,.)$ and $(.\,,.)_1$, where

$$(9.1) \qquad (x, y)_1 = (x, y)_A = (Ax, y) \qquad (x, y \in \mathfrak{E})\,,$$

we obtain the following two results.

Theorem 9.1. *If A is a Pesonen operator in $\mathfrak{E}$, then either $(Ax, x) > 0$ for every $x \in \mathfrak{P}^{00}$, or $(Ax, x) < 0$ for every $x \in \mathfrak{P}^{00}$.* □

Theorem 9.2. *If A is a Pesonen operator such that $x \in \mathfrak{P}^{00}$ implies $(Ax, x) > 0$, then either $(y, y) < 0$ implies $(Ay, y) > 0$, or $(z, z) > 0$ implies $(Az, z) > 0$.* □

Remark 9.3. It is not an essential restriction on the Pesonen operator A if we assume that $(Ax, x) > 0$ whenever $(x, x) \geqq 0$, $x \neq 0$. In fact, according to Theorems 9.1—9.2 this can always be achieved replacing, if necessary, the space $\mathfrak{E}$ by its anti-space $\mathfrak{E}'$ and the operator A by $-A$.

The next result is a consequence of Theorem 6.2.

Theorem 9.4. *If A is a Pesonen operator such that $(Ax, x) > 0$ for every $x \in \mathfrak{P}^{00}$, then*

$$(Ax, x) \geqq \mu_A(x, x) \qquad (x \in \mathfrak{E}),$$

where

$$\mu_A = \inf_{(x,x)=1} (Ax, x) > -\infty. \ \square$$

The analogy of Pesonen operators to symmetric operators in Hilbert space originates from the following fact.

Theorem 9.5. *Every invariant subspace of a Pesonen operator is non-degenerate.*

Proof. Let $\mathfrak{L}$ be an invariant subspace of the Pesonen operator A. Then $x \in \mathfrak{L}^0$ implies both $(x, x) = 0$ and $(Ax, x) = 0$; hence x must be zero. □

With the help of Lemma I.4.4 we obtain:

Corollary 9.6. *Every invariant semi-definite subspace of a Pesonen operator is definite.* □

Corollary 9.7. *A Pesonen operator has no neutral eigenvector.* □

Corollaries 9.6—9.7 and Lemma I.2.1 yield:

Corollary 9.8. *Every eigenspace of a Pesonen operator is definite.* □

Thus, on account of Lemma 3.8, we have:

Corollary 9.9. *Every eigenvalue of a Pesonen operator is real and semi-simple.* □

From Corollary 9.7, Corollary I.9.8 and Theorem 3.7 it follows that in a space of finite dimension every Pesonen operator is "diagonalizable":

Theorem 9.10. *If A is a Pesonen operator in the finite-dimensional inner product space $\mathfrak{E}$, then $\mathfrak{E}$ has a basis consisting of pairwise orthogonal eigenvectors of A.* □

10. Fundamental Projectors

Making use of certain linear operators arising from a fundamental decomposition, we pursue the study of decomposable spaces begun in Section I.11. We restrict our attention to non-degenerate spaces, i.e. (see Corollary I.11.2) to fundamental decompositions with $\mathfrak{E}^0 = 0$.

Let

(10.1) $$\mathfrak{E} = \mathfrak{E}^+ (\dotplus) \mathfrak{E}^-; \quad \mathfrak{E}^+ \subset \mathfrak{P}^{++}, \quad \mathfrak{E}^- \subset \mathfrak{P}^{--}$$

be a fundamental decomposition of the inner product space $\mathfrak{E}$. We define two orthogonal projectors P^+, P^- in $\mathfrak{E}$ by the relations

$$(10.2) \qquad P^+\mathfrak{E} = \mathfrak{E}^+, \quad P^-\mathfrak{E} = \mathfrak{E}^-$$

(cf. Theorem 3.10). In other words, we set

$$(10.3) \qquad P^+x = x^+, \quad P^-x = x^- \qquad (x \in \mathfrak{E}),$$

where

$$(10.4) \qquad x = x^+ + x^-; \qquad x^+ \in \mathfrak{E}^+, \quad x^- \in \mathfrak{E}^-$$

is the decomposition of the vector x corresponding to (10.1). The operators P^+, P^- will be called the *fundamental projectors* belonging to the fundamental decomposition (10.1).

Theorem 10.1. *If $\mathfrak{L}$ is a positive subspace of the decomposable, non-degenerate inner product space $\mathfrak{E}$, and P^+, P^- are the fundamental projectors belonging to a fundamental decomposition* (10.1) *of $\mathfrak{E}$, then the restriction $P^+|\mathfrak{L}$ is invertible. Similarly, if $\mathfrak{L}$ is negative, then $P^-|\mathfrak{L}$ is invertible.*

Proof. Let $\mathfrak{L}$ be positive. Suppose that for some $x \in \mathfrak{L}$ we have $P^+x = 0$. Then $x = P^-x$. Hence $x \in \mathfrak{L} \cap \mathfrak{E}^- \subset \mathfrak{P}^+ \cap \mathfrak{P}^{--} = 0$.

In the case of a negative $\mathfrak{L}$ a similar reasoning applies to $P^-|\mathfrak{L}$. □

The rest of the present section is devoted to quasi-positive and quasi-negative spaces. Recall that these spaces are decomposable (Theorem I.11.7).

Lemma 10.2. *If $\mathfrak{E}$ is a quasi-negative, non-degenerate inner product space, then for any maximal positive definite or maximal positive subspace $\mathfrak{L} \subset \mathfrak{E}$ and any fundamental decomposition* (10.1) *we have* $\dim \mathfrak{L} = \dim \mathfrak{E}^+$. *Analogously, if $\mathfrak{E}$ is a quasi-positive, non-denegerate space, then for any maximal negative definite or maximal negative subspace $\mathfrak{L}$ and any fundamental decomposition* (10.1) *we have* $\dim \mathfrak{L} = \dim \mathfrak{E}^-$.

Proof. Let $\mathfrak{L}$ be a positive (possibly positive definite) subspace of the quasi-negative, non-degenerate space $\mathfrak{E}$. Consider a fundamental decomposition (10.1) of $\mathfrak{E}$ and the corresponding fundamental projectors P^+, P^-.

If $P^+\mathfrak{L} \neq \mathfrak{E}^+$ then, $\mathfrak{E}^+$ being finite-dimensional and positive definite, there is a non-zero vector $y \in \mathfrak{E}^+$ orthogonal to $P^+\mathfrak{L}$. Since y is orthogonal to $\mathfrak{E}^-$ as well, for every $x \in \mathfrak{L}$ we have $(x, y) = (P^+x, y) + (P^-x, y) = 0$. Thus $y \perp \mathfrak{L}$. Therefore the proper extension $\langle y, \mathfrak{L}\rangle$ of $\mathfrak{L}$ is positive (cf. Remark I.3.2). Moreover, it is positive definite provided $\mathfrak{L}$ was so.

Consequently, for a maximal positive definite or maximal positive $\mathfrak{L}$ we have $P^+\mathfrak{L} = \mathfrak{E}^+$. Hence, in view of Theorem 10.1, $\mathfrak{L}$ is isomorphic to $\mathfrak{E}^+$. □

Corollary 10.3. *In a quasi-negative, non-degenerate space every maximal positive definite subspace is maximal positive. In a quasi-positive, non-degenerate space every maximal negative definite subspace is maximal negative.* □

Making use of Lemma I.11.4 we also find:

Corollary 10.4. *If*

$$\mathfrak{E} = \mathfrak{E}_j^+ \,(\dot{+})\, \mathfrak{E}_j^-; \quad \mathfrak{E}_j^+ \subset \mathfrak{P}^{++}, \quad \mathfrak{E}_j^- \subset \mathfrak{P}^{--} \quad (j = 1, 2) \tag{10.5}$$

are two fundamental decompositions of the quasi-negative, non-degenerate inner product space $\mathfrak{E}$, *then* $\dim \mathfrak{E}_1^+ = \dim \mathfrak{E}_2^+$. *For a quasi-positive* $\mathfrak{E}$ *we have* $\dim \mathfrak{E}_1^- = \dim \mathfrak{E}_2^-$. □

Lemma 10.2 justifies the following definitions.

Let $\mathfrak{E}$ be a quasi-negative, non-degenerate inner product space, and let $\mathfrak{L}$ be a maximal positive definite subspace of $\mathfrak{E}$. The non-negative integer $\dim \mathfrak{L}$ will be termed the *rank of positivity* of the space $\mathfrak{E}$ (or of the inner product $(.\,,.)$), and denoted by $\varkappa^+(\mathfrak{E})$. For every fundamental decomposition (10.1) we have

$$\varkappa^+(\mathfrak{E}) = \dim \mathfrak{E}^+ . \tag{10.6}$$

Next let $\mathfrak{E}$ be a quasi-positive, non-degenerate inner product space. Then the *rank of negativity* $\varkappa^-(\mathfrak{E})$ of $\mathfrak{E}$ can be defined as the dimension of a maximal negative definite subspace $\mathfrak{L} \subset \mathfrak{E}$. For every fundamental decomposition (10.1) of $\mathfrak{E}$ we have

$$\varkappa^-(\mathfrak{E}) = \dim \mathfrak{E}^- . \tag{10.7}$$

An extension of these definitions to more general spaces will be given in Chapter IV.

The following consequence of Corollary 10.3 is often useful.

Lemma 10.5. *Let* $\mathfrak{E}$ *be a quasi-negative (quasi-positive), non-degenerate inner product space. Then every subspace* $\mathfrak{L} \subset \mathfrak{E}$ *containing a maximal positive definite* (resp. *maximal negative definite*) *subspace of* $\mathfrak{E}$ *is non-degenerate.*

Proof. Let $\mathfrak{M} \subset \mathfrak{L} \subset \mathfrak{E}$, where $\mathfrak{E}$ is quasi-negative, $\mathfrak{E}^0 = 0$, and $\mathfrak{M}$ is maximal positive definite in $\mathfrak{E}$. If $x \in \mathfrak{L}^0$, then $(x, x) = 0$ and $x \perp \mathfrak{M}$, so that by Remark I.3.2 the subspace $\langle x, \mathfrak{M} \rangle$ is positive. But $\mathfrak{M}$ is maximal positive (Corollary 10.3). Therefore $x \in \mathfrak{M}$, i.e., $x = 0$. □

11. Fundamental Symmetries. Angular Operators

Let $\mathfrak{E}$ be a decomposable, non-degenerate inner product space. Consider the fundamental projectors P^+, P^- belonging to a fundamental decomposition (10.1) of $\mathfrak{E}$. We set

$$(11.1) \qquad\qquad J = P^+ - P^- ,$$

and say that J is the *fundamental symmetry* belonging to the fundamental decomposition (10.1).

The term "symmetry" for the linear operator J is justified by the fact that for every $x \in \mathfrak{E}$ the vectors x and Jx lie symmetrically with respect to the subspace $\mathfrak{E}^+$. Namely if

$$(11.2) \qquad\qquad x = x^+ + x^-; \qquad x^+ \in \mathfrak{E}^+ , \qquad x^- \in \mathfrak{E}^- ,$$

then

$$(11.3) \qquad\qquad Jx = x^+ - x^- .$$

Since the system (11.2)—(11.3) can be solved for x^+ and x^-, the fundamental symmetry J uniquely specifies the fundamental decomposition to which it belongs. This entitles us to speak of "the fundamental decomposition belonging to J".

Making use of relations (11.2)—(11.3) one easily verifies the following.

Lemma 11.1. *A fundamental symmetry J is completely invertible, and we have*

$$(11.4) \qquad\qquad J^{-1} = J .$$

Moreover, J is symmetric as well as isometric. $\qquad\square$

Let J be a fundamental symmetry on the decomposable, non-degenerate inner product space $\mathfrak{E}$. Since J is symmetric (Lemma 11.1), the formula

$$(11.5) \qquad\qquad (x, y)_J = (Jx, y) \qquad (x, y \in \mathfrak{E})$$

defines an inner product on $\mathfrak{E}$, the so-called *J-inner product*. (For the more general concept of *A*-inner product see Section 3 above.)

With the help of the fundamental decomposition (10.1) belonging to J, relation (11.5) can be written in the form

$$(11.6) \qquad\qquad (x, y)_J = (x^+, y^+) - (x^-, y^-) \qquad (x, y \in \mathfrak{E}) ,$$

where

$$(11.7) \quad x = x^+ + x^- , \quad y = y^+ + y^-; \quad x^+, y^+ \in \mathfrak{E}^+; \quad x^-, y^- \in \mathfrak{E}^- .$$

Note that different fundamental decompositions give rise to different *J*-inner products.

Lemma 11.2. *If* $\mathfrak{E} = \mathfrak{E}^+ (\dotplus) \mathfrak{E}^-$ *is a fundamental decomposition of the non-degenerate space* $\mathfrak{E}$, *and* J *is the corresponding fundamental symmetry, then* $\mathfrak{E}^+$ *is* J*-orthogonal to* $\mathfrak{E}^-$.

Proof. In (11.6) put $x^- = 0$, $y^+ = 0$. □

Lemma 11.3. *Every* J*-inner product is positive definite.*

Proof. From (11.6) we obtain

$$(x, x)_J = (x^+, x^+) - (x^-, x^-) \qquad (x \in \mathfrak{E}) .$$

The positive definiteness of $\mathfrak{E}^+$ implies $(x^+, x^+) \geqq 0$, where equality holds only if $x^+ = 0$. Similarly, the negative definiteness of $\mathfrak{E}^-$ yields $-(x^-, x^-) \geqq 0$ with equality holding if and only if $x^- = 0$. □

Let J be a fundamental symmetry on the non-degenerate inner product space $\mathfrak{E}$. Owing to Lemma 11.3 the formula

$$\|x\|_J = +\sqrt{(x, x)_J} \qquad (x \in \mathfrak{E}) \tag{11.8}$$

defines a norm on $\mathfrak{E}$. It will be called the *J-norm*. More explicitly:

$$\|x\|_J = ((x^+, x^+) - (x^-, x^-))^{1/2} \qquad (x \in \mathfrak{E}) , \tag{11.9}$$

where x^+ and x^- are the components of x with respect to the fundamental decomposition (10.1) belonging to J.

Different fundamental decompositions induce different J-norms.

Lemma 11.4. *For any* J*-norm we have the inequality*

$$|(x, y)| \leq \|x\|_J \|y\|_J \qquad (x, y \in \mathfrak{E}) .$$

Proof. Taking the fundamental decomposition (10.1) that belongs to the J-norm in question and applying the notation (11.7), Lemma I.2.2, the Cauchy inequality, and the definition (11.9) of the J-norm we find:

$$|(x, y)| = |(x^+, y^+) + (x^-, y^-)| \leqq$$
$$\leqq (x^+, x^+)^{1/2} (y^+, y^+)^{1/2} + |(x^-, x^-)|^{1/2} |(y^-, y^-)|^{1/2} \leqq$$
$$\leqq ((x^+, x^+) - (x^-, x^-))^{1/2} ((y^+, y^+) - (y^-, y^-))^{1/2} = \|x\|_J \|y\|_J . \ \square$$

The next result is an application of the concept of fundamental symmetry.

Theorem 11.5. *In a decomposable, non-degenerate inner product space* $\mathfrak{E}$ *every subspace has a dual companion. Moreover, one can choose the dual companion to be isometrically isomorphic to the subspace given.*

Proof. The assumption implies the existence of a fundamental symmetry J on $\mathfrak{E}$. For any subspace $\mathfrak{L} \subset \mathfrak{E}$ we set $\mathfrak{M} = J\mathfrak{L}$. If $x \in \mathfrak{L} \cap \mathfrak{M}^\perp$ then, in particular, $x \perp Jx$ i.e. $(x, x)_J = 0$, and by

Lemma 11.3 $x = 0$. The relation $\mathfrak{M} \cap \mathfrak{L}^{\perp} = 0$ follows in a similar way once we realize that $J\mathfrak{M} = J^2\mathfrak{L} = \mathfrak{L}$ (see (11.4)). Finally, $\mathfrak{M}$ is isometrically isomorphic to $\mathfrak{L}$, since J is an isometric operator (Lemma 11.1). □

By the aid of J-norms an extremely useful description of semi-definite subspaces can be given. For this purpose we need one more definition.

Let $\mathfrak{L}$ be a subspace of the decomposable, non-degenerate inner product space $\mathfrak{E}$. Consider a fundamental decomposition

$$\mathfrak{E} = \mathfrak{E}^+ \dotplus \mathfrak{E}^-; \qquad \mathfrak{E}^+ \subset \mathfrak{P}^{++}, \qquad \mathfrak{E}^- \subset \mathfrak{P}^{--},$$

and the corresponding fundamental projectors P^+, P^-. Suppose that $P^+|\mathfrak{L}$ is invertible. Then every element $x \in \mathfrak{L}$ is uniquely specified by its projection P^+x, so that the formula

$$K^+P^+x = P^-x \qquad (x \in \mathfrak{L}) \tag{11.10}$$

defines a linear operator $K^+ = K^+(\mathfrak{L})$ from $\mathfrak{E}^+$ to $\mathfrak{E}^-$ with domain $\mathfrak{D}(K^+) = P^+\mathfrak{L}$ and range $\mathfrak{R}(K^+) = P^-\mathfrak{L}$. We say, K^+ is the *angular operator* of the subspace $\mathfrak{L}$ with respect to $\mathfrak{E}^+$. Similarly, if $P^-|\mathfrak{L}$ is invertible, then the angular operator $K^- = K^-(\mathfrak{L})$ of $\mathfrak{L}$ with respect to $\mathfrak{E}^-$ can be defined by the relation

$$K^-P^-x = P^+x \qquad (x \in \mathfrak{L}); \tag{11.11}$$

we have $\mathfrak{D}(K^-) = P^-\mathfrak{L}$, $\mathfrak{R}(K^-) = P^+\mathfrak{L}$.

Theorem 11.6. *Suppose that*

$$\mathfrak{E} = \mathfrak{E}^+ \dotplus \mathfrak{E}^-; \qquad \mathfrak{E}^+ \subset \mathfrak{P}^{++}, \qquad \mathfrak{E}^- \subset \mathfrak{P}^{--}.$$

Let $\mathfrak{L}$ be a subspace of $\mathfrak{E}$. If $K^+(K^-)$ is the angular operator of $\mathfrak{L}$ with respect to $\mathfrak{E}^+(\mathfrak{E}^-)$, then

$$\mathfrak{L} = \{x^+ + K^+x^+ : \; x^+ \in \mathfrak{L}_1\} \tag{11.12}$$

(resp.

$$\mathfrak{L} = \{x^- + K^-x^- : \; x^- \in \mathfrak{L}_2\}) , \tag{11.13}$$

where $\mathfrak{L}_1(\mathfrak{L}_2)$ is a subspace of $\mathfrak{E}^+(\mathfrak{E}^-)$. Conversely, if $\mathfrak{L}$ can be written in the form (11.12) (resp. (11.13)), *where $\mathfrak{L}_1(\mathfrak{L}_2)$ is a subspace of $\mathfrak{E}^+(\mathfrak{E}^-)$ and $K^+(K^-)$ is a linear operator from $\mathfrak{L}_1$ to $\mathfrak{E}^-$ (from $\mathfrak{L}_2$ to $\mathfrak{E}^+$), then $K^+(K^-)$ is the angular operator of $\mathfrak{L}$ with respect to $\mathfrak{E}^+(\mathfrak{E}^-)$.*

The *proof* is straightforward. □

Theorem 11.7. *Suppose that*

$$\mathfrak{E} = \mathfrak{E}^+ \dotplus \mathfrak{E}^-; \qquad \mathfrak{E}^+ \subset \mathfrak{P}^{++}, \qquad \mathfrak{E}^- \subset \mathfrak{P}^{--},$$

and denote the corresponding fundamental symmetry by J. A subspace $\mathfrak{L} \subset \mathfrak{E}$ is positive if and only if the angular operator K^+ of $\mathfrak{L}$ with respect *to* $\mathfrak{H}^+$ *exists and satisfies the condition*

$$\|K^+x^+\|_J \leqq \|x^+\|_J \qquad (x^+ \in \mathfrak{D}(K^+)) .$$

In particular, positive definite subspaces are characterized by the property

$$\|K^+x^+\|_J < \|x^+\|_J \qquad (x^+ \in \mathfrak{D}(K^+), \quad x^+ \neq 0) ,$$

and neutral subspaces by

$$\|K^+x^+\|_J = \|x^+\|_J \qquad (x^+ \in \mathfrak{D}(K^+)) .$$

For negative subspaces similar statements, involving K^- *instead of* K^+, *are valid.*

Proof. If $\mathfrak{L}$ is positive, then the angular operator $K^+(\mathfrak{L})$ exists by Theorem 10.1. Moreover, in view of Theorem 11.6 for every $x \in \mathfrak{L}$ we have

$$(x, x) = (x^+, x^+) + (K^+x^+, K^+x^+) = \|x^+\|_J^2 - \|K^+x^+\|_J^2 .$$

Hence, under the assumption that $K^+(\mathfrak{L})$ exists, for $x \in \mathfrak{L}$ the relation $(x, x) > 0$ is equivalent to $\|K^+x^+\|_J < \|x^+\|_J$, while $(x, x) = 0$ is equivalent to $\|K^+x^+\|_J = \|x^+\|_J$. □

Notes to Chapter II

Most results of this chapter were originally obtained as tools for the study of linear operators in special kinds of inner product spaces, but it has been evident from the beginning that they remain valid under more general assumptions. Here we bring them together in order to make clear their respective ranges as well as to make subsequent proofs of deeper theorems more transparent.

Isometric and symmetric operators in inner product spaces of finite dimension have been studied for a long time. An account of the results appears in the book of Mal'cev [1].

Symmetric operators in certain quasi-positive inner product spaces of infinite dimension were introduced by Pontrjagin [1], in a wider class of spaces by Ginzburg [2] (cf. also Pesonen [1], Langer [1]), and in arbitrary non-degenerate spaces by Berezin [1].

More and more general types of isometric operators in indefinite inner product spaces of infinite dimension have been considered by Kreĭn and Rutman [1], Iohvidov [1], [2], Kreĭn [4], Iohvidov and Kreĭn [1], Ginzburg [1], [2], and again by Iohvidov [4], [11].

Corollary 2.2 was noted by Iohvidov and Kreĭn [1], and Lemma 2.1 by Iohvidov [4]. For the rest of the results of Section 2 see Mal'cev [1], Iohvidov and Kreĭn [1].

For finite-dimensional spaces a more detailed description of the structure of an isometric operator was given by Mal'cev [1]. Karrer [1] found essentially the same facts in a more algebraic language.

Theorem 3.10 stated for a smaller class of spaces is due to Pontrjagin [1], while Theorem 3.11 belongs to Langer [7], [12]. Other results of Section 3 have been presented by Mal'cev [1] for finite-dimensional spaces, and by Pontrjagin [1] and Langer [2] for certain classes of infinite-dimensional spaces (cf. also Iohvidov and Kreĭn [1]). A more complete structure theorem for symmetric operators of a finite-dimensional space is to be found in the book of Mal'cev [1].

Simple properties of a skew-symmetric operator defined in a quasi-positive space were employed by Lax and Phillips [1], [2; Chapter VI] to the study of a scattering problem.

Cayley transformations in indefinite inner product spaces of finite dimension have been considered by Mal'cev [1]. In the infinite-dimensional case the first application was made by Iohvidov [1]. In Sections 4—5 we follow the general and detailed treatment of Iohvidov and Kreĭn [1].

Several results of Sections 2—5, along with examples of symmetric and isometric operators having no Cayley transform, are contained in the expository paper of Azizov and Iohvidov [2].

A special case of the results of Section 6 is due to Kühne [1], while in the general case they belong to Kreĭn and Šmul'jan [1]. Cf. also Kühne [2].

Lemmas 7.1—7.3 were given by Pesonen [1] (cf. also Kühne [1]). Lemma 7.4 appears implicitly in the paper of Kreĭn and Šmul'jan [1].

The class of linear operators T with the property that $(Tx, Tx) \geqq (x, x)$ whenever $(x, x) \geqq 0$ was introduced by Kreĭn [4] (cf. also Iohvidov and Kreĭn [1]). Operators T satisfying $(Tx, Tx) \geqq (x, x)$ for every x in the space have first been investigated by Potapov [1] and Ginzburg [1], [2]; cf. also the Notes to Chapter VII. Both classes are contained in that of plus-operators.

The characteristic property of plus-operators first appears in a paper of M. L. Brodskiĭ [1]. It plays an important role in a basic result of Kreĭn [7], [8] and in its improvements (see Chapter VIII). The results of Section 8 belong to Kreĭn and Šmul'jan [1].

Most results of Section 9 have been obtained by Pesonen [1], but in terms of quadratic forms rather than operators. He established Theorem 9.10 and an extension of it to spaces of infinite dimension. Pesonen's results were improved and reformulated by Kühne [1]. In particular, the explicit forms and the elementary proofs of our propositions 9.5—9.9 belong to him. Theorem 9.10 was subsequently reproved by Kraljević [1].

Theorem 9.10 remains valid in real inner product spaces of dimension greater than 2. A proof given by J. Milnor is included in the book of Greub [1]. For more elementary proofs see Kraljević [1] and Wonenburger [1].

Theorem 10.1 was stated in this generality by Ginzburg [1]. In special cases it had earlier been applied by Pontrjagin [1] as well as by Iohvidov and Kreĭn [1]. The latter work also contains the rest of the material of Section 10. It should be noted that Corollary 10.4 is an extension of Sylvester's law of inertia.

Fundamental symmetries and J-norms have been introduced by Nevanlinna [1], [3]. Particular cases of Theorem 11.5 are due to Iohvidov and Kreĭn [1], and to Scheibe [1].

The concept of angular operator as well as Theorems 11.6—11.7 belong to Phillips [2]. An independent and more explicit treatment was given by Ginzburg [3] (cf. also Ginzburg and Iohvidov [1], Kreĭn [8]).

Chapter III. Partial Majorants and Admissible Topologies on Inner Product Spaces

Partial majorants (locally convex topologies in which the inner product is separately continuous; Sections 2—4) and admissible topologies (partial majorants with respect to which every continuous linear form is an inner multiplication; Sections 5—7), especially their relation to orthogonal companions and to the existence of projections are discussed. Sections 8—9 deal with the natural normed topology of definite subspaces. The list of more important results includes Corollary 2.5 as well as Theorems 3.3, 5.3, 6.1, 6.5, 7.1, and 9.2.

1. Locally Convex Topologies on Vector Spaces

In this section we list some definitions and facts related to topological spaces and, especially, topological vector spaces.

A function p defined on a vector space $\mathfrak{E}$ is said to be a *semi-norm* on $\mathfrak{E}$ if the following three requirements are fulfilled:

$$p(x) \geqq 0 \qquad (x \in \mathfrak{E}), \tag{1.1}$$

$$p(\alpha x) = |\alpha|\, p(x) \qquad (\alpha \in \boldsymbol{C}, \quad x \in \mathfrak{E}), \tag{1.2}$$

$$p(x_1 + x_2) \leqq p(x_1) + p(x_2) \qquad (x_1, x_2 \in \mathfrak{E}). \tag{1.3}$$

If, in addition, $p(x) = 0$ implies $x = 0$, then p is a norm.

The semi-norm p on the vector space $\mathfrak{E}$ is said to be *quadratic* if $p(x) = [x, x]^{1/2}$ $(x \in \mathfrak{E})$, where $[.\,,.]$ is a positive inner product on $\mathfrak{E}$.

A family $\{p_\gamma\}_{\gamma \in \Gamma}$ of semi-norms on the vector space $\mathfrak{E}$ *defines* a topology τ on $\mathfrak{E}$ through the following prescription: the set $\mathfrak{G} \subset \mathfrak{E}$ is τ-open if and only if to every $x \in \mathfrak{G}$ there are a finite number of indices $\gamma_1, \ldots, \gamma_n \in \Gamma$ and a positive number ε such that the relations $y \in \mathfrak{E}$, $p_{\gamma_j}(x - y) < \varepsilon$ $(j = 1, \ldots, n)$ imply $y \in \mathfrak{G}$. The topology τ is *locally convex*, which means that 1) the vector space operations are τ-continuous, and 2) any τ-neighbourhood of a point $x \in \mathfrak{E}$ contains a convex τ-neighbourhood of x.

Every locally convex topology is defined by some family of semi-norms.

A topology τ on $\mathfrak{E}$ is said to be *separated* if for any pair of distinct points $x_1, x_2 \in \mathfrak{E}$ there is a τ-neighbourhood $\mathfrak{U}_1$ of x_1 and a τ-neighbourhood $\mathfrak{U}_2$ of x_2 such that $\mathfrak{U}_1 \cap \mathfrak{U}_2 = \emptyset$.

The topology defined by the family $\{p_\gamma\}_{\gamma \in \Gamma}$ of semi-norms is separated if and only if the relations $p_\gamma(x) = 0 \quad (\gamma \in \Gamma)$ imply $x = 0$.

We shall mainly be concerned with separated locally convex topologies.

A separated locally convex topology is metrizable if and only if it can be defined by a countable family of semi-norms. A locally convex, metrizable, complete topology is called a *Fréchet topology.*

If the separated topology τ is defined by a finite number $p_1, \ldots, p_n$ of semi-norms on $\mathfrak{E}$, then τ is also defined by the single semi-norm $p(x) = \max_{1 \leqq j \leqq n} p_j(x) \quad (x \in \mathfrak{E})$, which is actually a norm. In this case τ is said to be a *normed topology*. A complete normed topology may be termed a *Banach topology.*

If the topology τ can be defined by a quadratic norm, we say τ is a *quadratic-normed topology*. A quadratic-normed, complete topology will be called a *Hilbert topology.*

In contrast to the notion of separated topology, a topology τ on the space $\mathfrak{E}$ is said to be *separable,* if $\mathfrak{E}$ contains a countable τ-dense subset.

Let τ be a locally convex topology defined by the semi-norms $p_\gamma \quad (\gamma \in \Gamma)$ on the vector space $\mathfrak{E}$. The restrictions $p_\gamma|\mathfrak{L}$ of the semi-norms $p_\gamma \ (\gamma \in \Gamma)$ to a subspace $\mathfrak{L} \subset \mathfrak{E}$ define a locally convex topology $\tau|\mathfrak{L}$, the topology *induced* by τ on $\mathfrak{L}$.

Let τ_1, τ_2 be two topologies on the same space $\mathfrak{E}$. We say that τ_1 is *weaker* than τ_2 (or that τ_2 is *stronger* than τ_1), and write $\tau_1 \leqq \tau_2$, if every τ_1-open set is τ_2-open or, equivalently, if every τ_1-neighbourhood of an arbitrary $x \in \mathfrak{E}$ contains a τ_2-neighbourhood of x.

If $\tau_1 \leqq \tau_2$ and $\tau_2 \leqq \tau_1$ at the same time, then τ_1 and τ_2 coincide: $\tau_1 = \tau_2$.

The semi-norm p_1 defines a weaker topology on $\mathfrak{E}$ than the semi-norm p_2 if and only if $p_1(x) \leqq \alpha p_2(x)$ for some $\alpha > 0$ and every $x \in \mathfrak{E}$.

If two semi-norms p_1, p_2 define the same topology, we say p_1 and p_2 are *equivalent.*

2. Partial Majorants. The Weak Topology

Let $\mathfrak{E}$ be a vector space with an inner product $(.\,,.)$, and let τ be a topology on $\mathfrak{E}$. We say that τ is a *partial majorant* of the inner product (or: a partial majorant on $\mathfrak{E}$) if 1) τ is locally convex, and 2) for any fixed $y \in \mathfrak{E}$ the function $\varphi_y(x) = (x, y)$ is τ-continuous on $\mathfrak{E}$.

Since $(x, y) = \overline{(y, x)}$ $(x, y \in \mathfrak{E})$, condition 2) is equivalent to the τ-continuity of $\psi_x(y) = (x, y)$ for fixed x. Therefore a locally convex topology τ on $\mathfrak{E}$ is a partial majorant if and only if the inner product is separately τ-continuous.

Let us mention an important example of a partial majorant.

The *weak topology* τ_0 of the inner product space $\mathfrak{E}$ is the locally convex topology defined by the family $\{p_y\}_{y\in\mathfrak{E}}$ of semi-norms, where

(2.1) $$p_y(x) = |(x, y)| \quad (x \in \mathfrak{E}) .$$

Theorem 2.1. *The weak topology τ_0 of the inner product space $\mathfrak{E}$ is a partial majorant on $\mathfrak{E}$. A locally convex topology τ on $\mathfrak{E}$ is a partial majorant if and only if $\tau \geqq \tau_0$.*

Proof. Let $x_0, y\in\mathfrak{E}$, $\varepsilon > 0$ be given. The relation $|(x, y) - (x_0, y)| < \varepsilon$ holds for every x belonging to the set $\{x\colon p_y(x - x_0) < \varepsilon\}$, which is a weak neighbourhood of x_0 (see the notation (2.1)). Hence τ_0 is a partial majorant. As a consequence, every locally convex topology $\tau \geqq \tau_0$ is also a partial majorant on $\mathfrak{E}$.

Let τ be a partial majorant on $\mathfrak{E}$. Then for any elements $x_0, y_j \in \mathfrak{E}$ $(j = 1, \ldots, n)$ and any number $\varepsilon > 0$ there are τ-neighbourhoods $\mathfrak{U}_j$ of x_0 such that $x \in \mathfrak{U}_j$ implies $|(x,y_j) - (x_0,y_j)| < \varepsilon$ i.e. $p_{y_j}(x - x_0) < \varepsilon$. Since also $\bigcap_{j=1}^{n} \mathfrak{U}_j$ is a τ-neighbourhood of x_0, it follows that every weak neighbourhood of x_0 contains a τ-neighbourhood of x_0. □

Corollary 2.2. *Every partial majorant of a non-degenerate inner product is separated.* □

Remark 2.3. A partial majorant τ on a degenerate inner product space $\mathfrak{E}$ is not necessarily separated. Replacing, however, each of the semi-norms p_y that define τ by $p_y + q$, where q is any norm on $\mathfrak{E}$, one obtains a separated partial majorant $\tau_1 \geqq \tau$.

Next we describe some simple properties of partial majorants and weak topologies involving the concept of orthogonality.

Lemma 2.4. *If τ is a partial majorant on the inner product space $\mathfrak{E}$, then for any subspace $\mathfrak{L} \subset \mathfrak{E}$ the orthogonal companion $\mathfrak{L}^\perp$ is τ-closed. Moreover, $\mathfrak{L}$ and its τ-closure $\overline{\mathfrak{L}}$ have the same orthogonal companion.*

Proof. If $x_0 \notin \mathfrak{L}^\perp$, then $(x_0, y) \neq 0$ for some $y \in \mathfrak{L}$; thus for every x in a suitable τ-neighbourhood of x_0 we have $(x, y) \neq 0$, $x \notin \mathfrak{L}^\perp$. In order to prove the second assertion suppose that the vector $x \in \mathfrak{E}$ is not orthogonal to $\overline{\mathfrak{L}}$, i.e., $(x, y_0) \neq 0$ for some $y_0 \in \overline{\mathfrak{L}}$. Then $(x, y) \neq 0$ for every y in a τ-neighbourhood $\mathfrak{U}$ of y_0. Choosing $y \in \mathfrak{L} \cap \mathfrak{U}$ it follows that $x \notin \mathfrak{L}^\perp$. □

Making use of Lemma I.9.6 we obtain:

Corollary 2.5. *Let τ be a partial majorant on the non-degenerate inner product space $\mathfrak{E}$. Then every ortho-complemented subspace of $\mathfrak{E}$ is τ-closed.* □

Further specialization yields:

Corollary 2.6. *Let τ be a partial majorant on the non-degenerate inner product space $\mathfrak{E}$, and let*

$$\mathfrak{E} = \mathfrak{E}^+ \dot{(+)} \mathfrak{E}^-; \qquad \mathfrak{E}^+ \subset \mathfrak{P}^{++}, \quad \mathfrak{E}^- \subset \mathfrak{P}^{--}$$

be a fundamental decomposition of $\mathfrak{E}$ (cf. Corollary I.11.2). *Then $\mathfrak{E}^+$ and $\mathfrak{E}^-$ are τ-closed.* □

3. Metrizable Partial Majorants

Theorem 2.1 (and Remark 2.3) establish the existence of (separated) partial majorants. One wonders if it is always possible to find a metrizable partial majorant. The answer turns out to be negative.

Theorem 3.1. *If an inner product has a metrizable partial majorant, then it has a normed partial majorant too.*

Proof. Let τ be a metrizable partial majorant on the inner product space $\mathfrak{E}$. Then τ is defined by a sequence $p_1, p_2, \ldots$ of semi-norms. Introduce the following notations:

$$(3.1)\quad \mathfrak{U}_n = \left\{x \in \mathfrak{E}\colon \sum_{j=1}^{n} p_j(x) < \frac{1}{n}\right\} \qquad (n = 1, 2, \ldots)\,,$$

$$(3.2)\quad \mathfrak{U}_n' = \{x \in \mathfrak{E}\colon |(x, y)| \leqq 1 \text{ for every } y \in \mathfrak{U}_n\} \quad (n = 1, 2, \ldots)\,,$$

$$(3.3)\quad \mathfrak{B} = \bigcup_{n=1}^{\infty} (\mathfrak{U}_n \cap \mathfrak{U}_n')\,,$$

$$(3.4)\quad \mathfrak{B}_1 = \left\{\sum_{j=1}^{n} \alpha_j x_j\colon\ n \geqq 1\,,\ \ \{x_j\}_1^n \subset \mathfrak{B}\,,\ \ \sum_{j=1}^{n} |\alpha_j| \leqq 1\right\},$$

$$(3.5)\quad p(x) = \inf\{\varrho > 0\colon\ x \in \varrho\, \mathfrak{B}_1\} \qquad (x \in \mathfrak{E})\,.$$

We shall prove that p is a semi-norm which defines a partial majorant on $\mathfrak{E}$. Replacing p by $p + q$, where q is an arbitrary norm on $\mathfrak{E}$, the theorem follows.

As τ is a partial majorant, and any τ-neighbourhood of 0 contains some of the sets $\mathfrak{U}_n$, for every $x \in \mathfrak{E}$ there exists an index n_0 such that $x \in \mathfrak{U}_{n_0}'$. On the other hand, by virtue of (3.1), for sufficiently small $\sigma > 0$ we have $\sigma x \in \mathfrak{U}_{n_0}$. Setting $\min\{1, \sigma\} = \sigma_1$ we find $\sigma_1 x \in \mathfrak{B}_1$ i.e. $x \in \frac{1}{\sigma_1} \mathfrak{B}_1$. Thus $p(x) < \infty$.

p clearly satisfies (1.1). The property (1.2) follows from the fact that for a positive number ϱ the relation $x \in \varrho\mathfrak{B}_1$ is equivalent to $|\alpha|\, x \in |\alpha|\, \varrho\, \mathfrak{B}_1$ or, in view of (3.4), to $\alpha x \in |\alpha|\, \varrho\, \mathfrak{B}_1$. In order to check (1.3) let $x_j \in \varrho_j \mathfrak{B}_1$ $(j = 1, 2)$. Then $\frac{1}{\varrho_j} x_j \in \mathfrak{B}_1$ and, by the convexity of $\mathfrak{B}_1$,

$$\frac{\varrho_1}{\varrho_1 + \varrho_2} \frac{1}{\varrho_1} x_1 + \frac{\varrho_2}{\varrho_1 + \varrho_2} \frac{1}{\varrho_2} x_2 \in \mathfrak{B}_1 ,$$

i.e. $x_1 + x_2 \in (\varrho_1 + \varrho_2)\, \mathfrak{B}_1$.

It remains to show that the semi-norm p defines a partial majorant on $\mathfrak{E}$.

First let $x, y \in \mathfrak{B}$. According to the definition (3.3) of $\mathfrak{B}$, there are indices m, n such that $x \in \mathfrak{U}_m \cap \mathfrak{U}'_m$, $y \in \mathfrak{U}_n \cap \mathfrak{U}'_n$. If $m \leqq n$, then from the relations $x \in \mathfrak{U}'_m$, $y \in \mathfrak{U}_n$, $\mathfrak{U}_n \subset \mathfrak{U}_m$ and (3.2) we infer that $|(x, y)| \leqq 1$. In the case $m > n$ we arrive at the same conclusion by interchanging the roles of x and y.

Next let x, y be a pair of elements in $\mathfrak{E}$ with $p(x) < 1$, $p(y) < 1$. On account of (3.5) and (3.4) we have $x, y \in \mathfrak{B}_1$, so that (3.4) yields the representations

$$x = \sum_{j=1}^{m} \alpha_j x_j ; \qquad \{x_j\}_1^m \subset \mathfrak{B} , \qquad \sum_{j=1}^{m} |\alpha_j| \leqq 1 ,$$

$$y = \sum_{k=1}^{n} \beta_k y_k ; \qquad \{y_k\}_1^n \subset \mathfrak{B} , \qquad \sum_{k=1}^{n} |\beta_k| \leqq 1 .$$

Applying the result of the preceding paragraph to x_j and y_k we obtain

$$|(x_j, y_k)| \leqq 1 \qquad (j = 1, \ldots, m;\ k = 1, \ldots, n) .$$

Therefore

$$(3.6) \qquad |(x, y)| \leqq 1 \qquad (p(x) < 1 , \quad p(y) < 1) .$$

Finally, if x, y are arbitrary elements of $\mathfrak{E}$, then replacing x by $\frac{n}{n+1} \frac{1}{p(x)} x$ or nx $(n = 1, 2, \ldots)$ according as $p(x) \neq 0$ or $p(x) = 0$, and proceeding analogously for y, from (3.6) we conclude that

$$(3.7) \qquad |(x, y)| \leqq p(x)\, p(y) \qquad (x, y \in \mathfrak{E}) .$$

In particular, the topology defined by p is a partial majorant of the inner product. □

Example 3.2. Let, as in Example I.11.3, $\mathfrak{E}$ be the vector space of those doubly infinite numerical sequences where only a finite number of terms with negative index is different from zero, and for $x = \{\xi_j\}_{-\infty}^{\infty} \in \mathfrak{E}$, $y = \{\eta_j\}_{-\infty}^{\infty} \in \mathfrak{E}$ let

$$(x, y) = \sum_{j=-\infty}^{\infty} \xi_j \bar{\eta}_{-j-1} \, . \tag{3.8}$$

Suppose that this inner product has a partial majorant defined by a norm $\|\cdot\|$. Then $|(x, y)| \leqq \beta(y) \|x\|$ for a suitable $\beta(y) > 0$ and every $x, y \in \mathfrak{E}$. Setting

$$y = e_{-n} = \{\delta_{j,-n}\}_{j=-\infty}^{\infty} \qquad (n = 1, 2, \ldots)$$

we obtain:

$$|(x, e_{-n})| = |\xi_{n-1}| \leqq \beta(e_{-n}) \|x\| \qquad (x \in \mathfrak{E}; \quad n = 1, 2, \ldots) \, .$$

The choice $x = \{\xi_j\}_{-\infty}^{\infty}$; $\xi_j = 0$ for $j < 0$, $\xi_j = (j+1)\, \beta(e_{-j-1})$ for $j \geqq 0$, yields $n\, \beta(e_{-n}) \leqq \beta(e_{-n}) \|x\|$ $(n = 1, 2, \ldots)$, which is impossible. Thus our inner product has no normed partial majorant; by Theorem 3.1 it cannot have a metrizable partial majorant either.

As to the uniqueness of metrizable partial majorants, we have the following:

Theorem 3.3. *Let $\mathfrak{E}$ be a non-degenerate inner product space. Suppose that the Fréchet topologies τ_1, τ_2 are partial majorants on $\mathfrak{E}$. Then $\tau_1 = \tau_2$.*

Proof. Each of τ_1, τ_2 is defined by a countable family of semi-norms. The union of these two families defines a metrizable locally convex topology τ.

Let $\{x_n\}_1^{\infty} \subset \mathfrak{E}$ be a Cauchy sequence with respect to τ. Since

$$\tau \geqq \tau_j \qquad (j = 1, 2) \, , \tag{3.9}$$

$\{x_n\}$ is a Cauchy sequence with respect to τ_j as well. Therefore $\{x_n\}$ converges to some element $y_j \in \mathfrak{E}$ with respect to τ_j $(j = 1, 2)$. Taking into account that τ_j is a partial majorant, for every $z \in \mathfrak{E}$ we obtain $(x_n, z) \to (y_j, z)$ $(n \to \infty; \ j = 1, 2)$. Thus $(y_1, z) = (y_2, z)$ $(z \in \mathfrak{E})$, i.e., $y_1 = y_2$. The τ_1-limit and the τ_2-limit of $\{x_n\}$ being equal, it is also a τ-limit. Consequently, τ is complete, hence a Fréchet topology.

Applying the closed graph principle to the identity mapping of the space $\mathfrak{E}$ topologized by τ onto the same space topologized by τ_j, from (3.9) we derive $\tau = \tau_j$ $(j = 1, 2)$. □

4. The Polar of a Normed Partial Majorant

In the following lines we show that every normed partial majorant of a non-degenerate inner product gives rise to another normed partial majorant. This fact, and the properties of the correspondence, will be applied in Chapter IV.

Consider a partial majorant τ defined by the single norm $\|\cdot\|$ on the non-degenerate inner product space $\mathfrak{E}$. Put

$$\|x\|' = \sup_{\|y\| \leqq 1} |(x, y)| \qquad (x \in \mathfrak{E}) . \tag{4.1}$$

It is easy to see that $\|\cdot\|'$ is a norm on $\mathfrak{E}$. We say, $\|\cdot\|'$ is the *polar* of the norm $\|\cdot\|$. The topology defined by $\|\cdot\|'$ will be denoted by τ' and called the *polar* of the topology τ.

τ' only depends on τ and not on the norm defining τ. Really, if $\|\cdot\|_1$ is another norm on $\mathfrak{E}$ such that

$$\alpha\|x\| \leqq \|x\|_1 \leqq \beta\|x\| \qquad (x \in \mathfrak{E}) ,$$

where $\alpha, \beta > 0$, then

$$\frac{1}{\beta} \sup_{\|y\| \leqq 1} |(x, y)| \leqq \sup_{\|y\|_1 \leqq 1} |(x, y)| \leqq \frac{1}{\alpha} \sup_{\|y\| \leqq 1} |(x, y)| \qquad (x \in \mathfrak{E}) .$$

Simultaneously we have proved the following:

Lemma 4.1. *Let τ_1, τ_2 be two normed partial majorants of a non-degenerate inner product. If $\tau_1 \leqq \tau_2$, then $\tau_1' \geqq \tau_2'$.* □

The polar of a partial majorant is also a partial majorant. Namely from (4.1) we obtain

$$|(x, y)| \leqq \|y\| \, \|x\|' \qquad (x, y \in \mathfrak{E}) . \tag{4.2}$$

Thus one can form the normed partial majorants $\tau'' = (\tau')'$, $\tau''' = (\tau'')'$ etc.

Lemma 4.2. *If τ is a normed partial majorant on the non-degenerate inner product space $\mathfrak{E}$, then so is its polar τ'. Moreover, we have*

$$\tau'' \leqq \tau , \tag{4.3}$$

$$\tau''' = \tau' . \tag{4.4}$$

Proof. We already know that τ' is a normed partial majorant. Interchanging x and y in (4.2) we find $\sup_{\|y\|' \leqq 1} |(x, y)| \leqq \|x\|$, which proves (4.3). From (4.3), replacing τ by τ', one obtains $\tau''' \leqq \tau'$. Finally, application of Lemma 4.1 to $\tau_1 = \tau''$, $\tau_2 = \tau$ yields $\tau''' \geqq \tau'$. □

Lemma 4.3. *If a norm $\|\cdot\|$ that defines a partial majorant on a non-degenerate space is quadratic, then so is the polar norm $\|\cdot\|'$.*

Proof. Let $\mathfrak{E}$ be a vector space with non-degenerate inner product $(.\,,.)$. Consider the quadratic norm $\|x\| = [x, x]^{1/2}$ $(x \in \mathfrak{E})$, where

$[.,.]$ is a positive definite inner product on $\mathfrak{E}$. The completion of $\mathfrak{E}$ with respect to $||\cdot||$ is a Hilbert space $\tilde{\mathfrak{E}}$. If $||\cdot||$ defines a partial majorant of $(.,.)$, then for any $y \in \mathfrak{E}$ the linear form $\varphi_y(x) = (x, y)$ $(x \in \mathfrak{E})$ is continuous with respect to $||\cdot||$, hence it admits a unique continuous extension to $\tilde{\mathfrak{E}}$. Consequently, by the Riesz representation theorem, there exists a $y' \in \tilde{\mathfrak{E}}$ such that

$$(x, y) = [x, y'] \qquad (x \in \mathfrak{E}) .$$

The correspondence $y \to y'$ is an isomorphism between the vector space $\mathfrak{E}$ and a subspace $\mathfrak{L} \subset \tilde{\mathfrak{E}}$. Moreover, $||y'|| = ||\varphi_y|| = \sup_{||x|| \leqq 1} |(x, y)| = ||y||'$. Since $||\cdot||$ is quadratic on $\mathfrak{L}$, $||\cdot||'$ is quadratic on $\mathfrak{E}$. □

A partial majorant τ satisfying the relation $\tau' = \tau$ is said to be *self-polar*.

5. Admissible Topologies

A topology τ on the inner product space $\mathfrak{E}$ is said to be *admissible*, if 1) τ is a partial majorant on $\mathfrak{E}$, and 2) for any τ-continuous linear form φ defined on $\mathfrak{E}$ there is an element $y_0 \in \mathfrak{E}$ such that $\varphi(x) = (x, y_0)$ $(x \in \mathfrak{E})$.

Theorem 5.1. *The weak topology of an inner product space is admissible.*

Proof. Consider the weak topology τ_0 of the inner product space $\mathfrak{E}$. We have already shown in Theorem 2.1 that τ_0 is a partial majorant on $\mathfrak{E}$. Now let φ be a weakly continuous linear form on $\mathfrak{E}$. Then we can find elements $y_1, \ldots, y_n \in \mathfrak{E}$ and a number $\delta > 0$ such that

$$|\varphi(x)| < 1 \quad \text{whenever} \quad |(x, y_j)| < \delta \qquad (j = 1, \ldots, n) . \tag{5.1}$$

If $y_1, \ldots, y_n$ are the fixed elements occurring in (5.1) and x runs over $\mathfrak{E}$, the n-tuples $\{(x, y_1), \ldots, (x, y_n)\}$ of complex numbers form a vector space of dimension not exceeding n. This space admits a basis

$$\{(x_1, y_1), \ldots, (x_1, y_n)\}, \ldots, \{(x_m, y_1), \ldots, (x_m, y_n)\} , \tag{5.2}$$

where $m \leqq n$, and $x_1, \ldots, x_m$ are suitable elements of $\mathfrak{E}$. Thus for every $x \in \mathfrak{E}$ there exist numbers $\alpha_1, \ldots, \alpha_m$ satisfying

$$\{(x, y_1), \ldots, (x, y_n)\} = \sum_{k=1}^{m} \alpha_k \{(x_k, y_1), \ldots, (x_k, y_n)\}$$

or, written by coordinates,

$$(5.3) \qquad (x, y_j) = (\sum_{k=1}^{m} \alpha_k x_k, y_j) \qquad (j = 1, \ldots, n).$$

Relations (5.1) and (5.3) imply

$$(5.4) \qquad \varphi(x) = \varphi(\sum_{k=1}^{m} \alpha_k x_k).$$

Really, setting $x' = x - \sum_{k=1}^{m} \alpha_k x_k$ we see from (5.3) that $(x', y_j) = 0$ $(j = 1, \ldots, n)$. Hence for every $\varrho > 0$ we find $(\varrho x', y_j) = 0$ $(j = 1, \ldots, n)$, and therefore, by (5.1), $|\varphi(\varrho x')| < 1$ $(0 < \varrho < \infty)$, i.e. $\varphi(x') = 0$.

As the system (5.2) is linearly independent, the equations

$$(5.5) \qquad (x_k, \sum_{j=1}^{n} \beta_j y_j) = \varphi(x_k) \qquad (k = 1, \ldots, m)$$

can be solved for $\beta_1, \ldots, \beta_n$. We set

$$(5.6) \qquad \sum_{j=1}^{n} \beta_j y_j = y_0.$$

Then (5.4), (5.5), (5.3) and (5.6) yield:

$$\varphi(x) = \varphi\left(\sum_{k=1}^{m} \alpha_k x_k\right) = \left(\sum_{k=1}^{m} \alpha_k x_k, \sum_{j=1}^{n} \beta_j y_j\right) = \left(x, \sum_{j=1}^{n} \beta_j y_j\right) =$$
$$= (x, y_0). \quad \square$$

It is clear that the weak topology is separated if and only if the inner product is non-degenerate. This remains true for every admissible topology.

Lemma 5.2. *If the inner product space $\mathfrak{E}$ is non-degenerate (degenerate), then every admissible topology of $\mathfrak{E}$ is separated (non-separated).*

Proof. In view of Corollary 2.2 it is sufficient to consider the degenerate case. So let x_0 be a non-zero isotropic vector of $\mathfrak{E}$. Furthermore, let τ be a separated locally convex topology on $\mathfrak{E}$ defined by the semi-norms p_γ $(\gamma \in \Gamma)$. Then for some $p = p_{\gamma_0}$ we have $p(x_0) \neq 0$. According to the Hahn-Banach extension principle there is a linear form φ on $\mathfrak{E}$ such that $\varphi(x_0) = p(x_0)$, $|\varphi(x)| \leqq p(x)$ $(x \in \mathfrak{E})$. In particular, φ is τ-continuous and $\varphi(x_0) \neq 0$. On the other hand, if τ is admissible, then for some $y_0 \in \mathfrak{E}$ we have $\varphi(x) = (x, y_0)$ $(x \in \mathfrak{E})$; hence $\varphi(x_0) = (x_0, y_0) = 0$. Contradiction. $\square$

Concerning the uniqueness of admissible topologies one can prove the following.

Theorem 5.3. *If τ_1 is an admissible topology defined by a countable family of semi-norms on the inner product space $\mathfrak{E}$, and τ is any admissible topology on $\mathfrak{E}$, then $\tau_1 \geqq \tau$. In particular, no more than one admissible topology of $\mathfrak{E}$ is metrizable.*

Proof. Let the topologies τ_1, τ be defined by the semi-norms p_j $(j = 1, 2, \ldots)$ and q_γ $(\gamma \in \Gamma)$, respectively. Assume that τ_1 is not stronger than τ. Then there exist indices $\gamma_1, \ldots, \gamma_m \in \Gamma$ and a number $\varepsilon > 0$ such that the set

$$\{x\colon\ q_{\gamma_k}(x) < \varepsilon \quad (k = 1, \ldots, m)\}$$

does not contain any τ_1-neighbourhood of the origin, in particular, any of the sets

$$\left\{x\colon\ p_j(x) < \frac{1}{n} \quad (j = 1, \ldots, n)\right\} \quad (n = 1, 2, \ldots)\,.$$

Consequently, there is a sequence $\{x_n\}_1^\infty \subset \mathfrak{E}$ with

$$p_j(x_n) < \frac{1}{n} \quad (j = 1, \ldots, n)\,, \qquad \max_{1 \leqq k \leqq m} q_{\gamma_k}(x_n) \geqq \varepsilon\,.$$

Since the maximum is taken over a finite set, it means no loss of generality if we write $q(x_n) \geqq \varepsilon$, where q denotes one of the semi-norms q_{γ_k}. Setting $n x_n = y_n$ we obtain:

$$\max_{1 \leqq j \leqq n} p_j(y_n) < 1 \quad (n = 1, 2, \ldots)\,, \tag{5.7}$$

$$q(y_n) \geqq n\varepsilon \quad (n = 1, 2, \ldots)\,. \tag{5.8}$$

The set $\mathfrak{L} = \{x\colon\ q(x) = 0\}$ is a subspace of $\mathfrak{E}$. We consider the quotient space $\hat{\mathfrak{E}} = \mathfrak{E}/\mathfrak{L}$ as a vector space with the (proper) norm $\|\hat{x}\| = q(x)$ $(\hat{x} \in \hat{\mathfrak{E}},\ x \in \hat{x})$. Let $\hat{\varphi}$ denote a linear form on $\hat{\mathfrak{E}}$ which is continuous:

$$|\hat{\varphi}(\hat{x})| \leqq \|\hat{\varphi}\|\,\|\hat{x}\| \quad (\hat{x} \in \hat{\mathfrak{E}})\,. \tag{5.9}$$

Then the formula

$$\varphi(x) = \hat{\varphi}(\hat{x}) \quad (x \in \mathfrak{E},\ x \in \hat{x})$$

defines a linear form on $\mathfrak{E}$ such that $|\varphi(x)| \leqq \|\hat{\varphi}\|\, q(x)$ $(x \in \mathfrak{E})$. In particular, φ is τ-continuous. As τ is admissible, $\varphi(x) = (x, z_0)$ $(x \in \mathfrak{E})$ with a suitable $z_0 \in \mathfrak{E}$. Hence φ is τ_1-continuous. Thus for some positive integer r and positive δ we have $|\varphi(x)| < 1$ whenever $\max_{1 \leqq j \leqq r} p_j(x) < \delta$ or, equivalently,

$$|\varphi(x)| \leqq \frac{1}{\delta} \max_{1 \leqq j \leqq r} p_j(x) \quad (x \in \mathfrak{E})\,. \tag{5.10}$$

From (5.7) and (5.10) we obtain

$$|\varphi(y_n)| < \frac{1}{\delta} \qquad (n \geqq r) .$$

Therefore the sequence $\{\hat{\varphi}(\hat{y}_n)\}_1^\infty$ is bounded for any fixed $\hat{\varphi}$ belonging to the conjugate space $\hat{\mathfrak{E}}^*$ of the normed space $\hat{\mathfrak{E}}$. But $\hat{\varphi}(\hat{y}_n)$ can be regarded as the value of a linear form $\hat{y}_n$ on the element $\hat{\varphi}$ of the Banach space $\hat{\mathfrak{E}}^*$. According to (5.9) this linear form is continuous; its norm is not greater than, and by the Hahn-Banach extension principle not less than, the value $\|\hat{y}_n\| = q(y_n)$. Thus (5.8) contradicts the principle of uniform boundedness. □

6. Orthogonal Companions and Admissible Topologies

The results to be presented in this and the next section show that admissible topologies are a natural tool in the study of orthogonal companions, the existence of projections etc. It also turns out that often in these applications the particular choice of the admissible topology is indifferent.

Theorem 6.1. *Let τ be an admissible topology on the inner product space $\mathfrak{E}$. Then the τ-closure of any subspace $\mathfrak{L} \subset \mathfrak{E}$ coincides with $\mathfrak{L}^{\perp\perp}$.*

Proof. Denoting the τ-closure of $\mathfrak{L}$ by $\bar{\mathfrak{L}}$ we have $\bar{\mathfrak{L}} \subset \mathfrak{L}^{\perp\perp}$, since $\mathfrak{L} \subset \mathfrak{L}^{\perp\perp}$ by definition and $\mathfrak{L}^{\perp\perp}$ is τ-closed by Lemma 2.4.

Now let $x_0 \notin \bar{\mathfrak{L}}$. Then x_0 has a τ-neighbourhood disjoint from $\mathfrak{L}$. This neighbourhood contains a set of the form

$$\{x\colon\, p_{\gamma_j}(x - x_0) < \varepsilon \qquad (j = 1, \ldots, n)\} ,$$

where $p_{\gamma_1}, \ldots, p_{\gamma_n}$ belong to the family of semi-norms defining τ, and ε is positive. Introducing the notations

$$p(x) = \max_{1 \leqq j \leqq n} p_{\gamma_j}(x) \qquad (x \in \mathfrak{E}) ,$$

$$q(x) = \inf_{x' \in \mathfrak{L}} p(x - x') \qquad (x \in \mathfrak{E}) \tag{6.1}$$

it is easy to see that q is a semi-norm and $q(x_0) \geqq \varepsilon$. Therefore the Hahn-Banach extension principle assures the existence of a linear form φ on $\mathfrak{E}$ with

$$\varphi(x_0) = \varepsilon ,$$

$$|\varphi(x)| \leqq q(x) \qquad (x \in \mathfrak{E}) . \tag{6.2}$$

Since, by virtue of (6.1), $q(x) \leqq p(x)$, relation (6.2) implies the τ-continuity of φ. But τ is an admissible topology. Consequently, there exists a $y_0 \in \mathfrak{E}$ such that

(6.3) $$\varphi(x) = (x, y_0) \quad (x \in \mathfrak{E}) .$$

If $x \in \mathfrak{L}$, then (6.1) yields $q(x) = 0$. Hence with the aid of (6.2) and (6.3) we obtain $y_0 \in \mathfrak{L}^\perp$. On the other hand, $(x_0, y_0) = \varphi(x_0) = \varepsilon \neq 0$. Thus $x_0 \notin \mathfrak{L}^{\perp\perp}$. □

Remark 6.2. Theorem 6.1 implies that the closure of a subspace with respect to any admissible topology is the same. Keeping this fact and Theorem 5.1 in mind, it will be convenient to speak of weakly closed and weakly dense subspaces instead of "subspaces which are closed (resp. dense) relative to some (and then every) admissible topology".

Corollary 6.3. *$\mathfrak{L}^{\perp\perp} = \mathfrak{L}$ if and only if the subspace $\mathfrak{L}$ is weakly closed.* □

Corollary 6.4. *A subspace $\mathfrak{L}$ of the non-degenerate space $\mathfrak{E}$ satisfies $\mathfrak{L}^\perp = 0$ if and only if $\mathfrak{L}$ is weakly dense in $\mathfrak{E}$.* □

In order to place Corollary 6.3 (above) and Theorem 6.5 (below) among previous results, recall that if $\mathfrak{L}$ is a subspace of the non-degenerate inner product space $\mathfrak{E}$ such that $\mathfrak{L} + \mathfrak{L}^\perp = \mathfrak{E}$, then a) $\mathfrak{L}$ is non-degenerate (Corollary I.9.5), b) $\mathfrak{L}^{\perp\perp} = \mathfrak{L}$ (Lemma I.9.6), and c) $\mathfrak{L}$ is weakly closed (Corollary 2.5 and Theorem 2.1). By Corollary 6.3 conditions b) and c) are equivalent. By Example I.9.7 conditions a) and b) do not imply $\mathfrak{L} + \mathfrak{L}^\perp = \mathfrak{E}$. The next result says that a weakening of "a) plus b)" is necessary and sufficient for $\mathfrak{L} + \mathfrak{L}^\perp$ to fill $\mathfrak{E}$ "approximately".

Theorem 6.5. *Let $\mathfrak{E}$ be a non-degenerate inner product space, and let $\mathfrak{L}$ be a subspace of $\mathfrak{E}$. The subspace $\mathfrak{L} + \mathfrak{L}^\perp$ is weakly dense in $\mathfrak{E}$ if and only if $\mathfrak{L}^{\perp\perp}$ is non-degenerate.*

Proof. For any subspace $\mathfrak{L}$, in view of (I.3.2) and the definition of orthogonal companion, we have $(\mathfrak{L} + \mathfrak{L}^\perp)^\perp = \mathfrak{L}^\perp \cap \mathfrak{L}^{\perp\perp} = \mathfrak{L}^{\perp\perp\perp} \cap \mathfrak{L}^{\perp\perp} = \mathfrak{L}^{\perp\perp} \cap (\mathfrak{L}^{\perp\perp})^\perp$. It remains to apply Corollary 6.4. □

Corollary 6.6. *For every subspace $\mathfrak{L}$ of a definite inner product space $\mathfrak{E}$ the subspace $\mathfrak{L} + \mathfrak{L}^\perp$ is weakly dense in $\mathfrak{E}$.* □

From Theorem 6.5 one can also derive the following extension of Corollary 6.6.

Lemma 6.7. *Suppose that the space $\mathfrak{E}$ as well as the subspace $\mathfrak{L} \subset \mathfrak{E}$ are non-degenerate and decomposable:*

(6.4) $$\mathfrak{E} = \mathfrak{E}^+ (\dotplus) \mathfrak{E}^-; \quad \mathfrak{E}^+ \subset \mathfrak{P}^{++}, \quad \mathfrak{E}^- \subset \mathfrak{P}^{--},$$

(6.5) $$\mathfrak{L} = \mathfrak{L}^+ (\dotplus) \mathfrak{L}^-; \quad \mathfrak{L}^+ \subset \mathfrak{P}^{++}, \quad \mathfrak{L}^- \subset \mathfrak{P}^{--}.$$

If the fundamental decompositions (6.4) *and* (6.5) *can be chosen in such a way that*

$$\mathfrak{L}^+ \subset \mathfrak{E}^+, \quad \mathfrak{L}^- \subset \mathfrak{E}^-, \tag{6.6}$$

then $\mathfrak{L} + \mathfrak{L}^\perp$ *is weakly dense in* $\mathfrak{E}$.

Proof. Let $x \in \mathfrak{L}^\perp$. Then $x \perp \mathfrak{L}^+$. Relation (6.4) yields a decomposition $x = x^+ + x^-$, where $x^+ \in \mathfrak{E}^+$, $x^- \in \mathfrak{E}^-$. Since (6.4) and (6.6) imply $x^- \perp \mathfrak{L}^+$, we obtain that the vector $x^+ = x - x^-$ is orthogonal to $\mathfrak{L}^+$. On the other hand, again by (6.4) and (6.6), $x^+ \perp \mathfrak{L}^-$. As a result, $x^+ \in \mathfrak{L}^\perp$, and $x^- = x - x^+ \in \mathfrak{L}^\perp$. Therefore

$$\mathfrak{L}^\perp = (\mathfrak{L}^\perp \cap \mathfrak{E}^+) \dotplus (\mathfrak{L}^\perp \cap \mathfrak{E}^-) .$$

In particular, $\mathfrak{L}^\perp$ fulfils the conditions postulated for $\mathfrak{L}$ (cf. Corollary I.4.3). Replacing in the above argument $\mathfrak{L}$ by $\mathfrak{L}^\perp$ we obtain:

$$\mathfrak{L}^{\perp\perp} = (\mathfrak{L}^{\perp\perp} \cap \mathfrak{E}^+) \dotplus (\mathfrak{L}^{\perp\perp} \cap \mathfrak{E}^-) .$$

Thus, in view of Corollary I.4.3, $\mathfrak{L}^{\perp\perp}$ is non-degenerate. By Theorem 6.5 $\mathfrak{L} + \mathfrak{L}^\perp$ is weakly dense in $\mathfrak{E}$. □

7. Projections and Admissible Topologies

In this section admissible topologies will be utilized for two purposes: a) to find all vectors admitting a projection on a fixed subspace, and b) to characterize ortho-complemented subspaces. Recall that problem a) was solved for decomposable subspaces in Theorems I.8.3 — I.8.5, while b) was not solved at all.

Theorem 7.1. *Let* $\mathfrak{L}$ *be a subspace of the inner product space* $\mathfrak{E}$. *Denote an admissible topology of the subspace* $\mathfrak{L}$ *by* τ. *The element* $y \in \mathfrak{E}$ *admits a projection on* $\mathfrak{L}$ *if and only if the linear form* $\varphi_y(x) = (x, y)$ $(x \in \mathfrak{L})$ *is* τ*-continuous.*

Proof. Since τ is admissible on $\mathfrak{L}$, the linear form φ_y (considered on $\mathfrak{L}$) is τ-continuous if and only if there exists an element $y_1 \in \mathfrak{L}$ satisfying $\varphi_y(x) = (x, y_1)$ $(x \in \mathfrak{L})$ i.e. $y - y_1 \perp \mathfrak{L}$. □

Theorem 7.2. *Let* $\mathfrak{L}$ *be a subspace of the inner product space* $\mathfrak{E}$. *Consider the weak topology* $\tau_0(\mathfrak{L})$ *of* $\mathfrak{L}$ *and the topology* $\tau_1 = \tau_0(\mathfrak{E})|\mathfrak{L}$ *induced on* $\mathfrak{L}$ *by the weak topology* $\tau_0(\mathfrak{E})$ *of* $\mathfrak{E}$. *We have* $\tau_0(\mathfrak{L}) \leqq \tau_1$; *equality holds if and only if* $\mathfrak{L}$ *is ortho-complemented.*

Proof. τ_1 is defined by the family $\{p_y|\mathfrak{L}\}_{y \in \mathfrak{E}}$ of semi-norms, where $p_y(x) = |(x, y)|$ $(x \in \mathfrak{E})$, while $\tau_0(\mathfrak{L})$ is defined by the subfamily $\{p_y|\mathfrak{L}\}_{y \in \mathfrak{L}}$. Hence $\tau_0(\mathfrak{L})$ is weaker than τ_1.

Let $\mathfrak{L}$ be ortho-complemented. Then for any $y \in \mathfrak{E}$ we have $p_y|\mathfrak{L} = p_{y_1}|\mathfrak{L}$, where y_1 denotes a projection of y on $\mathfrak{L}$. So in this case $\tau_0(\mathfrak{L})$ and τ_1 are defined by the same family of semi-norms.

Conversely, suppose that $\tau_0(\mathfrak{L}) = \tau_1$. For any $y \in \mathfrak{E}$, according to Theorem 2.1, the linear form $\varphi_y(x) = (x, y)$ is $\tau_0(\mathfrak{E})$-continuous on $\mathfrak{E}$. Therefore φ_y is $\tau_0(\mathfrak{L})$-continuous on $\mathfrak{L}$. Thus, in view of Theorems 7.1 and 5.1, y has a projection on $\mathfrak{L}$. □

8. Intrinsic Topology

Let $\mathfrak{L}$ be a definite subspace of the inner product space $\mathfrak{E}$. The relation

$$(8.1) \qquad |x|_{\mathfrak{L}} = |(x, x)|^{1/2} \qquad (x \in \mathfrak{L})$$

defines a quadratic norm on $\mathfrak{L}$. We call $|\cdot|_{\mathfrak{L}}$ the *intrinsic norm* on $\mathfrak{L}$, and the corresponding normed topology $\tau_{\text{int}}(\mathfrak{L})$ the *intrinsic topology* of $\mathfrak{L}$.

The adjective "intrinsic" will always refer to the topology $\tau_{\text{int}}(\mathfrak{L})$ of a definite subspace $\mathfrak{L}$. We say, for instance, that the definite subspace $\mathfrak{L}$ is *intrinsically complete*, if it is complete with respect to the norm $|\cdot|_{\mathfrak{L}}$. In other words, $\mathfrak{L}$ is intrinsically complete if $\mathfrak{L}$ or the anti-space of $\mathfrak{L}$ is a Hilbert space.

We define the *intrinsic completion* $\tilde{\mathfrak{L}}$ of a definite subspace $\mathfrak{L}$ as the completion of $\mathfrak{L}$ with respect to the intrinsic norm $|\cdot|_{\mathfrak{L}}$. The Hilbert space dimension of $\tilde{\mathfrak{L}}$ (or of the anti-space of $\tilde{\mathfrak{L}}$, if $\mathfrak{L}$ is negative definite) will be termed the *intrinsic dimension* of $\mathfrak{L}$, and denoted by $\dim_{\text{int}} \mathfrak{L}$. It is the minimal power of sets $\mathfrak{A} \subset \mathfrak{L}$ such that the span $\langle\mathfrak{A}\rangle$ is dense in $\mathfrak{L}$ with respect to $|\cdot|_{\mathfrak{L}}$.

Consider the following question. Given a vector $y \in \mathfrak{E}$ and a definite subspace $\mathfrak{L} \subset \mathfrak{E}$, is the linear form $\varphi_y(x) = (x, y)$ $(x \in \mathfrak{L})$ intrinsically continuous on $\mathfrak{L}$ (i.e., continuous with respect to $\tau_{\text{int}}(\mathfrak{L})$) or not? If $\mathfrak{L}$ has the property that φ_y is intrinsically continuous on $\mathfrak{L}$ for every $y \in \mathfrak{E}$, we say $\mathfrak{L}$ is *regular*. Otherwise $\mathfrak{L}$ is said to be *singular*.

Obviously, in a finite-dimensional or definite space $\mathfrak{E}$ every definite subspace is regular. On the other hand, owing to Theorem I.5.4 and Remark I.5.5, a definite subspace $\mathfrak{L}$ of an inner product space $\mathfrak{E}$ is regular if and only if it is regular when considered in a maximal non-degenerate subspace of $\mathfrak{E}$ containing $\mathfrak{L}$. Therefore the existence problem of singular subspaces is fully settled by the following theorem.

Theorem 8.1. *Every infinite-dimensional, non-degenerate, indefinite inner product space* $\mathfrak{E}$ *contains a singular subspace.*

Proof. Let e_1 denote any non-neutral element of $\mathfrak{E}$. The subspace $\langle e_1\rangle$ being ortho-complemented and $\mathfrak{E}$ being non-degenerate, there is a non-neutral $e_2 \in \mathfrak{E}$ orthogonal to e_1. Then $\langle e_1, e_2 \rangle$ is also ortho-complemented, and in its orthogonal companion one can find a non-neutral e_3. Pursuing the process we obtain a sequence $\{e_j\}_1^\infty \subset \mathfrak{E}$ of pairwise orthogonal, non-neutral elements. In view of the indefiniteness of $\mathfrak{E}$ we may also require $\operatorname{sign}(e_2, e_2) \neq \operatorname{sign}(e_1, e_1)$.

Suppose that $\{e_j\}$ contains an infinite number of positive elements. Then taking a subsequence and multiplying its members by suitable positive numbers we get a sequence $\{g_j\}_1^\infty \subset \mathfrak{E}$ which satisfies the conditions

$$(g_1, g_1) = -1\,, \quad (g_j, g_j) = 1 \qquad (j = 2, 3, \ldots)\,,$$
$$(g_j, g_k) = 0 \qquad (j \neq k;\ j,k = 1, 2, \ldots)\,.$$

The set $\mathfrak{L}$ of all finite sums $\sum_{j=1}^{\infty} \xi_j g_j$, where $\xi_1 = \xi_2 = \sum_{j=3}^{\infty}{}' \xi_j$ (cf. Example I.4.9), is easily seen to be a positive definite subspace of $\mathfrak{E}$. However, the linear form $\varphi_{g_1}(x) = (x, g_1)$ $(x \in \mathfrak{L})$ is not intrinsically continuous on $\mathfrak{L}$, since for the vectors

$$x_n = g_1 + g_2 + \frac{1}{n} \sum_{j=3}^{n+2}{}' g_j \in \mathfrak{L} \qquad (n = 1, 2, \ldots)$$

we have

$$|x_n|_{\mathfrak{L}}^2 = (x_n, x_n) = \frac{1}{n} \to 0 \qquad (n \to \infty)\,,$$

$$(x_n, g_1) = -1 \qquad (n = 1, 2, \ldots)\,.$$

If no more than a finite number of vectors e_j is positive, then the above construction applies to the anti-space of $\mathfrak{E}$, again providing a singular subspace in $\mathfrak{E}$. □

9. Projections and Intrinsic Topology

The following lemma is related to Theorem 7.1.

Lemma 9.1. *If the element $y \in \mathfrak{E}$ admits a projection on the definite subspace $\mathfrak{L}$, then the linear form $\varphi_y(x) = (x, y)$ $(x \in \mathfrak{L})$ is intrinsically continuous on $\mathfrak{L}$. Thus every ortho-complemented definite subspace is regular.*

Proof. Let y_1 be the projection of y on $\mathfrak{L}$. Then, according to Lemma I.2.2, for every $x \in \mathfrak{L}$ we have $|(x, y)| = |(x, y_1)| \leqq |y_1|_{\mathfrak{L}}\, |x|_{\mathfrak{L}}$. □

We note that a non-closed subspace of a Hilbert space is not ortho-complemented (cf. Corollary 2.5), though regular. In a definite, but (intrinsically) non-complete space a subspace $\mathfrak{L}$ can even be weakly closed without being ortho-complemented, and $\mathfrak{L}$ is automatically regular again. (Let $\mathfrak{E}$ be the vector space of continuous functions $f(t)$, $0 \leqq t \leqq 1$, with inner product

$$(f, g) = \int_0^1 f(t) \overline{g(t)} \, dt ,$$

and $\mathfrak{L}$ the set of those $f \in \mathfrak{E}$ which vanish for $0 \leqq t \leqq 1/2$.) If, however, a regular subspace $\mathfrak{L}$ of an inner product space is sequentially $\tau_0(\mathfrak{L})$-complete, then it is ortho-complemented. This turns out from the next result coupled with the well-known fact that for a definite inner product space weak sequential completeness is equivalent to (intrinsic) completeness.

Theorem 9.2. *If a definite subspace is regular and intrinsically complete, then it is ortho-complemented.*

Proof. Let $\mathfrak{L}$ be a positive definite, regular, intrinsically complete subspace of the inner product space $\mathfrak{E}$. Since $\mathfrak{L}$ is a Hilbert space, by the Riesz representation theorem $\tau_{\mathrm{int}}(\mathfrak{L})$ is an admissible topology of $\mathfrak{L}$. It remains to apply Theorem 7.1. If $\mathfrak{L}$ is negative definite, we pass to the anti-space of $\mathfrak{E}$. □

In certain cases intrinsic completeness is necessary, too, for ortho-complementedness.

Theorem 9.3. *If $\mathfrak{E}$ is a non-degenerate inner product space such that the relations*

$$(9.1) \qquad \lim_{m,n\to\infty} (x_m - x_n , y) = 0 \qquad (y \in \mathfrak{E})$$

imply the existence of an $x \in \mathfrak{E}$ satisfying

$$(9.2) \qquad \lim_{n\to\infty} (x_n - x, y) = 0 \qquad (y \in \mathfrak{E}) ,$$

then every ortho-complemented definite subspace $\mathfrak{L} \subset \mathfrak{E}$ is intrinsically complete.

Proof. Let $\{x_n\}_1^\infty \subset \mathfrak{L}$,

$$(9.3) \qquad \lim_{m,n\to\infty} |x_m - x_n|_{\mathfrak{L}} = 0 .$$

Then, by Lemma I.2.2,

$$(9.4) \qquad \lim_{m,n\to\infty} (x_m - x_n , y) = 0 \qquad (y \in \mathfrak{L}) .$$

But for any $y \in \mathfrak{E}$ we have $(x_m - x_n, y) = (x_m - x_n, y_0)$, where y_0 is the projection of y on $\mathfrak{L}$. Consequently, (9.4) implies (9.1). Thus for some $x \in \mathfrak{E}$ the relation (9.2) holds. Furthermore, in view of Corollary 2.5 and Theorem 2.1, $\mathfrak{L}$ is weakly closed. Hence $x \in \mathfrak{L}$.

On the other hand, owing to (9.3), there exists an element z belonging to the intrinsic completion $\tilde{\mathfrak{L}}$ of $\mathfrak{L}$ such that $|x_n - z|_{\tilde{\mathfrak{L}}} \to 0$ $(n \to \infty)$. Working in the space $\tilde{\mathfrak{L}}$, relation (9.2) and $(x_n - z, y) \to 0$ $(y \in \mathfrak{L})$ imply that the element $z - x \in \tilde{\mathfrak{L}}$ is orthogonal to $\mathfrak{L}$, a dense subset of $\tilde{\mathfrak{L}}$. Therefore $z = x$. In particular, $z \in \mathfrak{L}$. □

To finish with, we are going to give "pointwise" characterizations of regular and ortho-complemented definite subspaces, respectively. These results should be compared with Theorem I.8.4.

Theorem 9.4. *Consider an element y and a positive definite subspace $\mathfrak{L}$ of the inner product space $\mathfrak{E}$. The linear form $\varphi_y(x) = (x, y)$ $(x \in \mathfrak{L})$ is intrinsically continuous on $\mathfrak{L}$ if and only if*

$$(9.5) \qquad \inf_{x \in \mathfrak{L}} (y - x,\ y - x) > -\infty .$$

In case $\mathfrak{L}$ is negative definite, the respective condition reads

$$(9.6) \qquad \sup_{x \in \mathfrak{L}} (y - x,\ y - x) < \infty .$$

Proof. Let $|(x, y)| \leqq \alpha |x|_{\mathfrak{L}}$ $(x \in \mathfrak{L})$, where $\mathfrak{L}$ is a positive definite subspace. Then for every $x \in \mathfrak{L}$ we have:

$$(y - x,\ y - x) = (y, y) - (y, x) - (x, y) + (x, x) \geqq$$
$$\geqq (y, y) - 2\alpha |x|_{\mathfrak{L}} + |x|_{\mathfrak{L}}^2 .$$

The last expression is a quadratic function of $|x|_{\mathfrak{L}}$ which is bounded from below.

Let, conversely, $(y - x,\ y - x) \geqq \beta$ for a real number β and every vector x belonging to the positive definite subspace $\mathfrak{L}$. This can be written in the form

$$(9.7) \qquad (x, x) - (y, x) - (x, y) + \gamma \geqq 0 \quad (x \in \mathfrak{L}) ,$$

where $\gamma = (y, y) - \beta$. Fix the element x and choose a unimodular number ε such that $\varepsilon(x, y)$ is real. Replacing x in (9.7) by $\varepsilon\lambda x$, where $\lambda \in \boldsymbol{R}$, we find:

$$\lambda^2(x, x) - 2\lambda\varepsilon(x, y) + \gamma \geqq 0 \qquad (-\infty < \lambda < \infty) .$$

It follows that $4\varepsilon^2(x, y)^2 - 4\gamma(x, x) \leqq 0$, i.e., $|(x, y)| \leqq \gamma^{1/2} |x|_{\mathfrak{L}}$. Since γ is independent of x, the intrinsic continuity of $\varphi_y(x)$ $(x \in \mathfrak{L})$ is proved.

If $\mathfrak{L}$ is negative definite, the same argument can be applied to the anti-space of $\mathfrak{E}$. □

Theorem 9.5. *Let $\mathfrak{L}$ be a positive definite subspace of the inner product space $\mathfrak{E}$. The element $y \in \mathfrak{E}$ admits a projection on $\mathfrak{L}$ if and only if the following conditions are fulfilled:*

a) *The linear form* $\varphi_y(x) = (x, y)$ $(x \in \mathfrak{L})$ *is intrinsically continuous on* $\mathfrak{L}$.

b) *There exists an* $x_0 \in \mathfrak{L}$ *such that* $(x_0, x_0) = (x_0, y) = |\varphi_y|_{\mathfrak{L}}^2$, *where* $|\varphi_y|_{\mathfrak{L}} = \sup_{|x|_{\mathfrak{L}} \leqq 1} |\varphi_y(x)|$.

If $\mathfrak{L}$ *is negative definite, condition* b) *should be replaced by* $(x_0, x_0) = (x_0, y) = -|\varphi_y|_{\mathfrak{L}}^2$.

The element x_0 *appearing in* b) *is the projection of* y *on* $\mathfrak{L}$.

Proof. We may confine ourselves to the case where $\mathfrak{L}$ is positive definite.

If x_0 is the projection of y on $\mathfrak{L}$, then obviously $(x_0, y) = (x_0, x_0) = |x_0|_{\mathfrak{L}}^2$. It remains to prove that φ_y is intrinsically continuous on $\mathfrak{L}$, and $|\varphi_y|_{\mathfrak{L}} = |x_0|_{\mathfrak{L}}$. But this follows from the relations

$$|\varphi_y(x)| = |(x, y)| = |(x, x_0)| \leqq |x_0|_{\mathfrak{L}} \, |x|_{\mathfrak{L}} \qquad (x \in \mathfrak{L}),$$

$$\varphi_y(x_0) = (x_0, x_0) = |x_0|_{\mathfrak{L}}^2 .$$

Assume, conversely, that conditions a), b) hold. Owing to a) and the Riesz representation theorem, there is an element $\tilde{x}_0$ in the intrinsic completion $\tilde{\mathfrak{L}}$ of $\mathfrak{L}$ such that $\varphi_y(x) = (x, \tilde{x}_0)_{\tilde{\mathfrak{L}}}$ $(x \in \mathfrak{L})$; we have $|\varphi_y|_{\mathfrak{L}} = |\tilde{x}_0|_{\tilde{\mathfrak{L}}}$. Making use of these relations, and also of condition b), we find:

$$(x_0, y) = (x_0, \tilde{x}_0)_{\tilde{\mathfrak{L}}} ,$$

$$(x_0, x_0) = (x_0, y) = |\tilde{x}_0|_{\tilde{\mathfrak{L}}}^2 .$$

Therefore

$$|x - \tilde{x}_0|_{\tilde{\mathfrak{L}}}^2 = (x_0, x_0) - (x_0, \tilde{x}_0)_{\tilde{\mathfrak{L}}} - (\tilde{x}_0, x_0)_{\tilde{\mathfrak{L}}} + |\tilde{x}_0|_{\tilde{\mathfrak{L}}}^2 = 0 .$$

Hence $\tilde{x}_0 = x_0$, i.e. $\varphi_y(x) = (x, x_0)$ $(x \in \mathfrak{L})$, which shows that x_0 is the projection of y on $\mathfrak{L}$. □

Notes to Chapter III

Partial majorants and admissible topologies have been studied for some time in the duality theory of topological vector spaces; see e.g. Bourbaki [1], Köthe [1], Robertson and Robertson [1], Schaefer [1]. The application of the general theory to inner product spaces was initiated, in different directions and independently of each other, by Aronszajn [1] and Scheibe [1]. The investigations of Aronszajn were continued by Wittstock [2], [3], and those of Scheibe by Ginzburg and Iohvidov [1].

The mathematical treatment of the intrinsic topology of definite subspaces, a subject characteristic for inner product spaces, began with a short note of Bognár [2] in connection with a paper on quantum field theory by A. Uhlmann. The investigations were carried on by Iohvidov [5], [6], [8] and Ginzburg (see Ginzburg and Iohvidov [1]).

Theorem 3.1 is a weakened form of a result of Wittstock [2]. The idea of Example 3.2 belongs to M. L. Brodskiĭ as quoted by Ginzburg and Iohvidov [1]. Theorem 3.3 was proved by Scheibe [1] (cf. also Ginzburg and Iohvidov [1]).

For a more detailed discussion of polar topologies see Aronszajn [1] and Wittstock [3].

Theorems 5.1, 5.3 and 6.1 appear in the article of Ginzburg and Iohvidov [1], and in the books of Bourbaki [1], Köthe [1], Robertson and Robertson [1]. The three latter references also contain an extension (in the non-degenerate case) of Theorem 5.1, according to which a locally convex topology is admissible if and only if it is stronger than the weak topology and weaker than another particular topology called Mackey topology (theorem of Mackey and Arens); for a special case see Theorem IV.8.3 below.

Theorem 7.1 was proved by Ginzburg and Iohvidov [1], while Theorem 7.2 is essentially due to Scheibe [1].

Theorem 8.1 has first been proved by Bognár [2]. However, the idea of the present construction as well as characterizations of singular, resp. intrinsically complete singular, subspaces belong to Iohvidov [5], [6].

Theorem 9.3 seems to be new. All other results of Section 9 were obtained by Iohvidov [5], [6] (see also Ginzburg and Iohvidov [1]).

In connection with the intrinsic topology of definite subspaces Noël [1] introduced new topologies on inner product spaces.

Chapter IV. Majorant Topologies on Inner Product Spaces

Sections 1, 2, and 4 are concerned with the existence and general properties of majorants (roughly speaking, normed topologies in which the inner product is jointly continuous). Section 3 deals with orthonormal systems. Sections 5—7 discuss the existence and topological properties of fundamental decompositions in terms of majorants. In Section 8 the study of the projection problem is carried on. The following results should be specially mentioned: Theorems 3.3, 5.2, 6.4, 7.2, 8.6, Corollaries 6.3, 7.4, and Lemma 7.1.

1. Majorants

Let $\mathfrak{E}$ be a vector space with an inner product $(.\,,.)$. A topology τ on $\mathfrak{E}$ is said to be a *majorant* of the inner product (or: a majorant on $\mathfrak{E}$) if 1) τ is locally convex, and 2) the inner product is (jointly) τ-continuous.

It is evident that for a majorant τ the inner square (x, x) $(x \in \mathfrak{E})$ is τ-continuous. The next result says that for a topology τ defined by a single semi-norm the τ-continuity of the inner square is sufficient, too, in order that τ be a majorant.

Lemma 1.1. *Let p denote a semi-norm on the inner product space $\mathfrak{E}$. If*

$$|(x, x)| \leqq \alpha(p(x))^2 \qquad (x \in \mathfrak{E}), \tag{1.1}$$

where $\alpha \geqq 0$, then

$$|(x, y)| \leqq 2\alpha\, p(x)\, p(y) \qquad (x, y \in \mathfrak{E}). \tag{1.2}$$

Proof. Making use of inequality (1.1), for every $x, y \in \mathfrak{E}$ we find:

$$|(x, y) + (y, x)| = \left|\frac{1}{2}(x + y,\ x + y) - \frac{1}{2}(x - y,\ x - y)\right| \leqq$$

$$\leqq \frac{1}{2}\alpha(p(x + y))^2 + \frac{1}{2}\alpha(p(x - y))^2 \leqq \alpha(p(x) + p(y))^2 .$$

Application of this result to εx and y, where $|\varepsilon| = 1$, $\varepsilon(x, y) = |(x, y)|$, yields:

$$|(x, y)| \leqq \frac{1}{2}\alpha(p(x) + p(y))^2 \qquad (x, y \in \mathfrak{E}).$$

We finally replace x by $(1/p(x))x$ or nx ($n = 1, 2, \ldots$) according as $p(x) \neq 0$ or $p(x) = 0$, and do the same for y. □

Obviously, every majorant is a partial majorant. But a partial majorant need not be a majorant. Moreover, while the weak topology is always a partial majorant, there are inner products having no majorant at all. This will be seen with the help of the following result.

Lemma 1.2. *To every majorant there exists a weaker majorant defined by a single semi-norm.*

Proof. Let τ be a majorant defined by the semi-norms p_γ $(\gamma \in \Gamma)$. Since the inner product is τ-continuous, there are indices $\gamma_1, \ldots, \gamma_n \in \Gamma$ and a number $\varepsilon > 0$ such that $|(x, y)| < 1$ whenever $p_{\gamma_j}(x), p_{\gamma_j}(y) < \varepsilon$ $(j = 1, \ldots, n)$. The function

$$p(x) = \max_{1 \leqq j \leqq n} p_{\gamma_j}(x) \qquad (x \in \mathfrak{E})$$

is a semi-norm on $\mathfrak{E}$, and for $p(x), p(y) < \varepsilon$ we have $|(x, y)| < 1$. Thus the topology $\tau_1 \leqq \tau$ defined by p is a majorant. □

Remark 1.3. In Example III.3.2 we exhibited a non-degenerate inner product having no normed partial majorant, hence no normed majorant. From Lemma 1.2 and Corollary III.2.2 it follows that this inner product cannot have any kind of majorant.

Consequently, in the case of Example III.3.2 the weak topology is not a majorant. This fact is much more general.

Theorem 1.4. *The weak topology τ_0 of the non-degenerate inner product space $\mathfrak{E}$ is a majorant if and only if* $\dim \mathfrak{E} < \infty$.

Proof. Let $\dim \mathfrak{E} < \infty$. Then, owing to Corollary I.11.8, $\mathfrak{E}$ is decomposable. Therefore, by Lemma II.11.4, any J-norm defines a majorant τ_J on $\mathfrak{E}$. Since a finite-dimensional vector space has only one separated locally convex topology (the euclidean topology), τ_0 coincides with τ_J.

Suppose, conversely, that τ_0 is a majorant. Then there exist elements $z_1, \ldots, z_n \in \mathfrak{E}$ and a number $\varepsilon > 0$ such that $|(x, y)| < 1$ whenever $|(x, z_j)| < \varepsilon$, $|(y, z_j)| < \varepsilon$ $(j = 1, \ldots, n)$.

Denote the subspace $\langle z_1, \ldots, z_n \rangle$ by $\mathfrak{L}$. Let $x \in \mathfrak{L}^\perp$, $y \in \mathfrak{E}$. Then

$$|(mx, z_j)| = 0 < \varepsilon \qquad (j = 1, \ldots, n; \quad m = 1, 2, \ldots),$$

and for a suitable $\alpha > 0$ also

$$|(\alpha y, z_j)| < \varepsilon \qquad (j = 1, \ldots, n).$$

It follows that

$$|(mx, \alpha y)| < 1 \qquad (m = 1, 2, \ldots),$$

i.e., $(x, y) = 0$.

We have proved that $\mathfrak{L}^{\perp} \subset \mathfrak{E}^{\perp}$. Therefore $\mathfrak{L}^{\perp\perp} \supset \mathfrak{E}^{\perp\perp} = \mathfrak{E}$. On the other hand, according to Theorem I.10.10, $\mathfrak{L}^{\perp\perp} = \mathfrak{L}$. As a result, $\mathfrak{E} = \mathfrak{L}$. □

2. Majorants and Metrizable Partial Majorants

A necessary and sufficient condition for the existence of majorants reads as follows.

Theorem 2.1. *A non-degenerate inner product admits a majorant if and only if it admits a metrizable partial majorant.*

Proof. Let τ be a majorant of a non-degenerate inner product. Then Lemma 1.2 and Corollary III.2.2 assure the existence of a normed majorant τ_1; this is, all the more, a metrizable partial majorant.

The implication in the reverse direction has already been proved. Namely the normed partial majorant supplied by the proof of Theorem III.3.1 is, in fact, a majorant (cf. relation (III.3.7)). □

Theorem 2.1 is merely concerned with the existence of majorants. It cannot be strengthened by saying that every metrizable partial majorant, itself, is a majorant.

Example 2.2. Let $\mathfrak{E}$ be the vector space of finite sequences of complex numbers. Denote by e_k the sequence whose kth term is 1 and all others are 0. Introduce two inner products on $\mathfrak{E}$ with the help of the relations $(e_j, e_k) = k\,\delta_{jk}$, $[e_j, e_k] = \delta_{jk}$ $(j, k = 1, 2, \ldots)$. Set $\|x\| = [x, x]^{1/2}$ $(x \in \mathfrak{E})$. Then for any pair

$$x = \sum_{j=1}^{m} \alpha_j e_j, \qquad y = \sum_{k=1}^{n} \beta_k e_k$$

we have

$$|(x, y)| = \left| \sum_{k=1}^{m} k\,\alpha_k \bar{\beta}_k \right| \leqq m\,\|x\|\,\|y\| = \gamma(x)\,\|y\|\,.$$

On the other hand, $\|\cdot\|$ does not define a majorant of $(.\,,.)$, since $(e_k, e_k) \to \infty$ $(k \to \infty)$, whereas $\|e_k\| = 1$ $(k = 1, 2, \ldots)$.

If, however, the metrizable partial majorant is complete, this phenomenon cannot occur.

Theorem 2.3. *Every complete metrizable partial majorant is a majorant.*

Proof. Let τ be a complete metrizable partial majorant on the inner product space $\mathfrak{E}$. Consider two sequences $\{x_n\}_1^\infty$, $\{y_n\}_1^\infty \subset \mathfrak{E}$ such that

$\lim_{n\to\infty} x_n = \lim_{n\to\infty} y_n = 0$ with respect to τ. The linear forms φ_{y_n} ($n = 1, 2, \ldots$) defined by $\varphi_{y_n}(x) = (x, y_n)$ ($x \in \mathfrak{E}$) are τ-continuous, and for fixed x the numerical sequence $\{\varphi_{y_n}(x)\}_{n=1}^{\infty}$ is bounded (tends to 0). Thus, according to the uniform boundedness principle (valid in Fréchet spaces), given any neighbourhood $\mathfrak{U}$ of $0 \in \boldsymbol{C}$ there exists a τ-neighbourhood $\mathfrak{V}$ of $0 \in \mathfrak{E}$ so that $x \in \mathfrak{V}$ implies $\varphi_{y_n}(x) \in \mathfrak{U}$ ($n = 1, 2, \ldots$). In particular, $\varphi_{y_n}(x_n) \in \mathfrak{U}$ if n is sufficiently large. Hence $(x_n, y_n) \to 0$ ($n \to \infty$). □

If a non-complete, but normed partial majorant of a non-degenerate inner product is known, then the construction of a normed majorant (cf. the proof of Theorem III.3.1) can be simplified.

Lemma 2.4. *Assume that the norm* $\|\cdot\|$ *defines a partial majorant on the non-degenerate inner product space* $\mathfrak{E}$. *Then the norm*

(2.1) $$\|x\|_1 = \max\{\|x\|, \|x\|'\} \qquad (x \in \mathfrak{E})$$

defines a majorant on $\mathfrak{E}$.

Proof. From relation (III.4.2) we obtain $|(x, y)| \leqq \|x\|_1 \|y\|_1$ ($x, y \in \mathfrak{E}$). □

If $\|\cdot\|$ is quadratic, the norm (2.1) will not as a rule enjoy the same property. Nevertheless, in Lemma 2.4 we can replace $\|\cdot\|_1$ by another norm $\|\cdot\|_2$ that behaves well in this respect too (and is actually equivalent to $\|\cdot\|_1$).

Lemma 2.5. *Assume that the norm* $\|\cdot\|$ *defines a partial majorant on the non-degenerate inner product space* $\mathfrak{E}$. *Then the norm*

(2.2) $$\|x\|_2 = \sqrt{\|x\|^2 + \|x\|'^2} \qquad (x \in \mathfrak{E})$$

defines a majorant on $\mathfrak{E}$. *If* $\|\cdot\|$ *is quadratic, so is* $\|\cdot\|_2$.

Proof. Owing to relations (III.4.2) and (2.2) we have $|(x, y)| \leqq \leqq \|x\|' \|y\| \leqq \|x\|_2 \|y\|_2$ ($x, y \in \mathfrak{E}$). Furthermore, if there is an inner product $[.,.]$ on $\mathfrak{E}$ such that $\|x\|^2 = [x, x]$ ($x \in \mathfrak{E}$), then by Lemma III.4.3 we can find an inner product $[.,.]'$ with $\|x\|'^2 = [x, x]'$ ($x \in \mathfrak{E}$); the sum $[.,.]_2 = [.,.] + [.,.]'$ is clearly an inner product having the desired property $\|x\|_2^2 = [x, x]_2$ ($x \in \mathfrak{E}$). □

The polar topology can also be used for characterizing majorants among normed partial majorants.

Lemma 2.6. *A normed partial majorant* τ *on the non-degenerate inner product space* $\mathfrak{E}$ *is a majorant if and only if* $\tau' \leqq \tau$.

Proof. Let τ be defined by the norm $\|\cdot\|$. The relation $\tau' \leqq \tau$ holds if and only if there exists an $\alpha > 0$ such that

$$\|x\|' \leqq \alpha \|x\| \qquad (x \in \mathfrak{E}) .$$

By the definition of $||x||'$ this is equivalent to

$$|(x, y)| \leqq \alpha \, ||x|| \qquad (x \in \mathfrak{E}, \ ||y|| \leqq 1) \, ,$$

or

$$|(x, y)| \leqq \alpha \, ||x|| \, ||y|| \qquad (x, y \in \mathfrak{E}) \, . \quad \square$$

3. Orthonormal Systems

Let e_γ $(\gamma \in \Gamma)$ be elements of an inner product space $\mathfrak{E}$. If $|(e_\gamma, e_{\gamma'})| = \delta_{\gamma\gamma'}$ $(\gamma, \gamma' \in \Gamma)$, we say the system $\{e_\gamma\}_{\gamma \in \Gamma}$ is *orthonormal*. In other words, a system is orthonormal if its elements are pairwise orthogonal and have inner square $+1$ or -1.

It is easy to see that the elements of an orthonormal system are linearly independent and their span is non-degenerate.

Lemma 3.1. *If x is a non-zero neutral vector in the non-degenerate inner product space $\mathfrak{E}$, then there is an orthonormal system $\{e, f\} \subset \mathfrak{E}$ such that $x \in \langle e, f \rangle$.*

Proof. Let $y \in \mathfrak{E}$, $(x, y) \neq 0$. Then, by Corollary I.4.6, $\langle x, y \rangle$ is indefinite. Let $e \in \langle x, y \rangle$, $(e, e) = 1$. It follows from Lemma I.9.8 that $\langle x, y \rangle$ is spanned by e and a vector f orthogonal to e. Owing to Remark I.3.2 f must be negative. Hence we may require $(f, f) = -1$. $\square$

The next result relies on Lemma 3.1 and the idea of the Gram-Schmidt orthogonalization process.

Theorem 3.2. *Let $x_1, x_2, \ldots$ be a sequence of elements in the non-degenerate inner product space $\mathfrak{E}$. Then there exists a countable (finite or infinite) orthonormal system $\{e_j\} \subset \mathfrak{E}$ whose span contains each of the vectors $x_1, x_2, \ldots$.*

Proof. Let $x_{j_0} = y_0$ be the first non-zero element in the sequence $\{x_j\}$. If it is non-neutral, we set $e_1 = |(y_0, y_0)|^{-1/2} \, y_0$. In the opposite case we apply Lemma 3.1 to obtain an orthonormal system $\{e_1, e_2\}$ satisfying $y_0 \in \langle e_1, e_2 \rangle$.

Suppose that for some positive integer n we have already constructed a finite orthonormal system $\{e_1, \ldots, e_k\}$ whose span $\mathfrak{L}_k$ contains the vectors $x_1, \ldots, x_n$. Let j_n be the first index with $x_{j_n} \notin \mathfrak{L}_k$ (if such an index exists). Put

$$y_n = x_{j_n} - \sum_{j=1}^{k} (e_j, e_j) \, (x_{j_n}, e_j) \, e_j \, .$$

Then $y_n \neq 0$, $y_n \in \mathfrak{L}_k^\perp$.

According to Corollary I.11.9 and Lemma I.4.2 $\mathfrak{L}_k^\perp$ is non-degenerate. If $(y_n, y_n) \neq 0$, we define $e_{k+1} = |(y_n, y_n)|^{-1/2} y_n$. Otherwise we select an orthonormal system $\{e_{k+1}, e_{k+2}\} \subset \mathfrak{L}_k^\perp$ such that $y_n \in \langle e_{k+1}, e_{k+2}\rangle$ (cf. Lemma 3.1). In both cases we obtain a finite extension of the orthonormal system $\{e_1, \ldots, e_k\}$ with span containing x_{n+1} as well. □

Let τ be a locally convex topology on the inner product space $\mathfrak{E}$. We say that the system $\{x_\gamma\}_{\gamma \in \Gamma} \subset \mathfrak{E}$ is *complete* in $\mathfrak{E}$ with respect to τ (or: τ-complete in $\mathfrak{E}$), if the span $\langle x_\gamma \rangle_{\gamma \in \Gamma}$ is τ-dense in $\mathfrak{E}$.

Theorem 3.3. *Suppose that τ is a separable majorant on the non-degenerate inner product space $\mathfrak{E}$. Then every orthonormal system in $\mathfrak{E}$ is countable, and there is a τ-complete orthonormal system in $\mathfrak{E}$.*

Proof. Let $\tau_1 \leqq \tau$ be a normed majorant on $\mathfrak{E}$ (cf. Lemma 1.2 and Corollary III.2.2). Then also τ_1 is separable. Let $\mathfrak{L}_1$ be the subspace spanned by the positive elements of an orthonormal system $\{e_\gamma\}_{\gamma \in \Gamma}$. The topology $\tau_1|\mathfrak{L}_1$ induced by the separable normed topology τ_1 is separable. On the other hand, $\tau_1|\mathfrak{L}_1$ being a majorant on $\mathfrak{L}_1$, the intrinsic topology $\tau_{\text{int}}(\mathfrak{L}_1)$ is weaker than $\tau_1|\mathfrak{L}_1$. As a result, $\tau_{\text{int}}(\mathfrak{L}_1)$ is separable. Therefore, by the theory of Hilbert spaces, the set of positive elements e_γ is countable.

Passing to the anti-space of $\mathfrak{E}$ we conclude that the set of negative elements e_γ is also countable.

Finally, let $x_1, x_2, \ldots$ be a τ-dense sequence in $\mathfrak{E}$. According to Theorem 3.2 there exists a countable orthonormal system $\{e_j\} \subset \mathfrak{E}$ whose span contains each x_j. In particular, $\{e_j\}$ is τ-complete in $\mathfrak{E}$. □

The next result yields criteria for the validity of a natural generalization to indefinite inner product spaces of the well-known orthogonal expansion theorems.

Theorem 3.4. *Let $\{e_j\}_{j=1}^\infty$ be an orthonormal system in the non-degenerate inner product space $\mathfrak{E}$. The following conditions are equivalent:*

a) *$\mathfrak{E}$ admits a fundamental decomposition*

$$\mathfrak{E} = \mathfrak{E}^+ (\dot{+}) \mathfrak{E}^-; \quad \mathfrak{E}^+ \subset \mathfrak{P}^{++}, \quad \mathfrak{E}^- \subset \mathfrak{P}^{--} \tag{3.1}$$

such that the e_j's with $(e_j, e_j) = 1$ belong to $\mathfrak{E}^+$ and form there a $\tau_{\text{int}}(\mathfrak{E}^+)$-complete orthonormal system, whereas the e_j's with $(e_j, e_j) = -1$ belong to $\mathfrak{E}^-$ and form there a $\tau_{\text{int}}(\mathfrak{E}^-)$-complete orthonormal system.

b) *We have*

$$\sum_{j=1}^{\infty} |(x, e_j)|^2 < \infty \quad (x \in \mathfrak{E}) \tag{3.2}$$

and

$$-\sum_{(e_j, e_j) = -1} |(x, e_j)|^2 \leqq (x, x) \leqq \sum_{(e_j, e_j) = 1} |(x, e_j)|^2 \quad (x \in \mathfrak{E}). \tag{3.3}$$

c) *We have the relations* (3.2) *and*

$$(x, y) = \sum_{j=1}^{\infty}{}' (e_j, e_j)\,(x, e_j)\,(e_j, y) \qquad (x,y \in \mathfrak{E})\,. \tag{3.4}$$

d) *The function*

$$\|x\| = \left(\sum_{j=1}^{\infty}{}' |(x, e_j)|^2\right)^{1/2} \qquad (x \in \mathfrak{E}) \tag{3.5}$$

is a quadratic norm on $\mathfrak{E}$ *and it defines a majorant topology. With respect to this topology we have*

$$x = \sum_{j=1}^{\infty}{}' (e_j, e_j)\,(x, e_j)\,e_j \qquad (x \in \mathfrak{E})\,. \tag{3.6}$$

Proof. The following notation will be useful:

$$e_j^+ = \begin{cases} e_j & \text{if } (e_j, e_j) = 1\,, \\ 0 & \text{otherwise;} \end{cases} \qquad e_j^- = \begin{cases} e_j & \text{if } (e_j, e_j) = -1\,, \\ 0 & \text{otherwise.} \end{cases}$$

a) *implies* b). If $x = x^+ + x^-$, where $x^+ \in \mathfrak{E}^+$, $x^- \in \mathfrak{E}^-$, then by the theory of Hilbert spaces

$$\sum_{j=1}^{\infty}{}' |(x, e_j^+)|^2 = \sum_{j=1}^{\infty}{}' |(x^+, e_j^+)|^2 = (x^+, x^+)$$

and

$$\sum_{j=1}^{\infty}{}' |(x, e_j^-)|^2 = \sum_{j=1}^{\infty}{}' |(x^-, e_j^-)|^2 = -\,(x^-, x^-)\,.$$

Therefore (3.2) and (3.3) are valid.

b) *implies* c). We consider an $x \in \mathfrak{E}$ and put

$$x_n = \sum_{j=1}^{n}{}' (e_j\,, e_j)\,(x, e_j)\,e_j\,. \tag{3.7}$$

Then

$$(x_n\,, e_j) = \begin{cases} (x, e_j) & \text{for } j \leqq n\,, \\ 0 & \text{for } j > n\,. \end{cases}$$

Applying relation (3.3) to the element $x - x_n$ we obtain:

$$-\sum_{j=n+1}^{\infty}{}' |(x, e_j^-)|^2 \leqq (x, x) - \sum_{j=1}^{n}{}' (e_j\,, e_j)|(x,e_j)|^2 \leqq \sum_{j=n+1}^{\infty}{}' |(x, e_j^+)|^2\,.$$

In view of (3.2) the extreme members tend to 0 as $n \to \infty$. This proves (3.4) for $x = y$. Since the expressions on both sides of (3.4) represent inner products and the corresponding inner squares coincide, it follows from the polarization formula (I.2.3) that the inner products coincide themselves.

c) *implies* d). By (3.2), $||x||$ is finite for every x. It is also clear that $||x||^2 = [x, x]$ $(x \in \mathfrak{E})$, where

$$(3.8) \qquad [x, y] = \sum_{j=1}^{\infty} (x, e_j)(e_j, y) \qquad (x, y \in \mathfrak{E})$$

is a positive inner product on $\mathfrak{E}$. This inner product is even definite, because $[x, x] = 0$ implies $(x, e_j) = 0$ $(j = 1, 2, \ldots)$ and, owing to (3.4), $(x, y) = 0$ for every $y \in \mathfrak{E}$. Consequently, $||\cdot||$ is a quadratic norm. It defines a majorant of $(.,.)$, since in view of (3.4) and the Cauchy inequality we have

$$|(x, y)| \leqq \sum_{j=1}^{\infty} |(x, e_j)|\,|(e_j, y)| \leqq ||x||\,||y||\,.$$

Furthermore, considering an $x \in \mathfrak{E}$ and using the notation (3.7) as well as the assumption (3.2) we find:

$$||x - x_n||^2 = \sum_{j=1}^{\infty} |(x - x_n, e_j)|^2 = \sum_{j=n+1}^{\infty} |(x, e_j)|^2 \to 0 \qquad (n \to \infty)\,.$$

d) *implies* a). Set

$$(3.9) \qquad \mathfrak{E}^+ = \{x\colon (x, e_j^-) = 0 \quad (j = 1, 2, \ldots)\}\,,$$

$$(3.10) \qquad \mathfrak{E}^- = \{x\colon (x, e_j^+) = 0 \quad (j = 1, 2, \ldots)\}\,.$$

Then $e_j^+ \in \mathfrak{E}^+$, $e_j^- \in \mathfrak{E}^-$ $(j = 1, 2, \ldots)$. Moreover, on account of (3.6),

$$(3.11) \qquad x = \sum_{j=1}^{\infty} (x, e_j^+)\, e_j^+ \qquad (x \in \mathfrak{E}^+)\,,$$

$$(3.12) \qquad x = -\sum_{j=1}^{\infty} (x, e_j^-)\, e_j^- \qquad (x \in \mathfrak{E}^-)\,.$$

Since the topology involved in the convergence of the series (3.6), (3.11), (3.12) is a majorant, inner products can be calculated term by term. Making use of this remark it is easy to verify (3.1) and the relations

$$|x|_{\mathfrak{E}^+} = ||x|| \quad (x \in \mathfrak{E}^+), \qquad |x|_{\mathfrak{E}^-} = ||x|| \quad (x \in \mathfrak{E}^-)$$

(see (3.5)). In particular, (3.11) and (3.12) hold in the respective intrinsic topologies too, which proves the completeness properties required by condition a). □

4. Minimal Majorants

A majorant τ_* on the inner product space $\mathfrak{E}$ is said to be *minimal*, if no majorant $\tau \neq \tau_*$ is weaker than τ_*.

Lemma 4.1. *Let τ be a normed majorant on the non-degenerate inner product space $\mathfrak{E}$. There exists a normed majorant $\tau_\infty \leqq \tau$ which is self-polar.*

Proof. Let τ be defined by the norm $\|\cdot\|$. For some $\alpha > 0$ we have $|(x, y)| \leqq \alpha \|x\| \|y\| \quad (x,y \in \mathfrak{E})$. Setting

$$\|x\|_1 = \alpha^{1/2}\|x\| \qquad (x \in \mathfrak{E}) \tag{4.1}$$

we obtain:

$$|(x, y)| \leqq \|x\|_1 \|y\|_1 \qquad (x,y \in \mathfrak{E}) .$$

Let

$$\|x\|_2 = \sqrt{\frac{1}{2}(\|x\|_1^2 + \|x\|_1'^2)} \qquad (x \in \mathfrak{E}) .$$

It is easy to see that $\|\cdot\|_2$ is a norm on $\mathfrak{E}$. Furthermore, from the relation $|(x, y)| \leqq \|x\|_1' \|y\|_1$ (see (III.4.2)) and its symmetric counterpart $|(x, y)| \leqq \|x\|_1 \|y\|_1'$ we derive

$$|(x, y)| \leqq \frac{1}{2}(\|x\|_1' \|y\|_1 + \|x\|_1 \|y\|_1') \leqq \frac{1}{2}\sqrt{\|x\|_1^2 + \|x\|_1'^2}\sqrt{\|y\|_1^2 + \|y\|_1'^2}$$

i.e.

$$|(x, y)| \leqq \|x\|_2 \|y\|_2 \qquad (x,y \in \mathfrak{E}) .$$

Pursuing the process, we define the norms

$$\|x\|_{n+1} = \sqrt{\frac{1}{2}(\|x\|_n^2 + \|x\|_n'^2)} \qquad (x \in \mathfrak{E};\ n = 1, 2, \ldots) \tag{4.2}$$

and verify by recursion that

$$|(x, y)| \leqq \|x\|_n \|y\|_n \qquad (x,y \in \mathfrak{E};\quad n = 1, 2, \ldots) . \tag{4.3}$$

In view of (4.3) we have $\|x\|_n' \leqq \|x\|_n$. Thus, owing to relation (4.2),

$$\|x\|_{n+1} \leqq \|x\|_n \qquad (x \in \mathfrak{E};\ n = 1, 2, \ldots) . \tag{4.4}$$

Therefore the limit

$$\|x\|_\infty = \lim_{n\to\infty} \|x\|_n \qquad (x \in \mathfrak{E}) \tag{4.5}$$

exists.

$\|\cdot\|_\infty$ is clearly a semi-norm. But it is a norm as well. Namely (4.4) implies

$$\|x\|_{n+1}' \geqq \|x\|_n' \qquad (x \in \mathfrak{E};\ n = 1, 2, \ldots); \tag{4.6}$$

hence, taking the definition (4.2) into account,

$$||x||_{n+1} \geqq \sqrt{\frac{1}{2}||x||_n'^2} \geqq \frac{1}{\sqrt{2}}||x||_1' \qquad (n = 1,2, \ldots)$$

and, consequently, $||x||_\infty \geqq \frac{1}{\sqrt{2}} ||x||_1'$.

From (4.3) and (4.5) it follows that $|(x, y)| \leqq ||x||_\infty ||y||_\infty$ $(x,y \in \mathfrak{E})$. In particular, the topology τ_∞ defined by $||\cdot||_\infty$ is a majorant on $\mathfrak{E}$. Moreover, from (4.5), (4.4) and (4.1) we obtain the relation $||x||_\infty \leqq \leqq \alpha^{1/2}||x||$ $(x \in \mathfrak{E})$. Thus $\tau_\infty \leqq \tau$.

It remains to prove that $\tau_\infty' = \tau_\infty$. By (4.4) and (4.5) we have $||x||_n \geqq ||x||_\infty$. Therefore

$$(4.7) \qquad ||x||_n' \leqq ||x||_\infty' \qquad (x \in \mathfrak{E};\quad n = 1,2, \ldots).$$

In view of (4.6) and (4.7) the limit

$$(4.8) \qquad ||x||^\infty = \lim_{n\to\infty} ||x||_n' \qquad (x \in \mathfrak{E})$$

exists and satisfies the inequality

$$(4.9) \qquad ||x||^\infty \leqq ||x||_\infty' \qquad (x \in \mathfrak{E}).$$

On the other hand, relations (III.4.2), (4.6), (4.8) yield:

$$|(x, y)| \leqq ||x||_n' \, ||y||_n \leqq ||x||^\infty ||y||_n \qquad (x,y \in \mathfrak{E};\quad n = 1,2, \ldots).$$

Hence $|(x, y)| \leqq ||x||^\infty ||y||_\infty$, i.e.,

$$(4.10) \qquad ||x||_\infty' \leqq ||x||^\infty \qquad (x \in \mathfrak{E}).$$

According to (4.9), (4.10) and (4.8) we have:

$$(4.11) \qquad ||x||_\infty' = \lim_{n\to\infty} ||x||_n' \qquad (x \in \mathfrak{E}).$$

Passing in (4.2) to the limit with the help of formulas (4.5), (4.11), we conclude that $||x||_\infty = ||x||_\infty'$. □

Theorem 4.2. *A partial majorant τ on the non-degenerate inner product space $\mathfrak{E}$ is a minimal majorant if and only if it is normed and self-polar.*

Proof. Let τ be a minimal majorant on $\mathfrak{E}$. According to Lemma 1.2 and Corollary III.2.2 there exists a normed majorant $\tau_1 \leqq \tau$. Furthermore, by Lemma 4.1 there exists a normed self-polar majorant $\tau_\infty \leqq \tau_1$. The minimality of τ implies $\tau_\infty = \tau_1 = \tau$.

Let, conversely, τ be a normed self-polar partial majorant on $\mathfrak{E}$. By Lemma 2.6 τ is a majorant. Let $\tau_1 \leqq \tau$ be also a majorant. In view of Lemma 1.2 and Corollary III.2.2 we may assume that τ_1 is normed. Then Lemma 2.6 and Lemma III.4.1 yield: $\tau_1 \geqq \tau_1' \geqq \tau' = \tau$. □

As a consequence of Lemma 1.2, Corollary III.2.2, Lemma 4.1, and Theorem 4.2 we obtain the following.

Corollary 4.3. *To every majorant of a non-degenerate inner product a weaker minimal majorant can be found.* □

If there is a unique minimal majorant on the non-degenerate space $\mathfrak{E}$ then, according to Corollary 4.3, it is the *weakest majorant* on $\mathfrak{E}$ in the sense that it is weaker than any majorant. If there are several minimal majorants, then a weakest majorant does not exist.

Example 4.4. Let $\mathfrak{E}$ be the vector space of complex numerical sequences $x = \{\xi_j\}_1^\infty$ satisfying the relations

$$\sum_{j=1}^{\infty}{}' |\xi_j|^2 < \infty, \quad \sum_{j=1}^{\infty} (2j)^2 |\xi_{2j}|^2 < \infty, \quad \sum_{j=1}^{\infty} \frac{1}{(2j)^2} |\xi_{2j-1}|^2 < \infty.$$

For $x, y \in \mathfrak{E}$, $y = \{\eta_j\}_1^\infty$, put

$$(x, y) = \sum_{j=1}^{\infty}{}' (\xi_{2j-1}\bar{\eta}_{2j} + \xi_{2j}\bar{\eta}_{2j-1}).$$

Each of the norms

$$\|x\|_1 = \left(\sum_{j=1}^{\infty}{}' |\xi_j|^2\right)^{1/2}, \qquad \|x\|_2 = \left(\sum_{j=1}^{\infty} (2j)^2 |\xi_{2j}|^2 + \sum_{j=1}^{\infty} \frac{1}{(2j)^2} |\xi_{2j-1}|^2\right)^{1/2}$$

defines a majorant of this inner product, as one verifies most easily with the help of Lemma 1.1. Suppose $\|\cdot\|_3$ is a norm on $\mathfrak{E}$ that defines a weaker topology than any of $\|\cdot\|_1$ and $\|\cdot\|_2$. For the elements

$$x_n = \left\{\frac{1}{\sqrt{n}}\delta_{j,2n-1} + \frac{1}{\sqrt{n}}\delta_{j,2n}\right\}_{j=1}^{\infty},$$

$$y_n = \left\{2\sqrt{n}\,\delta_{j,2n-1} + \frac{1}{2\sqrt{n^3}}\delta_{j,2n}\right\}_{j=1}^{\infty}$$

we have $\|x_n\|_1 \to 0$, $\|y_n\|_2 \to 0$ $(n \to \infty)$. Therefore $\|x_n + y_n\|_3 \to 0$ $(n \to \infty)$. On the other hand, $(x_n + y_n, x_n + y_n) > 4$ $(n = 1, 2, \ldots)$. Thus the topology defined by $\|\cdot\|_3$ is not a majorant. It follows that our inner product has no weakest majorant.

In Chapter III we introduced admissible topologies, a subclass of partial majorants. Below we consider the special case where the admissible topology is even a majorant. (That this case is really a special one turns out from Theorem 1.4 and Theorem III.5.1.)

Theorem 4.5. *If the admissible topology τ on the non-degenerate inner product space $\mathfrak{E}$ is a majorant, then* a) *τ is a minimal majorant,* b) *τ is a Banach topology,* c) *τ is the only admissible majorant on $\mathfrak{E}$, and* d) *τ is stronger than any admissible topology on $\mathfrak{E}$.*

Proof. a) Let $\tau_1 \leqq \tau$ be a majorant. Making use of Lemma 1.2 and Corollary III.2.2 we find a normed majorant $\tau_2 \leqq \tau_1$. The majorant τ_2 being weaker than the admissible topology τ, it is necessarily admissible. Therefore, according to Theorem III.5.3, $\tau_2 \geqq \tau$. As a result, $\tau_1 = \tau$.

b) It follows from a) and Theorem 4.2 that τ is defined by a norm $\|\cdot\|$. So $\mathfrak{E}$ is a normed space, and its conjugate space $\mathfrak{E}^*$ is a Banach space. The admissibility of τ implies that the mapping $y \to \varphi_y$, where $\varphi_y(x) = (x, y)$ $(x \in \mathfrak{E})$, is an isomorphism between $\mathfrak{E}$ and $\mathfrak{E}^*$. Since $\|\varphi_y\| = \|y\|'$ and, by a) and Theorem 4.2, $\|\cdot\|'$ is equivalent to $\|\cdot\|$, this isomorphism is continuous in both directions. Thus, along with $\mathfrak{E}^*$, also $\mathfrak{E}$ must be complete.

Assertions c) — d) follow from b) and Theorem III.5.3. □

5. Majorants and Decomposability

Every fundamental decomposition of a decomposable non-degenerate inner product space $\mathfrak{E}$ gives rise to a quadratic norm, the corresponding J-norm. In view of Lemma II.11.4 the topology τ_J defined by this norm is a majorant on $\mathfrak{E}$; it will be called the *decomposition majorant* belonging to the fundamental decomposition in question.

A space having a quadratic-normed majorant need not be decomposable.

Example 5.1. Let $\mathfrak{E}$ denote the vector space of doubly infinite numerical sequences of the form $x = \{\xi_j\}_{j=-\infty}^{\infty}$, where $\xi_j = 0$ $(j \leqq j_0(x))$ and $\sum_{j=1}^{\infty}{}' |\xi_j|^2 < \infty$. (The first condition says that all but a finite number of terms with negative subscript vanish.) For $x, y \in \mathfrak{E}$, $y = \{\eta_j\}_{-\infty}^{\infty}$, define

$$(x, y) = \sum_{j=-\infty}^{\infty}{}' \xi_j \bar{\eta}_{-j-1} .$$

One verifies in just the same way as in Example I.11.3 that the non-degenerate inner product space $\mathfrak{E}$ so obtained is non-decomposable. Nevertheless, $\mathfrak{E}$ admits a majorant defined by the quadratic norm

$$\|x\| = \left(\sum_{j=-\infty}^{\infty}{}' |\xi_j|^2\right)^{1/2} \qquad (x \in \mathfrak{E}) .$$

Next let $\mathfrak{E}$ be an inner product space with a quadratic-normed and complete majorant: $|(x, y)| \leqq \alpha \|x\| \|y\|$ $(x, y \in \mathfrak{E})$, where $\alpha > 0$, $\|x\|^2 = [x, x]$ $(x \in \mathfrak{E})$, and $\mathfrak{E}$ is a Hilbert space with respect to the positive definite inner product $[.,.]$. Then there exists a linear operator G in $\mathfrak{E}$ such that

$$(5.1) \qquad (x, y) = [Gx, y] \qquad (x, y \in \mathfrak{E}) .$$

G is called the *Gram operator* of $(.,.)$ with respect to $[.,.]$. We have

$$(5.2) \qquad \|Gx\| \leqq \alpha \|x\| \qquad (x \in \mathfrak{E})$$

and

$$(5.3) \qquad [Gx, y] = [x, Gy] \qquad (x, y \in \mathfrak{E}) .$$

Theorem 5.2. *If the inner product space $\mathfrak{E}$ admits a Hilbert majorant τ, then $\mathfrak{E}$ is decomposable. Moreover, the fundamental decomposition of $\mathfrak{E}$ can be chosen so that its three components as well as the direct sum of any two of them are τ-closed.*

Proof. Consider a positive definite inner product $[.,.]$ such that the norm $\|x\|=[x, x]^{1/2}$ $(x \in \mathfrak{E})$ defines τ. Denote the Gram operator of the original inner product $(.,.)$ with respect to $[.,.]$ by G. On account of (5.2) and (5.3) G has a right-continuous spectral function $\{E(\lambda)\}_{\lambda=-\infty}^{\infty}$ satisfying $G(-\alpha-0) = 0$, $G(\alpha) = I$,

$$(5.4) \qquad Gx = \int_{-\alpha-0}^{\alpha} \lambda \, dE(\lambda)x \qquad (x \in \mathfrak{E}) ,$$

where the integrals exist in the τ-topology. It is easy to see that the subspaces

$$\mathfrak{E}^- = E(-0)\mathfrak{E} , \quad \mathfrak{E}^0 = (E(0) - E(-0))\mathfrak{E} , \quad \mathfrak{E}^+ = (I - E(0))\mathfrak{E}$$

form a fundamental decomposition of $\mathfrak{E}$; moreover, each of the subspaces $\mathfrak{E}^0$, $\mathfrak{E}^+$, $\mathfrak{E}^-$, $\mathfrak{E}^0 \dotplus \mathfrak{E}^+$, $\mathfrak{E}^0 \dotplus \mathfrak{E}^-$, $\mathfrak{E}^+ \dotplus \mathfrak{E}^-$ is τ-closed. □

Remark 5.3. For a non-degenerate $\mathfrak{E}$ the second half of Theorem 5.2 is weaker than Corollary III.2.6.

In the rest of this section we show that the existence of a Banach majorant is not sufficient for decomposability.

Lemma 5.4. *If τ_1 is a decomposition majorant on the non-degenerate inner product space $\mathfrak{E}$ and τ is a Banach majorant on $\mathfrak{E}$, then $\tau_1 \leqq \tau$.*

Proof. Let τ_1 belong to the decomposition

$$(5.5) \qquad \mathfrak{E} = \mathfrak{E}^+ (\dotplus) \mathfrak{E}^- ; \quad \mathfrak{E}^+ \subset \mathfrak{P}^{++} , \quad \mathfrak{E}^- \subset \mathfrak{P}^{--} .$$

Denote the corresponding fundamental projectors by P^+, P^-. From Corollary III.2.6 and the closed graph principle it follows that P^+, P^- are τ-continuous. For $J = P^+ - P^-$ and any norm $\| \cdot \|$ defining τ we have $\|x\|_J^2 = (Jx, x) \leqq \alpha \, \|Jx\| \, \|x\| \leqq \alpha_1 \|x\|^2$ $(x \in \mathfrak{E})$. □

Theorem 5.5. *If τ is an admissible majorant on a decomposable, non-degenerate inner product space, then τ is a Hilbert majorant.*

Proof. According to Theorem 4.5 τ is both a Banach majorant and a minimal majorant. Therefore, in view of Lemma 5.4, every decomposition majorant coincides with τ. Hence τ is quadratic-normed. □

Example 5.6. Let $\mathfrak{E}$ be a reflexive Banach space with norm $\|\cdot\|$ not equivalent to any quadratic norm. (We may choose e.g. $\mathfrak{E} = l^p$; $1 < p < \infty$, $p \neq 2$, since it contains closed subspaces without closed complementary subspaces, as first proved by Murray [1].) Consider the Banach space $\mathfrak{E} \times \mathfrak{E}^*$ of all pairs $\{x, \varphi\}$, where x belongs to $\mathfrak{E}$ and φ is a continuous linear form on $\mathfrak{E}$, linear operations and norm being defined by the relations

$$\alpha_1\{x_1, \varphi_1\} + \alpha_2\{x_2, \varphi_2\} = \{\alpha_1 x_1 + \alpha_2 x_2, \quad \alpha_1\varphi_1 + \alpha_2\varphi_2\}$$
$$(\alpha_1, \alpha_2 \in \boldsymbol{C}; \quad x_1, x_2 \in \mathfrak{E}; \quad \varphi_1, \varphi_2 \in \mathfrak{E}^*),$$

$$\|\{x, \varphi\}\| = (\|x\|^2 + \|\varphi\|^2)^{1/2} \qquad (x \in \mathfrak{E}, \quad \varphi \in \mathfrak{E}^*). \tag{5.6}$$

Introduce the non-degenerate inner product

$$(\{x, \varphi\}, \{y, \psi\}) = \psi(x) + \varphi(y) \qquad (x, y \in \mathfrak{E}; \quad \varphi, \psi \in \mathfrak{E}^*). \tag{5.7}$$

Obviously, the norm (5.6) defines a majorant of the inner product (5.7). We are going to prove that this majorant is an admissible topology.

Let Φ denote a linear form on $\mathfrak{E} \times \mathfrak{E}^*$ which is continuous relative to the norm (5.6). The restriction of Φ to $\mathfrak{E} \times 0$ can be looked upon as a continuous linear form on the Banach space $\mathfrak{E}$. Thus for a suitable $\varphi_0 \in \mathfrak{E}^*$ we have $\Phi(\{x, 0\}) = \varphi_0(x)$ $(x \in \mathfrak{E})$. Similarly, making use of the reflexivity of $\mathfrak{E}$, we find an $x_0 \in \mathfrak{E}$ with $\Phi(\{0, \varphi\}) = \varphi(x_0)$ $(\varphi \in \mathfrak{E}^*)$. As a result,

$$\Phi(\{x, \varphi\}) = \varphi_0(x) + \varphi(x_0) = (\{x, \varphi\}, \{x_0, \varphi_0\}) \qquad (\{x, \varphi\} \in \mathfrak{E} \times \mathfrak{E}^*).$$

Now suppose that $\mathfrak{E} \times \mathfrak{E}^*$ is decomposable. Then, in view of Theorem 5.5 and what we have just seen, the norm (5.6) is equivalent to a quadratic norm on $\mathfrak{E} \times \mathfrak{E}^*$. The restriction of the latter to $\mathfrak{E} \times 0$ yields a quadratic norm on $\mathfrak{E}$ equivalent to the original norm. This contradicts the choice of $\mathfrak{E}$.

6. Decomposition Majorants

In this section we consider decomposable, non-degenerate inner product spaces.

Theorem 6.1. *Every decomposition majorant is a minimal majorant.*

Proof. By Theorem 4.2 it is sufficient to prove that every J-norm is equivalent to its polar. But the latter property follows from Lemma II.11.1 and the theory of Hilbert spaces:

$$\|x\|_J' = \sup_{\|y\|_J \leqq 1} |(x, y)| = \sup_{\|z\|_J \leqq 1} |(x, Jz)| = \sup_{\|z\|_J \leqq 1} |(x, z)_J| = \|x\|_J . \quad \square$$

In certain cases decomposition majorants are not only minimal, but also weakest majorants.

Theorem 6.2. *Suppose that the inner product space* $\mathfrak{E}$ *has a fundamental decomposition of the form*

$$\mathfrak{E} = \mathfrak{E}^+ (\dot{+}) \mathfrak{E}^-; \quad \mathfrak{E}^+ \subset \mathfrak{P}^{++}, \quad \mathfrak{E}^- \subset \mathfrak{P}^{--}, \tag{6.1}$$

where one of the components $\mathfrak{E}^+$, $\mathfrak{E}^-$ *is intrinsically complete. Then there is only one minimal majorant on* $\mathfrak{E}$.

Proof. Denote the fundamental projectors belonging to (6.1) by P^+, P^-. Consider the fundamental symmetry $J = P^+ - P^-$, the corresponding decomposition majorant τ_J, and a minimal majorant τ on $\mathfrak{E}$. Suppose that, say, $\mathfrak{E}^+$ is intrinsically complete. The proof will advance through the following stages: 1) $\tau|\mathfrak{E}^+ \leqq \tau_J|\mathfrak{E}^+$, 2) P^+ is τ-continuous, 3) $\tau_J \leqq \tau$.

Owing to Theorem 4.2, τ is defined by a norm $\|\cdot\|$, and for some $\alpha > 0$ we have

$$\|x^+\| \leqq \alpha \|x^+\|' = \alpha \sup_{\|y\| \leqq 1} |(x^+, y)| \qquad (x^+ \in \mathfrak{E}^+) . \tag{6.2}$$

In order to establish the τ_J-continuity of the right-hand expression, we are going to examine the linear forms φ_y $(y \in \mathfrak{E},\ \|y\| \leqq 1)$ defined on $\mathfrak{E}^+$ by the relation

$$\varphi_y(x^+) = (x^+, y) \qquad (x^+ \in \mathfrak{E}^+) .$$

Since τ_J is a majorant (cf. Lemma II.11.4), each of the forms φ_y is continuous on the Hilbert space $\mathfrak{E}^+$:

$$|\varphi_y(x^+)| \leqq \|y\|_J \|x^+\|_J \qquad (x^+ \in \mathfrak{E}^+) .$$

Moreover, as τ is also a majorant, at every fixed x^+ the values $\varphi_y(x^+)$ $(\|y\| \leqq 1)$ constitute a bounded set:

$$|\varphi_y(x^+)| \leqq \beta \|x^+\| \|y\| \leqq \beta \|x^+\| = \beta_1(x^+) \qquad (y \in \mathfrak{E},\ \|y\| \leqq 1) .$$

Thus, according to the principle of uniform boundedness, for a suitable $\gamma > 0$ we have

$$|(x^+, y)| \leqq \gamma \|x^+\|_J \qquad (x^+ \in \mathfrak{E}^+,\ y \in \mathfrak{E},\ \|y\| \leqq 1) .$$

Consequently, from (6.2),

$$\|x^+\| \leqq \delta \|x^+\|_J \qquad (x^+ \in \mathfrak{E}^+) , \tag{6.3}$$

where $\delta = \alpha\gamma$.

Now let $x \in \mathfrak{E}$ be arbitrary. Setting $P^+x = x^+$, with the help of (6.3) we find:

$$||x^+||^2 \leqq \delta^2 ||x^+||_J^2 = \delta^2 (x^+, x^+) = \delta^2 (x^+, x) \leqq \beta\delta^2 \, ||x^+|| \, ||x|| \, .$$

Hence $||x^+|| \leqq \beta\delta^2 ||x||$, so that

$$||x||_J^2 = (Jx, x) \leqq \beta \, ||Jx|| \, ||x|| = \beta \, ||2x^+ - x|| \, ||x|| \leqq$$
$$\leqq \beta (2\beta\delta^2 + 1) \, ||x||^2 \, .$$

Therefore the majorant τ_J is weaker than the minimal majorant τ. As a result, $\tau_J = \tau$. □

From Theorems 6.1—6.2 we obtain:

Corollary 6.3. *If $\mathfrak{E}$ has a fundamental decomposition of the form* (6.1), *where one of the subspaces $\mathfrak{E}^+$, $\mathfrak{E}^-$ is intrinsically complete, then there is only one decomposition majorant on $\mathfrak{E}$.* □

There are also other cases where the decomposition majorants coincide.

Theorem 6.4. *If the non-degenerate inner product* (. , .) *has a Hilbert majorant, then* (. , .) *has only one decomposition majorant.*

Proof. Let τ be a Hilbert majorant on our non-degenerate space $\mathfrak{E}$. By Theorem 5.2 $\mathfrak{E}$ is decomposable. Consider two fundamental symmetries J_1, J_2 on $\mathfrak{E}$, and denote the respective decomposition majorants by τ_1, τ_2.

From Corollary III.2.6 and the closed graph principle it follows that every fundamental projector of $\mathfrak{E}$ is τ-continuous. In particular, the operators J_1, J_2 and $T = J_1 J_2$ are τ-continuous. Moreover, owing to Lemma II.11.1, T is J_1-symmetric:

$$(Tx, y)_{J_1} = (J_1 Tx, y) = (J_2 x, y) =$$
$$= (x, J_2 y) = (x, J_1 Ty) = (x, Ty)_{J_1} \quad (x, y \in \mathfrak{E}) \, .$$

Finally, since τ is a majorant and J_1 is τ-continuous, the J_1-norm is τ-continuous:

$$||y||_{J_1} \leqq \alpha ||y|| \quad (y \in \mathfrak{E}) \, , \tag{6.4}$$

where $|| \cdot ||$ is a norm defining τ, and α is a suitable positive number.

Let us show that the facts enumerated so far imply the τ_1-continuity of T.

In view of Lemma I.2.2 for every vector x and non-negative integer n we have

$$||T^{2^n} x||_{J_1} = (T^{2^n} x, T^{2^n} x)_{J_1}^{1/2} = (T^{2^{n+1}} x, x)_{J_1}^{1/2} \leqq ||T^{2^{n+1}} x||_{J_1}^{1/2} \, ||x||_{J_1}^{1/2} \, .$$

Hence, by iteration,

$$||Tx||_{J_1} \leqq ||T^{2^n} x||_{J_1}^{2^{-n}} \, ||x||_{J_1}^{1-2^{-n}} \qquad (n = 0, 1, 2, \ldots) \, .$$

Therefore, applying the inequality (6.4) to $y = T^{2^n}x$,

$$||Tx||_{J_1} \leqq \alpha^{2^{-n}} ||T|| \, ||x||^{2^{-n}} ||x||_{J_1}^{1-2^{-n}}.$$

For $n \to \infty$ this yields:

(6.5) $||Tx||_J \leqq ||T|| \, ||x||_{J_1}$ $(x \in \mathfrak{E})$.

Consequently, T is τ_1-continuous indeed.

With the help of (6.5) we find:

$$||x||_{J_2}^2 = (J_2x, x) = (J_1 Tx, x) = (Tx, x)_{J_1} \leqq ||Tx||_{J_1} ||x||_{J_1} \leqq$$
$$\leqq ||T|| \, ||x||_{J_1}^2 \, .$$

Thus $\tau_2 \leqq \tau_1$. The relation $\tau_1 \leqq \tau_2$ can be proved similarly. $\square$

7. Invariant Properties of $\mathfrak{E}^+$ and $\mathfrak{E}^-$

In the previous section we have seen that in some instances the decomposition majorant does not depend on how the fundamental decomposition is chosen. In order to prove other invariant properties of fundamental decompositions, we need the following improvement of Theorem II.10.1.

Lemma 7.1. *Suppose that the inner product space $\mathfrak{E}$ admits a fundamental decomposition of the form*

(7.1) $\mathfrak{E} = \mathfrak{E}^+ (+) \mathfrak{E}^-$; $\mathfrak{E}^+ \subset \mathfrak{P}^{++}$, $\mathfrak{E}^- \subset \mathfrak{P}^{--}$.

Consider the corresponding fundamental projectors P^+, P^-, fundamental symmetry $J = P^+ - P^-$, and decomposition majorant τ_J. If $\mathfrak{L} \subset \mathfrak{E}$ is a positive subspace, then the linear operator $P^+|\mathfrak{L}$ and its inverse are τ_J-continuous. Similarly, if $\mathfrak{L}$ is negative, then $P^-|\mathfrak{L}$ and its inverse are τ_J-continuous.

Proof. For every $x \in \mathfrak{E}$ Lemma II.11.2 yields $||P^+x||_J \leqq ||x||_J$. If $\mathfrak{L}$ is positive, then for $x \in \mathfrak{L}$ we also have

$$||x||_J^2 = ||P^+x||_J^2 + ||P^-x||_J^2 = 2||P^+x||_J^2 - (x, x) \leqq 2||P^+x||_J^2 .$$

Analogously, for x in a negative subspace $\mathfrak{L}$ we find

$$||x||_J^2 = 2||P^-x||_J^2 + (x, x) \leqq 2||P^-x||_J^2 . \quad \square$$

Now, making use of the conclusion of Corollary 6.3, we can establish the invariant character of the condition appearing in the same corollary (and in Theorem 6.2).

Theorem 7.2. *Let*

(7.1) $\mathfrak{E} = \mathfrak{E}^+ (+) \mathfrak{E}^-$; $\mathfrak{E}^+ \subset \mathfrak{P}^{++}$, $\mathfrak{E}^- \subset \mathfrak{P}^{--}$

and

$$(7.2) \qquad \mathfrak{G} = \mathfrak{G}^{(+)}(\dot{+})\mathfrak{G}^{(-)}; \qquad \mathfrak{G}^{(+)} \subset \mathfrak{P}^{++}, \qquad \mathfrak{G}^{(-)} \subset \mathfrak{P}^{--}$$

be two fundamental decompositions of the inner product space $\mathfrak{G}$. *If* $\mathfrak{G}^{(+)}$ *is intrinsically complete, so is* $\mathfrak{G}^{+}$. *Similarly, if* $\mathfrak{G}^{(-)}$ *is intrinsically complete, so is* $\mathfrak{G}^{-}$.

Proof. Denote the fundamental projectors belonging to (7.1) by P^+ and P^-. Let $\mathfrak{G}^{(+)}$ be intrinsically complete. Then $\mathfrak{G}^{(+)}$ is complete with respect to the decomposition majorant corresponding to (7.2). In view of Corollary 6.3, $\mathfrak{G}^{(+)}$ is complete with respect to the decomposition majorant corresponding to (7.1) as well. On account of Lemma 7.1 the latter property is shared by the subspace $P^+\mathfrak{G}^{(+)}$. In other words, $P^+\mathfrak{G}^{(+)}$ is intrinsically complete.

Suppose that $P^+\mathfrak{G}^{(+)} \neq \mathfrak{G}^+$. As the orthogonal decomposition theorem known from Hilbert space theory makes use of the completeness of the subspace, but not of the completeness of the whole space, we can find an element $x_0 \neq 0$ in $\mathfrak{G}^+$ such that $x_0 \perp P^+\mathfrak{G}^{(+)}$. Then $x_0 \perp \mathfrak{G}^{(+)}$ and, owing to Remark I.3.2, $\langle x_0, \mathfrak{G}^{(+)}\rangle$ is a positive definite proper extension of $\mathfrak{G}^{(+)}$. This contradicts Lemma I.11.4. □

Lemma 7.3. *Suppose that the inner product space* $\mathfrak{G}$ *admits a fundamental decomposition of the form* (7.1). *Let* $\mathfrak{L}$ *be a positive definite subspace of* $\mathfrak{G}$. *Then* $\dim_{\text{int}} \mathfrak{L} \leqq \dim_{\text{int}} \mathfrak{G}^+$. *Similarly, if* $\mathfrak{L} \subset \mathfrak{G}$ *is negative definite, then* $\dim_{\text{int}} \mathfrak{L} \leqq \dim_{\text{int}} \mathfrak{G}^-$.

Proof. Consider the fundamental symmetry J and the decomposition majorant τ_J belonging to (7.1). For any subspace $\mathfrak{M} \subset \mathfrak{G}$ denote by $\dim_J \mathfrak{M}$ the τ_J-dimension of $\mathfrak{M}$, i.e., the minimal power of sets $\mathfrak{A} \subset \mathfrak{M}$ whose span $\langle\mathfrak{A}\rangle$ is τ_J-dense in $\mathfrak{M}$. Let $\mathfrak{L}$ be positive definite. Since τ_J is a majorant, $\tau_{\text{int}}(\mathfrak{L}) \leqq \tau_J|\mathfrak{L}$; hence $\dim_{\text{int}} \mathfrak{L} \leqq \dim_J \mathfrak{L}$. Further, by Lemma 7.1, $\dim_J \mathfrak{L} \leqq \dim_J \mathfrak{G}^+$. Finally, $\dim_J \mathfrak{G}^+ = \dim_{\text{int}} \mathfrak{G}^+$ by definition. □

Corollary 7.4. *If* (7.1), (7.2) *are two fundamental decompositions of* $\mathfrak{G}$, *then*

$$(7.3) \qquad \dim_{\text{int}} \mathfrak{G}^+ = \dim_{\text{int}} \mathfrak{G}^{(+)}, \qquad \dim_{\text{int}} \mathfrak{G}^- = \dim_{\text{int}} \mathfrak{G}^{(-)}. \quad \square$$

Remark 7.5. If the inner product space $\mathfrak{G}$ is allowed to be degenerate, and (cf. Lemma I.11.1)

$$(7.4) \qquad \mathfrak{G} = \mathfrak{G}^0 (\dot{+}) \mathfrak{G}^+ (\dot{+}) \mathfrak{G}^-; \qquad \mathfrak{G}^+ \subset \mathfrak{P}^{++}, \qquad \mathfrak{G}^- \subset \mathfrak{P}^{--},$$

$$(7.5) \qquad \mathfrak{G} = \mathfrak{G}^0 (\dot{+}) \mathfrak{G}^{(+)} (\dot{+}) \mathfrak{G}^{(-)}; \qquad \mathfrak{G}^{(+)} \subset \mathfrak{P}^{++}, \qquad \mathfrak{G}^{(-)} \subset \mathfrak{P}^{--}$$

are two fundamental decompositions, then in view of Corollary I.5.3 and Lemma I.5.2 the subspaces $\mathfrak{G}^+ (\dot{+}) \mathfrak{G}^-$ and $\mathfrak{G}^{(+)} (\dot{+}) \mathfrak{G}^{(-)}$ are non-degenerate and isometrically isomorphic to each other. It follows that in this case the relations (7.3) still hold.

On the basis of Corollary 7.4 and Remark 7.5 the rank of positivity and the rank of negativity, introduced in Section II.10 for non-degenerate quasi-negative resp. quasi-positive spaces, can be defined for every decomposable space. Namely, if (7.4) is a fundamental decomposition of the inner product space $\mathfrak{E}$, the intrinsic dimension of $\mathfrak{E}^+$ ($\mathfrak{E}^-$) will be called the *rank of positivity* (*rank of negativity*) of $\mathfrak{E}$, and denoted by $\varkappa^+(\mathfrak{E})$ (resp. $\varkappa^-(\mathfrak{E})$). The cardinal number

$$\varkappa(\mathfrak{E}) = \min\{\varkappa^+(\mathfrak{E}),\ \varkappa^-(\mathfrak{E})\}$$

will be termed the *rank of indefiniteness* of $\mathfrak{E}$.

8. Subspaces of Spaces with a Hilbert Majorant

Let the norm $\|\cdot\|$ define a Hilbert majorant τ of the inner product $(.\,,.)$ on the space $\mathfrak{E}$. Then a) $\|x\| = [x, x]^{1/2}$ $(x \in \mathfrak{E})$, where $[.\,,.]$ is a positive definite inner product on $\mathfrak{E}$; b) $\mathfrak{E}$ is complete with respect to $\|\cdot\|$; c) there is an $\alpha > 0$ such that

$$|(x, y)| \leqq \alpha\, \|x\|\, \|y\| \qquad (x, y \in \mathfrak{E})\,. \tag{8.1}$$

Let G denote the Gram operator of $(.\,,.)$ with respect to $[.\,,.]$:

$$(x, y) = [Gx, y] \qquad (x, y \in \mathfrak{E})\,. \tag{8.2}$$

Consider a τ-closed subspace $\mathfrak{L} \subset \mathfrak{E}$ and the corresponding $[.\,,.]$ orthogonal projector $Q_{\mathfrak{L}}$ characterized by the properties

$$\mathfrak{R}(Q_{\mathfrak{L}}) = \mathfrak{L}\,, \quad [x - Q_{\mathfrak{L}}x,\ y] = 0 \qquad (x \in \mathfrak{E}\,,\quad y \in \mathfrak{L})\,. \tag{8.3}$$

Set

$$G_{\mathfrak{L}} = Q_{\mathfrak{L}} G | \mathfrak{L}\,. \tag{8.4}$$

Obviously, $G_{\mathfrak{L}}$ is an operator in $\mathfrak{L}$ satisfying the equation

$$(x, y) = [G_{\mathfrak{L}}x, y] \qquad (x, y \in \mathfrak{L})\,. \tag{8.5}$$

Therefore $G_{\mathfrak{L}}$ is the Gram operator of $(.\,,.)$ with respect to $[.\,,.]$ when both inner products are restricted to $\mathfrak{L}$.

Now the following result can be added to our list of projection theorems (see Chapters I and III).

Lemma 8.1. *Let $\mathfrak{E}$ be an inner product space with a Hilbert majorant τ, and let $\mathfrak{L}$ be a τ-closed subspace of $\mathfrak{E}$. The element $x \in \mathfrak{E}$ admits a projection on $\mathfrak{L}$ if and only if, with the notations introduced above, $Q_{\mathfrak{L}}G\,x \in \mathfrak{R}(G_{\mathfrak{L}})$.*

Proof. Let x_0 be a projection of x on $\mathfrak{L}$. Then for every $y \in \mathfrak{L}$ we have $(x, y) = (x_0, y)$ i.e. $[Gx, y] = [G_{\mathfrak{L}}x_0, y]$ or, equivalently, $[Q_{\mathfrak{L}}Gx, y] = [G_{\mathfrak{L}}x_0, y]$. Hence $Q_{\mathfrak{L}}Gx = G_{\mathfrak{L}}x_0$. The reasoning applies in the reverse direction too. □

Corollary 8.2. *$\mathfrak{L}$ is ortho-complemented if and only if $\mathfrak{R}(Q_{\mathfrak{L}}G) = \mathfrak{R}(G_{\mathfrak{L}})$.* □

Another criterion for ortho-complementedness in a space with a Hilbert majorant can be formulated by the aid of a new topology.

Let $\mathfrak{E}$ be a vector space with inner product $(.\,,.)$. Let, at the same time, $\mathfrak{E}$ be a Hilbert space with respect to the positive definite inner product $[.\,,.]$. Suppose that the norm $||x|| = [x, x]^{1/2}$ $(x \in \mathfrak{E})$ defines a majorant of $(.\,,.)$. Set

$$p(x) = ||Gx|| \qquad (x \in \mathfrak{E}) , \tag{8.6}$$

where G is the Gram operator of $(.\,,.)$ relative to $[.\,,.]$. The semi-norm p defines a topology τ_M, called the *Mackey topology* of the inner product space $\mathfrak{E}$.

From Theorem 8.3 below and Theorem III.5.3 it follows that the Mackey topology does not depend on the choice of $[.\,,.]$.

Theorem 8.3. *The Mackey topology is admissible.*

Proof. Since $|(x, y)| = |[Gx, y]| \leqq ||Gx||\, ||y||$, the topology τ_M is a partial majorant. Let the linear form φ be τ_M-continuous on the space $\mathfrak{E}$:

$$|\varphi(x)| \leqq \gamma ||Gx|| \qquad (x \in \mathfrak{E}) .$$

For $Gx = 0$ we have $\varphi(x) = 0$. Thus we can define a linear form Φ on $\mathfrak{R}(G)$ setting

$$\Phi(Gx) = \varphi(x) \qquad (x \in \mathfrak{E}) .$$

Then $|\Phi(Gx)| = |\varphi(x)| \leqq \gamma ||Gx||$ $(x \in \mathfrak{E})$. Therefore Φ can be extended to a continuous linear form on the Hilbert space $\mathfrak{E}$. Consequently,

$$\Phi(z) = [z, y_0] \qquad (z \in \mathfrak{R}(G)) ,$$

where y_0 is a suitable element of $\mathfrak{E}$. Hence $\varphi(x) = \Phi(Gx) = [Gx, y_0] = (x, y_0)$ $(x \in \mathfrak{E})$. □

Making use of Theorem III.5.3 we obtain:

Corollary 8.4. *The Mackey topology is stronger than any other admissible topology.* □

Theorem 8.5. *Let $\mathfrak{E}$ be an inner product space with a Hilbert majorant τ. The τ-closed subspace $\mathfrak{L} \subset \mathfrak{E}$ is ortho-complemented if and only if its Mackey topology $\tau_M(\mathfrak{L})$ coincides with the induced topology $\tau_1 = \tau_M(\mathfrak{E})|\mathfrak{L}$.*

Proof. Let $\mathfrak{L}$ be ortho-complemented. According to Theorem 8.3 $\tau_M(\mathfrak{L})$ is admissible on $\mathfrak{L}$. In view of Theorem III.5.3 it is sufficient to show that τ_1 has the same property. With the notations constantly used in this section $|(x, y)| = |[Gx, y]| \leqq \|Gx\| \, \|y\|$ $(x, y \in \mathfrak{L})$, hence τ_1 is a partial majorant on $\mathfrak{L}$. Consider a τ_1-continuous linear form φ on $\mathfrak{L}$: $|\varphi(x)| \leqq \gamma \|Gx\|$ $(x \in \mathfrak{L})$. By the Hahn-Banach extension principle φ admits a $\tau_M(\mathfrak{E})$-continuous linear extension defined on $\mathfrak{E}$. Thus, owing to Theorem 8.3, there exists an $y \in \mathfrak{E}$ such that $\varphi(x) = (x, y)$ $(x \in \mathfrak{L})$. Denoting a projection of y on $\mathfrak{L}$ by y_0, we have $\varphi(x) = (x, y_0)$ $(x \in \mathfrak{L})$.

Let, conversely, $\tau_1 = \tau_M(\mathfrak{L})$. For any fixed $y \in \mathfrak{E}$ the linear form $\varphi_y(x) = (x, y)$ $(x \in \mathfrak{L})$ is τ_1-continuous, hence $\tau_M(\mathfrak{L})$-continuous. On account of Theorems 8.3 and III.7.1, $\mathfrak{L}$ is ortho-complemented. □

Let us give a sufficient condition of ortho-complementedness in spaces with a Hilbert majorant τ. By Corollary III.2.5, Theorem 5.2, Lemma I.9.2 and Lemma I.9.3 we may restrict our attention to τ-closed definite subspaces.

Theorem 8.6. *Let $\mathfrak{E}$ be an inner product space having a Hilbert majorant τ. If the definite subspace $\mathfrak{L} \subset \mathfrak{E}$ is τ-closed and intrinsically complete, then it is ortho-complemented.*

Proof. The intrinsic topology $\tau_{\text{int}}(\mathfrak{L})$ as well as the induced topology $\tau|\mathfrak{L}$ are Hilbert topologies on $\mathfrak{L}$. The first being weaker than the second one, the closed graph principle implies that they coincide. Therefore, on account of the Riesz representation theorem, $\tau|\mathfrak{L}$ is admissible on $\mathfrak{L}$. It remains to apply Theorem III.7.1. □

The following modification of Lemma I.5.1 is contained in Theorem 5.2. We treat it separately in order to stress its more elementary nature and facilitate future reference.

Lemma 8.7. *Let $\mathfrak{E}$ be an inner product space with a Hilbert majorant τ, and let $\mathfrak{L}$ be a τ-closed subspace of $\mathfrak{E}$. Then the isotropic part $\mathfrak{L}^0$ of $\mathfrak{L}$ is τ-closed. Moreover, $\mathfrak{L}$ is the orthogonal direct sum of $\mathfrak{L}^0$ and a τ-closed non-degenerate subspace $\mathfrak{L}^1$.*

Proof. $\mathfrak{L}^\perp$ is τ-closed by Lemma III.2.4. Hence $\mathfrak{L}^0 = \mathfrak{L} \cap \mathfrak{L}^\perp$ is τ-closed. According to the orthogonal decomposition theorem of Hilbert spaces, $\mathfrak{L}^0$ has a τ-closed complementary subspace $\mathfrak{L}^1$ in $\mathfrak{L}$. Clearly, $\mathfrak{L}^1$ is non-degenerate and orthogonal to $\mathfrak{L}^0$ (cf. Lemma I.5.1). □

Notes to Chapter IV

Similarly to Chapter III, a considerable part of the present chapter is closely related to the duality theory of topological vector spaces (see e.g. Bourbaki [1], Köthe [1], Robertson and Robertson [1], Schaefer [1]).

Theorem 2.1 belongs to Wittstock [2]. Conditions for the existence of majorants have also been studied by Jarchow [1], [2], who showed that in a class of structures wider than that of locally convex topologies every inner product has a "majorant".

For the present proof of Theorem 2.3 see Ginzburg and Iohvidov [1]. Lemmas 2.4—2.6 and related results can be found in the papers of Aronszajn [1] and Wittstock [3].

Theorem 3.2 was proved by Savage [1]. For spaces with a separable Hilbert majorant Theorem 3.3 is due to Nevanlinna [3], and for those with an arbitrary separable majorant to Jelínek and Virsik [1]. Theorem 3.4 was obtained by Nevanlinna [3] and reformulated by Ginzburg and Iohvidov [1].

Lemma 4.1 is a result of Aronszajn [1]. Variants of its proof have been given by Wittstock [3], [5]. Theorem 4.2 and Corollary 4.3 belong to Wittstock [3]. Example 4.4 and Theorem 4.5 are borrowed from the paper of Jelínek and Virsik [1] (for Theorem 4.5 cf. also Jarchow [1]).

An analogue of Example 5.1 appears in a short note of Ovčinnikov [1] as a modification of Example I.11.3. The first formulations of Theorem 5.2 were given by Hestenes [1] and, independently, Nevanlinna [3]. The proof only requires a part of the spectral theorem that can easily be reduced (see Lemma VI.8.3 below) to the more elementary subject of positive square roots in Hilbert space. Example 5.6 is due to Wittstock (personal communication).

A special case of Theorem 6.1 was obtained by Nevanlinna [3], [4]; the general case is due to Wittstock [2]. Criteria for the existence of a weakest majorant, also Theorem 6.2 in a stronger form, are contained in the works of Wittstock [1], [3], [5]. A special case of Corollary 6.3 was earlier established by Iohvidov and Kreĭn [1].

In Theorem 6.4 the existence of a Banach majorant would suffice. The original proof belongs to Wittstock [2], and the present one to Bognár [7]. In the text of the proof we reproduce an argument of Reid [1] rather than refer to the result we need (the latter has first been obtained by Kreĭn [2]). An example showing that Theorem 6.4 cannot be extended to spaces with a Fréchet majorant was given by Wittstock [2].

Lemma 7.1 goes back to Ginzburg [3].

Theorem 7.2 is due to Wittstock [4]. For special cases see Iohvidov and Kreĭn [1], Scheibe [1], Ginzburg [3], Bognár and Krámli [1]. All of these papers contain special cases of Corollary 7.4 too.

The first steps towards Lemma 8.1 were made by Nevanlinna [4] and Louhivaara [4]. The lemma itself belongs to Browder [1] (cf. also Louhivaara [5]). Related facts for non-symmetric bilinear forms have been obtained by Littman [1] and Browder [2] (cf. also Louhivaara [6]), and in a special case by Wittstock [1]. An account of all these results appears in the survey article of Louhivaara [7]. The most complete study of the problem, however, is due to Ginzburg [4] (cf. also Ginzburg and Iohvidov [1]).

On the initiation of Nevanlinna [1] the theory of projections in an inner product space with a Hilbert majorant has been utilized for extending the variational method to general elliptic or even non-elliptic boundary value problems (Louhivaara [1], [2], [3], Browder [1], [2], Littman [1], Hildebrandt [1]; for a summary cf. also Dolph [1] and Louhivaara [7]).

For the definition and properties of the Mackey topology when the existence of a Hilbert majorant is not assumed see e.g. Robertson and Robertson [1].

Theorem 8.5 was announced by Ginzburg [4]. The present proof of this theorem as well as that of Theorem 8.3 is taken from the paper of Ginzburg and Iohvidov [1], where the case of a possibly non-symmetric bilinear form with a reflexive Banach majorant is considered. Theorem 8.6 appears in the same paper.

Non-symmetric bilinear forms with a Banach or Hilbert majorant were also investigated by Jalava [2].

Iohvidov [10] made a study of non-closed maximal definite subspaces in an inner product space having a Hilbert majorant.

Chapter V. The Geometry of Krein Spaces

Krein spaces, the most important type of inner product spaces, can roughly be characterized as non-degenerate, decomposable, complete spaces. Sections 1—2 deal with their global properties. In Sections 4, 7, 9, 10 maximal semi-definite subspaces of Krein spaces are investigated, while Sections 3, 5, 6, 8 are essentially concerned with ortho-complemented subspaces. Theorems 1.3, 3.4, 3.5, 4.2, 5.3, 5.7, 6.3, 7.1 and 9.1 give an idea of the contents of the chapter.

1. Krein Spaces

If an inner product space $\mathfrak{H}$ admits a fundamental decomposition of the form

$$(1.1) \qquad \mathfrak{H} = \mathfrak{H}^+ (\dot{+}) \mathfrak{H}^-; \qquad \mathfrak{H}^+ \subset \mathfrak{P}^{++}, \qquad \mathfrak{H}^- \subset \mathfrak{P}^{--},$$

where the subspaces $\mathfrak{H}^+$ and $\mathfrak{H}^-$ are intrinsically complete, then we shall say that $\mathfrak{H}$ is a *Krein space*.

Lemma I.11.1 and Theorem IV.7.2 yield:

Theorem 1.1. *A Krein space $\mathfrak{H}$ is non-degenerate and decomposable. Every fundamental decomposition of $\mathfrak{H}$ looks like* (1.1), *and the components $\mathfrak{H}^+$, $\mathfrak{H}^-$ are intrinsically complete.* □

Loosely speaking, a Krein space is a non-degenerate, decomposable, complete inner product space.

According to our definition, the class of Krein spaces includes Hilbert spaces ($\mathfrak{H}^- = 0$) as well as anti-spaces of Hilbert spaces ($\mathfrak{H}^+ = 0$). Furthermore, on account of Corollary I.11.8 and Lemma I.11.1, every finite-dimensional non-degenerate inner product space is a Krein space.

A connection between Hilbert spaces and arbitrary Krein spaces is given by the following consequence of Lemmas II.11.3, II.11.2, and Theorem 1.1.

Corollary 1.2. *The decomposable, non-degenerate inner product space $\mathfrak{E}$ is a Krein space if and only if, for every fundamental symmetry J, the J-inner product turns $\mathfrak{E}$ into a Hilbert space.* □

The rest of this book is devoted to Krein spaces and their linear operators. The letter $\mathfrak{H}$, when used without explanation, will always denote a Krein space.

Theorem 1.3. *A vector space* $\mathfrak{E}$ *with an inner product* $(.\,,.)$ *is a Krein space if and only if* 1) $(.\,,.)$ *has a Hilbert majorant* τ *and* 2) *the Gram operator of* $(.\,,.)$ *with respect to any positive definite inner product* $[.\,,.]$ *such that the norm*

$$(1.2) \qquad \|x\| = [x, x]^{1/2} \qquad (x \in \mathfrak{E})$$

defines τ *is completely invertible.*

Proof. Suppose that

$$(1.3) \qquad \mathfrak{E} = \mathfrak{E}^+ (\dot{+}) \mathfrak{E}^-; \qquad \mathfrak{E}^+ \subset \mathfrak{P}^{++}, \quad \mathfrak{E}^- \subset \mathfrak{P}^{--},$$

where $\mathfrak{E}^+$ and $\mathfrak{E}^-$ are intrinsically complete. Denote the fundamental symmetry belonging to (1.3) by J. The decomposition majorant τ_J defined by $\|\cdot\|_J$ is a Hilbert majorant on $\mathfrak{E}$. Since, in view of Lemma II.11.1, $(x, y) = (J^2x, y) = (Jx, y)_J$, the Gram operator of $(.\,,.)$ with respect to $(.\,,.)_J$ is J, a completely invertible operator.

Owing to Theorem III.3.3 $(.\,,.)$ has only one Hilbert majorant. Let $[.\,,.]_j$ $(j = 1,2)$ be two positive definite inner products on $\mathfrak{E}$ such that the norms $\|x\|_j = [x, x]_j^{1/2}$ $(x \in \mathfrak{E})$ define the Hilbert majorant of $(.\,,.)$. Since $\|\cdot\|_1$ is equivalent to $\|\cdot\|_2$, there exist linear operators G_{12}, G_{21} on $\mathfrak{E}$ satisfying the relations

$$[x, y]_2 = [G_{12}x, y]_1, \qquad [x, y]_1 = [G_{21}x, y]_2 \qquad (x,y \in \mathfrak{E}).$$

We have $G_{12}G_{21} = G_{21}G_{12} = I$; in particular, G_{12} and G_{21} are completely invertible. Now if G_j denotes the Gram operator of $(.\,,.)$ relative to $[.\,,.]_j$, i.e.

$$(x, y) = [G_jx, y]_j \qquad (x,y \in \mathfrak{E}; \quad j = 1,2),$$

then $G_2 = G_{21}G_1$, $G_1 = G_{12}G_2$; hence G_2 is completely invertible if and only if G_1 is so.

Let, conversely, $(x, y) = [Gx, y]$ $(x,y \in \mathfrak{E})$, where the norm $\|\cdot\|$ given by (1.2) defines a Hilbert majorant τ and the τ-continuous linear operator G is completely invertible. Then $\mathfrak{E}$ is non-degenerate and decomposable (cf. Theorem IV.5.2). Thus, according to Lemma I.11.1, $\mathfrak{E}$ has a fundamental decomposition of the form (1.3). On account of Corollary 1.2 it is sufficient to show that the corresponding decomposition majorant τ_J coincides with τ.

The complete invertibility of G and the closed graph principle imply $\tau_M = \tau$. Therefore, by Theorem IV.8.3, the majorant τ is an admissible topology on $\mathfrak{E}$. Hence, in view of Theorem IV.4.5, τ is a minimal majorant. On the other hand, Lemma IV.5.4 yields $\tau_J \leqq \tau$. Consequently, $\tau_J = \tau$. □

From the well-known fact that a Hilbert space is completely characterized by its dimension one readily obtains the classification of Krein spaces by means of their ranks of positivity and negativity.

Theorem 1.4. *Two Krein spaces* $\mathfrak{H}_1$, $\mathfrak{H}_2$ *are isometrically isomorphic if and only if* $\varkappa^+(\mathfrak{H}_1) = \varkappa^+(\mathfrak{H}_2)$ *and* $\varkappa^-(\mathfrak{H}_1) = \varkappa^-(\mathfrak{H}_2)$. □

Owing to Corollary IV.6.3 all J-norms on a Krein space $\mathfrak{H}$ are equivalent. They will be called *natural norms* on $\mathfrak{H}$, and the Hilbert majorant they define, the *strong topology* of $\mathfrak{H}$.

In view of Theorem 1.3 the strong topology of $\mathfrak{H}$ is equal to the Mackey topology. Therefore it may be denoted by $\tau_M(\mathfrak{H})$.

As a corollary of Theorem IV.8.3 and Lemma II.11.4 we have:

Theorem 1.5. *The strong topology of* $\mathfrak{H}$ *is an admissible majorant.* □

In a Krein space $\mathfrak{H}$ all notions involving some kind of topology such as *convergent, closed, dense, continuous, homeomorphic,* (infinite) *dimension* etc. will be understood, unless otherwise stated, to refer to the strong topology $\tau_M(\mathfrak{H})$. The *closure* of a set $\mathfrak{A} \subset \mathfrak{H}$ will be denoted by $\overline{\mathfrak{A}}$.

One more definition. Let $\mathfrak{L} \dotplus \mathfrak{M} = \mathfrak{L}_1$, where $\mathfrak{L}$, $\mathfrak{M}$ and $\mathfrak{L}_1$ are closed subspaces of $\mathfrak{H}$. In this case the cardinal number $\dim \mathfrak{M}$ is called the *codimension* of $\mathfrak{L}$ with respect to $\mathfrak{L}_1$, and denoted by $\operatorname{codim}_{\mathfrak{L}_1} \mathfrak{L}$. On account of the closed graph principle, the definition is correct. Furthermore, according to Corollary I.1.2, it is compatible with the definition of codimension we gave in Section I.1.

2. Krein Spaces as Completions

It is well known that every positive definite inner product space $\mathfrak{E}$ can be embedded in a Hilbert space $\mathfrak{H}$ as a dense subspace, and that this procedure is essentially unique. The corresponding result for more general inner product spaces reads as follows.

Theorem 2.1. *Let*

(2.1) $$\mathfrak{E} = \mathfrak{E}^+ (\dotplus) \mathfrak{E}^-; \quad \mathfrak{E}^+ \subset \mathfrak{P}^{++}, \quad \mathfrak{E}^- \subset \mathfrak{P}^{--}$$

be a fundamental decomposition of the decomposable, non-degenerate inner product space $\mathfrak{E}$. *Denote the intrinsic completion of* $\mathfrak{E}^+$ *and* $\mathfrak{E}^-$ *by* $\mathfrak{H}^+$ *and* $\mathfrak{H}^-$, *respectively. Then the cartesian product* $\mathfrak{H} = \mathfrak{H}^+ \times \mathfrak{H}^-$ *is a Krein space, and* $\mathfrak{E}$ *is isometrically isomorphic to a dense subspace of* $\mathfrak{H}$.

Let

(2.2) $$\mathfrak{E} = \mathfrak{E}_j^+ (\dotplus) \mathfrak{E}_j^-; \quad \mathfrak{E}_j^+ \subset \mathfrak{P}^{++}, \quad \mathfrak{E}_j^- \subset \mathfrak{P}^{--} \quad (j = 1, 2)$$

be two fundamental decompositions of $\mathfrak{E}$, *with corresponding decomposition majorants* τ_j. *Denote the intrinsic completion of* $\mathfrak{E}_j^{\pm}$ *by* $\mathfrak{H}_j^{\pm}$. *If* $\tau_1 = \tau_2$, *and only then, there exists an isometrical isomorphism between* $\mathfrak{H}_1 = \mathfrak{H}_1^+ \times \mathfrak{H}_1^-$ and $\mathfrak{H}_2 = \mathfrak{H}_2^+ \times \mathfrak{H}_2^-$ *which leaves the elements of* $\mathfrak{E}$ *unchanged.*

Proof. That $\mathfrak{H}$ is a Krein space and $\mathfrak{E}$ can be identified with a dense subspace of $\mathfrak{H}$ is clear. Let $\tau_1 = \tau_2$. If $x_1 \in \mathfrak{H}_1$, there exists a sequence $\{y^{(n)}\}_{n=1}^{\infty} \subset \mathfrak{E}$ converging to x_1 in the strong topology $\tau_M(\mathfrak{H}_1)$. Then $\{y^{(n)}\}$ is a Cauchy sequence relative to $\tau_M(\mathfrak{H}_1)|\mathfrak{E} = \tau_1 = \tau_2 = \tau_M(\mathfrak{H}_2)|\mathfrak{E}$, so that it has a $\tau_M(\mathfrak{H}_2)$-limit $x_2 \in \mathfrak{H}_2$. The mapping T uniquely defined by the relation $Tx_1 = x_2$ is an isomorphism between $\mathfrak{H}_1$ and $\mathfrak{H}_2$ which leaves the elements of $\mathfrak{E}$ unchanged. Since $\tau_M(\mathfrak{H}_1)$ and $\tau_M(\mathfrak{H}_2)$ are majorants, T preserves inner products too.

Suppose, conversely, that there exists an isometrical isomorphism between $\mathfrak{H}_1$ and $\mathfrak{H}_2$ which leaves the elements of $\mathfrak{E}$ unchanged. Then we may regard τ_1 and τ_2 as induced on $\mathfrak{E}$ by two decomposition majorants of the same Krein space. Consequently, $\tau_1 = \tau_2$. □

Remark 2.2. Comparing Theorem 2.1 with Corollary IV.6.3 and Theorem IV.6.4, we obtain sufficient conditions on $\mathfrak{E}$ in order that its "completion" $\mathfrak{H}$ be essentially unique.

3. Subspaces

Corollary 1.2 and Theorem IV.5.2 yield:

Theorem 3.1. *If* $\mathfrak{L}$ *is a closed subspace of the Krein space* $\mathfrak{H}$, *then* $\mathfrak{L}$ *admits a fundamental decomposition whose components, as well as the direct sum of any two of them, are closed.* □

On the other hand, a non-closed subspace of $\mathfrak{H}$ need not be decomposable.

Example 3.2. The non-decomposable space $\mathfrak{E}$ of Example IV.5.1 can be embedded in a Krein space $\mathfrak{H}$ as follows. Let $\mathfrak{H}$ consist of all doubly infinite numerical sequences $\{\xi_j\}_{j=-\infty}^{\infty}$ satisfying $\sum_{j=-\infty}^{\infty} |\xi_j|^2 < \infty$. Set

$$(\{\xi_j\}, \{\eta_j\}) = \sum_{j=-\infty}^{\infty} \xi_j \bar{\eta}_{-j-1} .$$

The relations

$$[\{\xi_j\}, \{\eta_j\}] = \sum_{j=-\infty}^{\infty} \xi_j \bar{\eta}_j , \qquad \|\{\xi_j\}\| = [\{\xi_j\}, \{\xi_j\}]^{1/2}$$

define a Hilbert majorant of $(.\,,.)$, and the respective Gram operator G is completely invertible: $G\{\xi_j\}_{j=-\infty}^{\infty} = \{\xi_{-j-1}\}_{j=-\infty}^{\infty}$. By Theorem 1.3 $\mathfrak{H}$ is a Krein space.

A closed, non-degenerate subspace of a Krein space, though decomposable, need not be a Krein space.

Example 3.3. In Example I.9.7 $\mathfrak{E}$ is a Krein space and $\mathfrak{L}$ is a closed positive definite subspace of $\mathfrak{E}$. The elements

$$x^{(n)} = \{\xi_j^{(n)}\}_{j=1}^{\infty} \qquad (n = 1, 2, \ldots)$$

defined by the relations

$$\xi_{2j-1}^{(n)} = (2j-1)^{-1/2} \qquad (j \leqq n),$$
$$\xi_{2j}^{(n)} = 2j(2j-1)^{-3/2} \qquad (j \leqq n),$$
$$\xi_j^{(n)} = 0 \qquad (j > 2n)$$

belong to $\mathfrak{L}$ and we have

$$|x^{(n)} - x^{(m)}|_{\mathfrak{L}} \to 0 \qquad (m, n \to \infty).$$

If $x^{(n)}$ had an intrinsic limit $x^{(0)} \in \mathfrak{L}$, then the coordinates of the latter would be

$$\xi_{2j-1}^{(0)} = (2j-1)^{-1/2}, \qquad \xi_{2j}^{(0)} = 2j(2j-1)^{-3/2} \qquad (j = 1, 2, \ldots),$$

so that $\sum_{j=1}^{\infty} |\xi_j^{(0)}|^2 = \infty$. Hence $\mathfrak{L}$ is not intrinsically complete.

According to the next theorem, a closed subspace of $\mathfrak{H}$ is a Krein space if and only if it is ortho-complemented.

Theorem 3.4. *A subspace $\mathfrak{L}$ of the Krein space $\mathfrak{H}$ is ortho-complemented if and only if* 1) *$\mathfrak{L}$ is closed and* 2) *$\mathfrak{L}$ is a Krein space.*

Proof. If $\mathfrak{L}$ is ortho-complemented, then it is closed by Corollary III.2.5. On account of Corollary IV.8.2 and Theorem 1.3, a closed subspace $\mathfrak{L} \subset \mathfrak{H}$ is ortho-complemented if and only if

$$\mathfrak{R}(G_{\mathfrak{L}}) = \mathfrak{L}, \tag{3.1}$$

where $G_{\mathfrak{L}}$ is an operator in the subspace $\mathfrak{L}$ determined by the relation $(x, y) = (G_{\mathfrak{L}}x, y)_J$ $(x, y \in \mathfrak{L})$ with a fixed fundamental symmetry J of $\mathfrak{H}$. Since $G_{\mathfrak{L}}$ is J-symmetric, (3.1) is equivalent to the complete invertibility of $G_{\mathfrak{L}}$, i.e. (see Theorem 1.3), to $\mathfrak{L}$ being a Krein space. □

Let us give another characterization of ortho-complemented subspaces in $\mathfrak{H}$.

Theorem 3.5. *A subspace $\mathfrak{L}$ of the Krein space $\mathfrak{H}$ is ortho-complemented if and only if* 1) *$\mathfrak{L}$ is closed,* 2) *$\mathfrak{L}$ is non-degenerate, and* 3) *for any fundamental decomposition*

$$\mathfrak{L} = \mathfrak{L}^{+}(\dotplus)\mathfrak{L}^{-}; \qquad \mathfrak{L}^{+} \subset \mathfrak{P}^{++}, \qquad \mathfrak{L}^{-} \subset \mathfrak{P}^{--} \tag{3.2}$$

of $\mathfrak{L}$ *a fundamental decomposition*

$$(3.3)\qquad \mathfrak{H} = \mathfrak{H}^+(\dot{+})\mathfrak{H}^-; \qquad \mathfrak{H}^+ \subset \mathfrak{P}^{++}, \qquad \mathfrak{H}^- \subset \mathfrak{P}^{--}$$

of $\mathfrak{H}$ *can be found so that*

$$(3.4)\qquad \mathfrak{L}^+ \subset \mathfrak{H}^+, \quad \mathfrak{L}^- \subset \mathfrak{H}^-.$$

Proof. Let $\mathfrak{L}$ be ortho-complemented. Then $\mathfrak{L}$ and $\mathfrak{L}^\perp$ are closed and non-degenerate (cf. Lemma I.9.1, Corollary I.9.5, and Corollary III.2.5). In view of Corollary 1.2, Theorem IV.5.2 and Lemma I.11.1 $\mathfrak{L}$ and $\mathfrak{L}^\perp$ admit fundamental decompositions (3.2) and

$$\mathfrak{L}^\perp = \mathfrak{L}_1^+(\dot{+})\mathfrak{L}_1^-; \qquad \mathfrak{L}_1^+ \subset \mathfrak{P}^{++}, \qquad \mathfrak{L}_1^- \subset \mathfrak{P}^{--},$$

respectively. The subspaces

$$(3.5)\qquad \mathfrak{H}^+ = \mathfrak{L}^+(\dot{+})\mathfrak{L}_1^+, \qquad \mathfrak{H}^- = \mathfrak{L}^-(\dot{+})\mathfrak{L}_1^-$$

satisfy (3.3) and (3.4).

Suppose, conversely, that $\mathfrak{L}$ is closed and non-degenerate, and that for suitable decompositions (3.2), (3.3) the inclusions (3.4) hold. By Corollary III.2.6 $\mathfrak{L}^+$ is closed in $\mathfrak{L}$. Consequently, $\mathfrak{L}^+$ is closed in $\mathfrak{H}$. In particular, $\mathfrak{L}^+$ is ortho-complemented in the Hilbert space $\mathfrak{H}^+$. Analogously, $\mathfrak{L}^-$ is ortho-complemented in $\mathfrak{H}^-$. Thus for certain subspaces $\mathfrak{L}_1^+$, $\mathfrak{L}_1^-$ we have (3.5). The subspace $\mathfrak{L}_1 = \mathfrak{L}_1^+(\dot{+})\mathfrak{L}_1^-$ is orthogonal to $\mathfrak{L}$ and satisfies the relation $\mathfrak{L}\,(\dot{+})\,\mathfrak{L}_1 = \mathfrak{H}$. □

With a view to later application we mention the following consequence of Theorem 1.5 and Theorem III.6.1:

Theorem 3.6. *If* $\mathfrak{L}$ *is a subspace of* $\mathfrak{H}$, *then* $\mathfrak{L}^{\perp\perp} = \mathfrak{L}$. □

Corollary 3.7. *If* $\mathfrak{L}$ *is a closed subspace of* $\mathfrak{H}$, *then the isotropic part of* $\mathfrak{L}^\perp$ *coincides with that of* $\mathfrak{L}$. □

4. Maximal Semi-definite Subspaces

The first three results presented below establish conditions (one necessary, one necessary and sufficient, and one sufficient) on a positive (negative) subspace of $\mathfrak{H}$ for being maximal positive (maximal negative).

Theorem 4.1. *Every maximal positive or maximal negative subspace of a Krein space is closed.*

Proof. Since the strong topology is a majorant, the closure of a positive (negative) subspace is positive (negative). □

Theorem 4.2. *Let*

$$(4.1)\qquad \mathfrak{H} = \mathfrak{H}^+(\dot{+})\mathfrak{H}^-; \qquad \mathfrak{H}^+ \subset \mathfrak{P}^{++}, \qquad \mathfrak{H}^- \subset \mathfrak{P}^{--}$$

be a fundamental decomposition of the Krein space $\mathfrak{H}$. *Denote the corresponding fundamental projectors by* P^+, P^-. *The positive (negative) subspace* $\mathfrak{L} \subset \mathfrak{H}$ *is maximal positive (maximal negative) if and only if* $P^+\mathfrak{L} = \mathfrak{H}^+$ *(resp.* $P^-\mathfrak{L} = \mathfrak{H}^-$*).*

Proof. Let $\mathfrak{L}$ be positive and $P^+\mathfrak{L} = \mathfrak{H}^+$. Then, owing to Theorem II.10.1, $\mathfrak{L}$ is maximal positive.

Suppose, conversely, that $\mathfrak{L}$ is maximal positive. By Theorem 4.1 and Lemma IV.7.1 $P^+\mathfrak{L}$ is closed. The subspace $\mathfrak{H}^+$ being intrinsically complete (cf. Theorem 1.1), so is $P^+\mathfrak{L}$. If $P^+\mathfrak{L} \neq \mathfrak{H}^+$ then, according to the orthogonal decomposition theorem of Hilbert spaces, there exists a vector x satisfying the relations $x \in \mathfrak{H}^+$, $x \perp P^+\mathfrak{L}$, $x \neq 0$; in view of Remark I.3.2, $\langle \mathfrak{L}, x \rangle$ is a proper positive extension of $\mathfrak{L}$. Contradiction. □

Lemma 4.3. *Every closed, maximal positive definite (maximal negative definite) subspace of* $\mathfrak{H}$ *is maximal positive (maximal negative).*

Proof. Let $\mathfrak{L} \subset \mathfrak{H}$ be closed and positive definite. Following the notations and proof of Theorem 4.2 we find that in the case $P^+\mathfrak{L} = \mathfrak{H}^+$ the subspace $\mathfrak{L}$ is maximal positive, while in the opposite case $\mathfrak{L}$ admits a positive definite proper extension. □

Lemma I.6.3 says that the orthogonal companion of a maximal positive (maximal negative) subspace is negative (resp. positive). In Krein spaces we can prove more.

Theorem 4.4. *If* $\mathfrak{L}$ *is a maximal positive (maximal negative) subspace of* $\mathfrak{H}$, *then* $\mathfrak{L}^\perp$ *is maximal negative (maximal positive).*

Proof. Let $\mathfrak{L}$ be maximal positive. On account of Lemmas I.6.3 and III.2.4 $\mathfrak{L}^\perp$ is negative and closed. Consider a fundamental decomposition (4.1) with corresponding projectors P^+, P^-. By Lemma IV.7.1 and Theorem 1.1 $P^-\mathfrak{L}^\perp$ is an intrinsically complete subspace of the intrinsically complete space $\mathfrak{H}^-$.

Suppose that $\mathfrak{L}^\perp$ is not maximal negative. Then, according to Theorem 4.2, $P^-\mathfrak{L}^\perp \neq \mathfrak{H}^-$. Consequently, there exists an $x \in \mathfrak{H}^-$ satisfying the relations $x \perp P^-\mathfrak{L}^\perp$, $x \neq 0$. In particular, $(x, x) < 0$ and $x \in \mathfrak{L}^{\perp\perp}$. But, in view of Theorems 3.6 and 4.1, $\mathfrak{L}^{\perp\perp}$ coincides with the positive subspace $\mathfrak{L}$. Contradiction. □

In Krein spaces also a converse of Lemma I.6.3 holds.

Lemma 4.5. *If* $\mathfrak{L}$ *is a positive (negative) subspace with negative (positive) orthogonal companion in* $\mathfrak{H}$, *then the closure of* $\mathfrak{L}$ *is maximal positive (maximal negative).*

Proof. Suppose that $\mathfrak{L}$ is positive, but its closure is not maximal positive. Consider a fundamental decomposition (4.1) and the corresponding projectors P^+, P^-. On account of Theorem 4.2 and Lemma IV.7.1 there exists an $x \in \mathfrak{H}^+$ with the properties $x \perp P^+\bar{\mathfrak{L}}$, $x \neq 0$. In particular, x is a positive element of $\mathfrak{L}^\perp$. □

We end this section by introducing a quantity which measures the non-maximality of positive subspaces. It will next appear in Chapter VII.

Let $\mathfrak{L}$ be a closed positive subspace of the Krein space $\mathfrak{H}$. Let $\mathfrak{L}_1 \supset \mathfrak{L}$ be a maximal positive subspace of $\mathfrak{H}$ (cf. Section I.6). The codimension of $\mathfrak{L}$ relative to $\mathfrak{L}_1$ will be called the *plus-defect* of $\mathfrak{L}$ and denoted by $d^+(\mathfrak{L})$.

From the next result it follows that the cardinal number $d^+(\mathfrak{L})$ does not depend on the choice of the maximal positive subspace $\mathfrak{L}_1 \supset \mathfrak{L}$.

Lemma 4.6. *If $\mathfrak{L} \subset \mathfrak{L}_1 \subset \mathfrak{H}$, where $\mathfrak{L}_1$ is maximal positive, then for any fundamental decomposition* (4.1) *and respective fundamental projectors P^+, P^- we have*

$$\operatorname{codim}_{\mathfrak{L}_1} \mathfrak{L} = \operatorname{codim}_{\mathfrak{H}^+} P^+\mathfrak{L} .$$

Proof. Let $\mathfrak{M}$ be a closed complementary subspace to $\mathfrak{L}$ in $\mathfrak{L}_1$. Then, by Theorem 4.2 and Lemma IV.7.1, $P^+\mathfrak{M}$ is a closed complementary subspace to $P^+\mathfrak{L}$ in $\mathfrak{H}^+$, and $\dim P^+\mathfrak{M} = \dim \mathfrak{M}$. □

5. Uniformly Definite Subspaces

Consider a semi-definite subspace $\mathfrak{L}$ in the Krein space $\mathfrak{H}$. According to Theorems 3.4 and 1.1 $\mathfrak{L}$ is ortho-complemented if and only if it is closed, definite, and intrinsically complete. On the other hand, the intrinsic topology $\tau_{\text{int}}(\mathfrak{L})$ of a definite $\mathfrak{L}$ being weaker than the topology $\tau_M(\mathfrak{H})|\mathfrak{L}$ induced on $\mathfrak{L}$ by the strong topology of $\mathfrak{H}$, Corollary 1.2 and the closed graph principle imply that a closed definite subspace $\mathfrak{L}$ is intrinsically complete if and only if $\tau_{\text{int}}(\mathfrak{L}) = \tau_M(\mathfrak{H})|\mathfrak{L}$. This motivates the following definition.

A (closed or non-closed) subspace $\mathfrak{L}$ of the Krein space $\mathfrak{H}$ is said to be *uniformly positive* (*uniformly negative*), if 1) $\mathfrak{L}$ is positive definite (negative definite) and 2) $\tau_{\text{int}}(\mathfrak{L}) = \tau_M(\mathfrak{H})|\mathfrak{L}$.

In other words, $\mathfrak{L} \subset \mathfrak{H}$ is uniformly positive if

$$(5.1) \qquad (x, x) \geqq \alpha \|x\|_J^2 \qquad (x \in \mathfrak{L}) ,$$

and uniformly negative if

$$(5.2) \qquad (x, x) \leqq -\alpha \|x\|_J^2 \qquad (x \in \mathfrak{L}) ,$$

where $\|\cdot\|_J$ is any fixed natural norm and α is a positive number depending on $\mathfrak{L}$.

Uniformly positive and uniformly negative subspaces may be termed *uniformly definite.*

The above arguments and definitions lead to the following two results.

Lemma 5.1. *A closed definite subspace of $\mathfrak{H}$ is intrinsically complete if and only if it is uniformly definite.* □

Theorem 5.2. *A semi-definite subspace of $\mathfrak{H}$ is ortho-complemented if and only if it is closed and uniformly definite.* □

For a possibly indefinite subspace $\mathfrak{L} \subset \mathfrak{H}$ we have:

Theorem 5.3. *A subspace $\mathfrak{L}$ of the Krein space $\mathfrak{H}$ is ortho-complemented if and only if $\mathfrak{L}$ is the orthogonal direct sum of a closed uniformly positive and a closed uniformly negative subspace.*

Proof. By Theorem 5.2 and Lemma I.9.2 the condition is sufficient. To prove necessity, suppose that $\mathfrak{L}$ is ortho-complemented. Then, on account of Theorem 3.4, $\mathfrak{L}$ is the orthogonal direct sum of a positive definite and a negative definite subspace. Application of Lemma I.9.2 and Theorem 5.2 yields the desired conclusion. □

Having made clear the basic importance of uniformly definite subspaces, we are going to present further characterizations of them, valuable partly from the intuitive, partly from the technical point of view.

Lemma 5.4. *A subspace of $\mathfrak{H}$ is uniformly positive (uniformly negative) if and only if its closure is so.*

Proof. Employ the description (5.1)—(5.2) of uniformly definite subspaces. □

From Lemma 5.4 and Theorem 5.2 one obtains:

Corollary 5.5. *A semi-definite subspace of $\mathfrak{H}$ is uniformly definite if and only if its closure is ortho-complemented.* □

Lemma 5.4 combined with Corollary 5.5 and Theorem 3.5 yields:

Theorem 5.6. *The subspace $\mathfrak{L} \subset \mathfrak{H}$ is uniformly positive (uniformly negative) if and only if $\mathfrak{H}$ admits a fundamental decomposition*

$$\mathfrak{H} = \mathfrak{H}^+ (\dot{+}) \mathfrak{H}^-; \qquad \mathfrak{H}^+ \subset \mathfrak{P}^{++}, \quad \mathfrak{H}^- \subset \mathfrak{P}^{--} \tag{5.3}$$

such that $\mathfrak{L} \subset \mathfrak{H}^+$ $(\mathfrak{L} \subset \mathfrak{H}^-)$. □

In contrast to Theorem 5.6, the following characterization of uniformly definite subspaces is useful in cases where a fundamental decomposition of $\mathfrak{H}$ is fixed in advance.

Theorem 5.7. *Consider a fundamental decomposition* (5.3) *of the Krein space* $\mathfrak{H}$. *Denote the corresponding fundamental symmetry by* J. *A subspace* $\mathfrak{L} \subset \mathfrak{H}$ *is uniformly positive if and only if the angular operator* $K^+(\mathfrak{L})$ *of* $\mathfrak{L}$ *with respect to* $\mathfrak{H}^+$ *exists and satisfies the inequality* $\|K^+(\mathfrak{L})\|_J < 1$. *Analogously,* $\mathfrak{L}$ *is uniformly negative if and only if* $K^-(\mathfrak{L})$, *the angular operator of* $\mathfrak{L}$ *relative to* $\mathfrak{H}^-$, *exists and satisfies the condition* $\|K^-(\mathfrak{L})\|_J < 1$.

Proof. We restrict ourselves to the positive case. Let $\mathfrak{L}$ be uniformly positive. Then, according to Theorem II.11.7, $K^+(\mathfrak{L})$ exists. Moreover, for

$$x^+ \in \mathfrak{H}^+, \quad x^- \in \mathfrak{H}^-, \quad x^+ + x^- \in \mathfrak{L} \tag{5.4}$$

relation (5.1) yields

$$\|x^+\|_J^2 - \|x^-\|_J^2 \geqq \alpha\,(\|x^+\|_J^2 + \|x^-\|_J^2) \tag{5.5}$$

i.e.

$$\|x^-\|_J^2 \leqq \frac{1-\alpha}{1+\alpha}\,\|x^+\|_J^2, \tag{5.6}$$

where $\alpha > 0$. Thus

$$\|K^+(\mathfrak{L})\|_J \leqq \frac{1-\alpha}{1+\alpha} < 1 .$$

Let, conversely, $\mathfrak{L}$ be a subspace such that $K^+(\mathfrak{L})$ exists and $\|K^+(\mathfrak{L})\|_J = \varrho < 1$. Then for the positive number $\alpha = \dfrac{1-\varrho}{1+\varrho}$ and any vectors (5.4) the relation (5.6) holds. From (5.6) we obtain (5.5) and, finally, (5.1). □

6. Non-uniformly Definite Subspaces

By Theorem 5.6 every non-zero Krein space contains uniformly definite non-zero subspaces. Next we investigate the existence of definite subspaces which are not uniformly definite (*non-uniformly definite subspaces* for short).

Lemma 6.1. *Let* $\mathfrak{H}$ *be a Krein space with rank of positivity* $\varkappa^+(\mathfrak{H})$ *and rank of negativity* $\varkappa^-(\mathfrak{H})$. *In order that every positive definite subspace* $\mathfrak{L} \subset \mathfrak{H}$ *would have a positive definite closure it is necessary and sufficient that either* $\varkappa^+(\mathfrak{H}) < \infty$ *or* $\varkappa^-(\mathfrak{H}) = 0$. *Similarly, the closure of every negative definite* $\mathfrak{L} \subset \mathfrak{H}$ *is negative definite if and only if at least one of the conditions* $\varkappa^-(\mathfrak{H}) < \infty$, $\varkappa^+(\mathfrak{H}) = 0$ *is satisfied.*

Proof. Owing to Theorem II.10.1 the conditions are sufficient. Next suppose that $\varkappa^+(\mathfrak{H}) = \infty$, $\varkappa^-(\mathfrak{H}) > 0$. Let $\{e_j\}_1^\infty$ and $\{f_1\}$, respectively, be orthonormal systems contained in the components $\mathfrak{H}^+$ and $\mathfrak{H}^-$ of a fundamental decomposition of $\mathfrak{H}$. Denote by $\mathfrak{L}$ the set of all elements of the form $\sum_1^\infty \alpha_j e_j + \beta_1 f_1$, where $\alpha_j \neq 0$ only for a finite number of indices and, moreover, $\sum_2^\infty \alpha_j = \alpha_1 = \beta_1$ (cf. also Example I.4.9 and the proof of Theorem III.8.1). Then $\mathfrak{L}$ is a positive definite subspace, but the vectors

$$e_1 + \frac{1}{n} \sum_2^{n+1} e_j + f_1 \in \mathfrak{L} \qquad (n = 1, 2, \ldots)$$

converge to the neutral non-zero vector $e_1 + f_1$. □

From Lemmas 5.4 and 6.1 we obtain:

Corollary 6.2. *Every indefinite Krein space of infinite dimension contains non-closed, non-uniformly definite subspaces.* □

As to the existence of closed, non-uniformly definite subspaces, we have the following.

Theorem 6.3. *Let $\mathfrak{H}$ be a Krein space with rank of indefiniteness $\varkappa(\mathfrak{H})$. In order that every closed, positive definite (negative definite) subspace of $\mathfrak{H}$ would be uniformly positive (uniformly negative), it is necessary and sufficient that $\varkappa(\mathfrak{H}) < \infty$.*

Proof. Consider a fundamental decomposition

$$\mathfrak{H} = \mathfrak{H}^+ (\dot{+}) \mathfrak{H}^-; \qquad \mathfrak{H}^+ \subset \mathfrak{P}^{++}, \qquad \mathfrak{H}^- \subset \mathfrak{P}^{--} \tag{6.1}$$

and a closed, positive definite subspace $\mathfrak{L} \subset \mathfrak{H}$. On account of Theorem II.11.7 we have the inequality

$$\|K^+ x^+\|_J < 1 \qquad (x^+ \in \mathfrak{D}(K^+), \quad \|x^+\|_J \leqq 1), \tag{6.2}$$

where K^+ is the angular operator of $\mathfrak{L}$ with respect to $\mathfrak{H}^+$ and J is the fundamental symmetry belonging to (6.1). Moreover, owing to Lemma IV.7.1, $\mathfrak{D}(K^+)$ is closed.

Let $\varkappa(\mathfrak{H}) < \infty$. Then either the domain $\mathfrak{D}(K^+) \subset \mathfrak{H}^+$ or the range $\mathfrak{R}(K^+) \subset \mathfrak{H}^-$ of K^+ has finite dimension. In the first case (6.2) implies that

$$\|K^+ x^+\|_J \leqq \gamma \qquad (x^+ \in \mathfrak{D}(K^+), \quad \|x^+\|_J \leqq 1), \tag{6.3}$$

where $\gamma < 1$. In the second case the same conclusion can be drawn by splitting $\mathfrak{D}(K^+)$ into the orthogonal direct sum of $\mathfrak{R}(K^+)$ and its finite-dimensional orthogonal companion in $\mathfrak{D}(K^+)$. Therefore, in view of Theorem 5.7, $\mathfrak{L}$ is uniformly positive.

Now let $\varkappa(\mathfrak{H}) = \infty$. Choose two orthonormal systems $\{e_j\}_1^\infty \subset \mathfrak{H}^+$, $\{f_j\}_1^\infty \subset \mathfrak{H}^-$. Define a linear operator K^+ in the following way:

$$\mathfrak{D}(K^+) = \left\{ \sum_{j=1}^{\infty}{}' \alpha_j e_j : \quad \sum_{j=1}^{\infty}{}' |\alpha_j|^2 < \infty \right\},$$

$$K^+\left(\sum_{j=1}^{\infty}{}' \alpha_j e_j \right) = \sum_{j=1}^{\infty}{}' \frac{j}{j+1} \alpha_j f_j \qquad \left(\sum_{j=1}^{\infty}{}' |\alpha_j|^2 < \infty \right).$$

Then $\mathfrak{D}(K^+)$ is closed, (6.2) is fulfilled, but (6.3) does not hold for any $\gamma < 1$. On account of Lemma IV.7.1, Theorem II.11.7 and Theorem 5.7, respectively, the subspace

$$\mathfrak{L} = \{x^+ + K^+x^+ : \quad x^+ \in \mathfrak{D}(K^+)\}$$

is closed, positive definite, but not uniformly positive.

The case of negative definite subspaces can be treated by passing to the anti-space of $\mathfrak{H}$. □

Remark 6.4. For uniformly definite subspaces, according to their definition and Lemma 5.1, the properties of being closed or intrinsically complete imply each other. However, by the same lemma, a non-uniformly definite subspace cannot have more than one of these properties. The existence of closed, non-uniformly definite subspaces has just been discussed in Theorem 6.3. For intrinsically complete, non-uniformly definite subspaces the existence problem is settled by the next example.

Example 6.5. Consider a Krein space $\mathfrak{H}$ with $\varkappa^+(\mathfrak{H}) = \infty$, $0 < \varkappa^-(\mathfrak{H}) < \infty$. Let $\mathfrak{L}$ be a positive definite subspace of $\mathfrak{H}$ such that the closure $\mathfrak{L}_1$ of $\mathfrak{L}$ is degenerate (cf. Lemmas 6.1 and I.4.4). Applying Remark I.5.5 and Theorem I.5.4 to the inner product space $\mathfrak{L}_1$ we find a positive definite subspace $\mathfrak{L}_2 \supset \mathfrak{L}$ with the property

$$\mathfrak{L}_2 \dotplus \mathfrak{L}_1^0 = \mathfrak{L}_1\,,$$

where $\mathfrak{L}_1^0$ stands for the isotropic part of $\mathfrak{L}_1$. On the other hand, by Lemmas IV.8.7 and I.4.4, there exists a closed positive definite subspace $\mathfrak{L}_3$ such that

$$\mathfrak{L}_3 \dotplus \mathfrak{L}_1^0 = \mathfrak{L}_1\,.$$

In view of Theorem 6.3 $\mathfrak{L}_3$ is uniformly positive, hence intrinsically complete. But, according to Lemma I.5.2, $\mathfrak{L}_2$ is isometrically isomorphic to $\mathfrak{L}_3$. In particular, $\mathfrak{L}_2$ is intrinsically complete. At the same time, having a degenerate closure $\mathfrak{L}_1$, the subspace $\mathfrak{L}_2$ is not uniformly positive (cf. Lemma 5.4).

7. Maximal Uniformly Definite Subspaces

A uniformly positive (uniformly negative) subspace of the Krein space $\mathfrak{H}$ is said to be *maximal uniformly positive* (*maximal uniformly negative*) if it is not contained in any other uniformly positive (uniformly negative) subspace of $\mathfrak{H}$.

From Theorem 5.6 and Lemma I.11.4 we obtain:

Theorem 7.1. *The subspace $\mathfrak{L} \subset \mathfrak{H}$ is maximal uniformly positive (maximal uniformly negative) if and only if $\mathfrak{H}$ has a fundamental decomposition*

$$\mathfrak{H} = \mathfrak{H}^+ (\dotplus) \mathfrak{H}^-; \qquad \mathfrak{H}^+ \subset \mathfrak{P}^{++}, \qquad \mathfrak{H}^- \subset \mathfrak{P}^{--} \tag{7.1}$$

such that $\mathfrak{L} = \mathfrak{H}^+$ $(\mathfrak{L} = \mathfrak{H}^-)$. □

Theorem 7.1 and Remark I.11.6 yield:

Corollary 7.2. *Every maximal uniformly positive (maximal uniformly negative) subspace of $\mathfrak{H}$ is maximal positive (maximal negative).* □

Remark 7.3. Corollary 7.2 should be compared with Lemma 6.1 which, in view of the existence of maximal definite subspaces containing a given definite subspace, implies that in "most" Krein spaces a maximal positive definite subspace $\mathfrak{L}$ need not be maximal positive. (For a closed $\mathfrak{L}$, however, see Lemma 4.3.)

The following consequence of Theorem 7.1 should also be mentioned.

Corollary 7.4. *If $\mathfrak{L}$ is a maximal uniformly positive (maximal uniformly negative) subspace of $\mathfrak{H}$, then $\mathfrak{L}^\perp$ is maximal uniformly negative (maximal uniformly positive).* □

The next two results are concerned with complementary subspaces.

Lemma 7.5. *Suppose that the subspace $\mathfrak{L} \subset \mathfrak{H}$ contains a maximal uniformly positive (maximal uniformly negative) subspace. Then $\bar{\mathfrak{L}}$ is ortho-complemented and $\mathfrak{L}^\perp$ is uniformly negative (uniformly positive).*

Proof. Let $\mathfrak{L}_1 \subset \mathfrak{L}$ be maximal uniformly positive. On account of Corollary 7.4 $\mathfrak{L}_1^\perp$ is uniformly negative. Therefore the subspace $\mathfrak{L}^\perp \subset \mathfrak{L}_1^\perp$ is uniformly negative. Moreover, owing to Lemma III.2.4, $\mathfrak{L}^\perp$ is closed. By Theorem 5.2 $\mathfrak{L}^\perp$ is ortho-complemented. Hence, in view of Lemma I.9.1, also $\mathfrak{L}^{\perp\perp}$ is ortho-complemented. It remains to observe that $\mathfrak{L}^{\perp\perp} = \bar{\mathfrak{L}}$ (cf. Theorem 3.6). □

Lemma 7.6. *Let $\mathfrak{L}_1$ be a maximal uniformly positive and $\mathfrak{L}_2$ a maximal negative subspace of the Krein space $\mathfrak{H}$. Then $\mathfrak{L}_1 \dotplus \mathfrak{L}_2 = \mathfrak{H}$. The same conclusion holds if $\mathfrak{L}_1$ is maximal uniformly negative and $\mathfrak{L}_2$ is maximal positive.*

Proof. According to Theorem 7.1 $\mathfrak{H}$ has a fundamental decomposition (7.1) such that $\mathfrak{L}_1 = \mathfrak{H}^+$. On the other hand, Theorem 4.2 yields $P^-\mathfrak{L}_2 = \mathfrak{H}^-$, where P^- is the fundamental projector with $\mathfrak{R}(P^-) = \mathfrak{H}^-$. Thus

$$\mathfrak{H}^+ = \mathfrak{L}_1 \subset \mathfrak{L}_1 \dotplus \mathfrak{L}_2 ,$$

$$\mathfrak{H}^- = P^-\mathfrak{L}_2 \subset P^+\mathfrak{L}_2 \dotplus \mathfrak{L}_2 \subset \mathfrak{H}^+ \dotplus \mathfrak{L}_2 = \mathfrak{L}_1 \dotplus \mathfrak{L}_2 . \quad \square$$

8. Regular and Singular Subspaces

In Section III.8 definite subspaces were divided into two classes: those of regular and singular subspaces. In a Krein space these classes can easily be characterized.

We shall make use of the following lemma.

Lemma 8.1. *Let $\mathfrak{L}$ be a subspace with positive definite closure in the Krein space $\mathfrak{H}$. Suppose that $\mathfrak{L}$ admits a decomposition $\mathfrak{L} = \mathfrak{L}_1 \dotplus \mathfrak{L}_2$, where* $\dim \mathfrak{L}_1 = 1$ *and $\mathfrak{L}_2$ is uniformly positive. Then $\mathfrak{L}$ is uniformly positive.*

Proof. According to Lemma 5.4 the closure $\bar{\mathfrak{L}}_2$ of $\mathfrak{L}_2$ is uniformly positive. Hence for $\mathfrak{L}_1 \subset \bar{\mathfrak{L}}_2$ i.e. $\mathfrak{L} \subset \bar{\mathfrak{L}}_2$ the assertion follows at once. Next consider the case $\mathfrak{L}_1 \cap \bar{\mathfrak{L}}_2 = 0$. Set

$$\mathfrak{L}' = \mathfrak{L}_1 \dotplus \bar{\mathfrak{L}}_2 , \quad \mathfrak{L}_1' = (\bar{\mathfrak{L}}_2)^{\perp} \cap \mathfrak{L}' .$$

In view of Theorem 5.2 we have

$$\mathfrak{L}' = \mathfrak{L}_1' \,(\dotplus)\, \bar{\mathfrak{L}}_2 .$$

Corollary I.1.2 and the required positive definiteness of $\bar{\mathfrak{L}}$, respectively, yield that $\mathfrak{L}_1'$ is 1-dimensional and positive definite. Being the orthogonal direct sum of two ortho-complemented subspaces, $\mathfrak{L}'$ is ortho-complemented (cf. Lemma I.9.2). Thus, by Theorem 5.2, $\mathfrak{L}'$ is uniformly positive, and therefore so is the subspace $\mathfrak{L} \subset \mathfrak{L}'$. $\square$

Theorem 8.2. *A definite subspace of the Krein space $\mathfrak{H}$ is regular if and only if it is uniformly definite.*

Proof. Let $\mathfrak{L} \subset \mathfrak{H}$ be uniformly definite. Since, for every $y \in \mathfrak{H}$, the linear form $\varphi_y(x) = (x, y)$ $(x \in \mathfrak{L})$ is continuous relative to $\tau_M(\mathfrak{H})|\mathfrak{L}$, it is continuous relative to $\tau_{\text{int}}(\mathfrak{L})$. So $\mathfrak{L}$ is regular.

Suppose, conversely, that the subspace $\mathfrak{L} \subset \mathfrak{H}$ is non-uniformly definite, say, positive definite. We shall distinguish between two cases.

1. $\bar{\mathfrak{L}}$ contains a neutral non-zero vector x_0. In this case, taking an element $y \in \mathfrak{H}$ with $(x_0, y) \neq 0$ and a sequence $\{x_n\}_1^\infty \subset \mathfrak{L}$ with $x_n \to x_0$ $(n \to \infty)$, we find:

$$\lim_{n\to\infty} |x_n|_{\mathfrak{L}} = 0\,, \quad \lim_{n\to\infty} (x_n, y) \neq 0\,.$$

Hence $\mathfrak{L}$ is singular.

2. $\bar{\mathfrak{L}}$ is positive definite. In this case we proceed as follows.

We consider a fundamental decomposition

$$\mathfrak{H} = \mathfrak{H}^+ (\dot{+}) \mathfrak{H}^-; \qquad \mathfrak{H}^+ \subset \mathfrak{P}^{++}, \quad \mathfrak{H}^- \subset \mathfrak{P}^{--} \tag{8.1}$$

with respective fundamental projectors P^+, P^- and fundamental symmetry $J = P^+ - P^-$. Instead of P^+x, as usual, we shall write x^+.

The subspace $\mathfrak{L}$ being non-uniformly definite, there exists an $x_1 \in \mathfrak{L}$ such that $(x_1, x_1) < ||x_1||_J^2$. By Theorem II.10.1 we may require $(x_1^+, x_1^+) = 1$.

Set

$$\mathfrak{L}_1 = \langle x_1 \rangle, \qquad \mathfrak{L}_2 = \{x \in \mathfrak{L}\colon \; x^+ \perp x_1^+\}\,. \tag{8.2}$$

Then $\mathfrak{L}_1 \dot{+} \mathfrak{L}_2 = \mathfrak{L}$. Thus, on account of Lemma 8.1, $\mathfrak{L}_2$ is non-uniformly definite. Consequently, there exists an $x_2 \in \mathfrak{L}_2$ satisfying the conditions $(x_2, x_2) < \frac{1}{8}||x_2||_J^2$, $(x_2^+, x_2^+) = 1$. In view of (8.2) we also have $(x_1^+, x_2^+) = 0$.

Pursuing the process, we construct a sequence $\{x_n\}_1^\infty \subset \mathfrak{L}$ such that

$$(x_n, x_n) < \frac{1}{n^3}||x_n||_J^2\,, \qquad (x_m^+, x_n^+) = \delta_{mn} \qquad (m,n = 1, 2, \ldots)\,.$$

Consider the elements

$$n x_n = y_n \qquad (n = 1, 2, \ldots)\,,$$

$$\sum_{n=1}^{\infty} \frac{1}{n} x_n^+ = x^+.$$

The positive definiteness of $\bar{\mathfrak{L}}$ implies $||x_n||_J^2 < 2||x_n^+||_J^2$. Therefore

$$(y_n, y_n) < \frac{2}{n} \qquad (n = 1,2, \ldots)\,.$$

On the other hand, $(y_n, x^+) = 1$ for every n. So $\mathfrak{L}$ is singular. □

9. Alternating Pairs

An ordered pair $\{\mathfrak{L}_1, \mathfrak{L}_2\}$ of subspaces of the Krein space $\mathfrak{H}$ will be called an *alternating pair* provided $\mathfrak{L}_1$ is positive, $\mathfrak{L}_2$ is negative, and

$\mathfrak{L}_1 \perp \mathfrak{L}_2$. If, in addition, $\mathfrak{L}_1$ is maximal positive and $\mathfrak{L}_2$ is maximal negative, we say that $\{\mathfrak{L}_1, \mathfrak{L}_2\}$ is an *alternating maximal pair*.

By an *alternating extension* of the alternating pair $\{\mathfrak{L}_1, \mathfrak{L}_2\}$ we mean an alternating pair $\{\mathfrak{L}_1', \mathfrak{L}_2'\}$ such that $\mathfrak{L}_1 \subset \mathfrak{L}_1'$, $\mathfrak{L}_2 \subset \mathfrak{L}_2'$.

Theorem 9.1. *Every alternating pair in the Krein space $\mathfrak{H}$ can be extended to an alternating maximal pair.*

Proof. Consider a fundamental decomposition

$$\mathfrak{H} = \mathfrak{H}^+ (\dotplus) \mathfrak{H}^-; \qquad \mathfrak{H}^+ \subset \mathfrak{P}^{++}, \quad \mathfrak{H}^- \subset \mathfrak{P}^{--}, \tag{9.1}$$

the corresponding fundamental projectors P^+, P^-, and the fundamental symmetry $J = P^+ - P^-$.

Let $\mathfrak{L}_1 \subset \mathfrak{P}^+$, $\mathfrak{L}_2 \subset \mathfrak{P}^-$, $\mathfrak{L}_1 \perp \mathfrak{L}_2$. Since the inner product is continuous on $\mathfrak{H}$, we may assume that $\mathfrak{L}_1$ and $\mathfrak{L}_2$ are closed. Then, according to Lemma IV.7.1, $P^+\mathfrak{L}_1$ and $P^-\mathfrak{L}_2$ are also closed. The orthogonal (and J-orthogonal) projectors with range $P^+\mathfrak{L}_1$ resp. $P^-\mathfrak{L}_2$ (cf. Theorem 5.2 and Theorem II.3.10) will be denoted by P_1 and P_2.

In view of Theorem II.11.7 the angular operators $K^+(\mathfrak{L}_1)$, $K^-(\mathfrak{L}_2)$ exist and their J-norms do not exceed 1. For technical reasons we set

$$K_1 = K^+(\mathfrak{L}_1), \qquad K_2 = K^-(\mathfrak{L}_2) P_2 . \tag{9.2}$$

Then K_2 is defined throughout $\mathfrak{H}$ and we have

$$\|K_1\|_J \leqq 1, \quad \|K_2\|_J \leqq 1 . \tag{9.3}$$

Moreover, the assumption $\mathfrak{L}_1 \perp \mathfrak{L}_2$ yields

$$(x + K_1 x, \; P_2 y + K_2 y) = 0 \qquad (x \in P^+\mathfrak{L}_1, \quad y \in \mathfrak{H})$$

or, equivalently,

$$P_2 K_1 x = K_2^{[*]} x \qquad (x \in P^+\mathfrak{L}_1), \tag{9.4}$$

where $K_2^{[*]}$ is the Hilbert space adjoint of K_2 specified by the relation

$$(K_2^{[*]} x, y)_J = (x, K_2 y)_J \qquad (x, y \in \mathfrak{H}) . \tag{9.5}$$

On account of Theorem 4.4 it is sufficient to find a maximal positive subspace $\mathfrak{L}_1^{(0)}$ containing $\mathfrak{L}_1$ and orthogonal to $\mathfrak{L}_2$. By Theorem II.11.7 it will do if we construct the angular operator $K = K^+(\mathfrak{L}_1^{(0)})$, i.e., a linear operator K which satisfies the conditions $K_1 \subset K$, $\mathfrak{D}(K) = \mathfrak{H}^+$ (cf. Theorem 4.2), $\mathfrak{R}(K) \subset \mathfrak{H}^-$, $\|K\|_J \leqq 1$ and, in analogy with (9.4),

$$P_2 K x = K_2^{[*]} x \qquad (x \in \mathfrak{H}^+) . \tag{9.6}$$

Since (9.6) can be regarded as a definition of the first term in the representation $K = P_2 K + (I - P_2) K$, it remains to choose $(I - P_2) K$ to be an extension of $(I - P_2) K_1$ with domain $\mathfrak{H}^+$ and range contained in $(I - P_2) \mathfrak{H}^-$ such that

$$\|(I - P_2) K x\|_J^2 \leqq \|x\|_J^2 - \|K_2^{[*]} x\|_J^2 \qquad (x \in \mathfrak{H}^+) . \tag{9.7}$$

Consider the inner product

$$(x, y)_2 = (x, y) + (K_2^{[*]}x, K_2^{[*]}y) \qquad (x, y \in \mathfrak{H}^+).$$

By (9.2), (9.3) and (9.5) we have

$$(x, x)_2 = \|x\|_J^2 - \|K_2^{[*]} x\|_J^2 \geqq 0 \qquad (x \in \mathfrak{H}^+).$$

In terms of the semi-norm $p_2(x) = (x, x)_2^{1/2}$ $(x \in \mathfrak{H}^+)$ condition (9.7) takes the form

$$\|(I - P_2)Kx\|_J \leqq p_2(x) \qquad (x \in \mathfrak{H}^+). \tag{9.8}$$

For $x \in P^+\mathfrak{L}_1$ the respective inequality

$$\|(I - P_2)K_1x\|_J \leqq p_2(x) \qquad (x \in P^+\mathfrak{L}_1) \tag{9.9}$$

is guaranteed by (9.4) and (9.3).

Put

$$\mathfrak{H}_0^+ = \{x \in \mathfrak{H}^+ : \ p_2(x) = 0\}.$$

On the quotient vector space $\mathfrak{H}^+/\mathfrak{H}_0^+$ the positive inner product $(.\,,\,.)_2$ induces (cf. Section I.5, especially Corollary I.5.3) a positive definite inner product $(.\,,\,.)_2^\wedge$. With respect to the corresponding norm $\|\cdot\|_2^\wedge$ the space $\mathfrak{H}^+/\mathfrak{H}_0^+$ can be completed to a Hilbert space $\mathfrak{E}^+$.

In view of (9.9) the formula

$$T_1\hat{x} = (I - P_2)K_1x \qquad (x \in \hat{x}, \quad x \in P^+\mathfrak{L}_1)$$

uniquely defines a linear operator T_1 on a subspace of $\mathfrak{E}^+$; this operator has its values in $(I - P_2)\,\mathfrak{H}^-$ and, owing to (9.9), is contractive:

$$\|T_1\hat{x}\|_J \leqq \|\hat{x}\|_2^\wedge \qquad (\hat{x} \in \mathfrak{D}(T_1)).$$

We extend T_1 to the whole of the Hilbert space $\mathfrak{E}^+$ as a contractive operator T with values in $(I - P_2)\,\mathfrak{H}^-$. Finally, we set

$$(I - P_2)Kx = T\hat{x} \qquad (x \in \mathfrak{H}^+).$$

Then (9.8) and all other relevant conditions will be fulfilled. □

10. Dissipative Operators in Hilbert Space

The purpose of this section is to give an idea of how one applies alternating pairs.

Let $\mathfrak{H}_0$ be a Hilbert space with inner product $[.\,,\,.]_0$ and norm $\|\cdot\|_0$. Let T_1, T_2 be *dissipative* operators in $\mathfrak{H}_0$; by this we mean that T_1, T_2 are linear operators with dense domains $\mathfrak{D}(T_1)$, $\mathfrak{D}(T_2)$ satisfying

$$[T_jx_j, x_j]_0 + [x_j, T_jx_j]_0 \leqq 0 \qquad (x_j \in \mathfrak{D}(T_j);\ j = 1,2). \tag{10.1}$$

Suppose also that

$$[T_1x_1, x_2]_0 = [x_1, T_2x_2]_0 \qquad (x_1 \in \mathfrak{D}(T_1), \quad x_2 \in \mathfrak{D}(T_2)). \tag{10.2}$$

The question is whether there exist operators $T_1^{(0)} \supset T_1$, $T_2^{(0)} \supset T_2$ in $\mathfrak{H}_0$ which are *maximal dissipative* (i.e., dissipative operators having no dissipative proper extensions) and satisfy the analogue of (10.2):

$$(10.3)\quad [T_1^{(0)}x_1, x_2]_0 = [x_1, T_2^{(0)}x_2]_0 \qquad (x_1 \in \mathfrak{D}(T_1^{(0)}),\quad x_2 \in \mathfrak{D}(T_2^{(0)}))\,.$$

In order to solve the problem we consider the cartesian product $\mathfrak{H} = \mathfrak{H}_0 \times \mathfrak{H}_0$, i.e. the Hilbert space of all pairs $\{x, y\}$ $(x,y \in \mathfrak{H}_0)$ with inner product

$$(10.4)\qquad [\{x_1, y_1\}, \{x_2, y_2\}] = [x_1, x_2]_0 + [y_1, y_2]_0$$

and norm

$$(10.5)\qquad \|\{x, y\}\| = (\|x\|_0^2 + \|y\|_0^2)^{1/2}\,.$$

We introduce an indefinite inner product on $\mathfrak{H}$ setting

$$(10.6)\qquad (\{x_1, y_1\}, \{x_2, y_2\}) = [x_1, y_2]_0 + [y_1, x_2]_0\,.$$

The Gram operator G of $(.\,,.)$ with respect to $[.\,,.]$ exists, namely $G\{x, y\} = \{y, x\}$. Moreover, G is continuous relative to $\|\cdot\|$, and completely invertible. Thus, according to Theorem 1.3, $\mathfrak{H}$ equipped with the inner product (10.6) is a Krein space.

From (10.1) it follows that the graph

$$\mathfrak{G}(-T_1) = \{\{x_1, -Tx_1\}:\ x_1 \in \mathfrak{D}(T_1)\}$$

is a positive subspace of $\mathfrak{H}$. Similarly, the graph

$$\mathfrak{G}(T_2) = \{\{x_2, T_2x_2\}:\ x_2 \in \mathfrak{D}(T_2)\}$$

is a negative subspace of $\mathfrak{H}$. Further, by (10.2), $\mathfrak{G}(-T_1) \perp \mathfrak{G}(T_2)$ (in the sense of (10.6)). So $\{\mathfrak{G}(-T_1), \mathfrak{G}(T_2)\}$ is an alternating pair in $\mathfrak{H}$.

Let $\{\mathfrak{L}_1', \mathfrak{L}_2'\}$ be an alternating extension of $\{\mathfrak{G}(-T_1), \mathfrak{G}(T_2)\}$. Suppose that $\mathfrak{L}_1'$ contains an element of the form $\{0, y\}$. By (10.6) such an element is neutral. Therefore, on account of Lemma I.4.4, $\{0, y\} \perp \mathfrak{L}_1'$. In particular,

$$(\{0, y\},\ \{x_1, -T_1x_1\}) = 0$$

i.e. $[y, x_1]_0 = 0$ for every $x_1 \in \mathfrak{D}(T_1)$. Hence $y = 0$.

Consequently, $\mathfrak{L}_1'$ is the graph of an operator $-T_1'$. The positivity of $\mathfrak{L}_1'$ implies that T_1' is dissipative. In the same manner one verifies that $\mathfrak{L}_2'$ is the graph of a dissipative operator T_2'. Also the properties $T_1 \subset T_1'$, $T_2 \subset T_2'$,

$$(10.7)\qquad [T_1'x_1, x_2]_0 = [x_1, T_2'x_2]_0 \qquad (x_1 \in \mathfrak{D}(T_1'),\quad x_2 \in \mathfrak{D}(T_2'))$$

follow immediately.

Conversely, if T_1', T_2' are dissipative extensions of T_1 and T_2, respectively, such that (10.7) holds, then $\{\mathfrak{G}(-T_1'), \mathfrak{G}(T_2')\}$ is an alternating extension of $\{\mathfrak{G}(-T_1), \mathfrak{G}(T_2)\}$.

We are led to the following result. If $\{\mathfrak{L}_1^{(0)}, \mathfrak{L}_2^{(0)}\}$ is an alternating maximal pair extending $\{\mathfrak{G}(-T_1), \mathfrak{G}(T_2)\}$, then the operators $T_1^{(0)}$, $T_2^{(0)}$ defined by the relations $\mathfrak{G}(-T_1^{(0)}) = \mathfrak{L}_1^{(0)}$, $\mathfrak{G}(T_2^{(0)}) = \mathfrak{L}_2^{(0)}$ yield a solution of our problem, and every solution can be obtained in this way. Thus the existence of a solution is guaranteed by Theorem 9.1.

Notes to Chapter V

The objects called here Krein spaces for the first time occurred implicitly in papers of Nevanlinna [4] and Pesonen [1]. Formally they were introduced by Ginzburg [1] in 1957 and investigated thereupon by several authors including Iohvidov, Langer, Phillips, and Scheibe. Nonetheless, the new term seems to be justified by the fact that the theory of these spaces grew out, for the most part, of the school of M. G. Kreĭn, and has largely been influenced by Kreĭn's personality and contributions.

The present definition of Krein spaces was formulated by Scheibe [1]. Most authors start from a Hilbert space and introduce the indefinite inner product as a derived concept. In the latter treatment the indefinite rather than the definite inner product is called J-inner product.

The last assertion of Theorem 1.1 was proved by Iohvidov and Kreĭn [1] in a special case. In the general case it is due to Scheibe [1]. It follows also from independent investigations of Ginzburg [3].

Theorem 1.3 has first been obtained by Nevanlinna [4] and independently reproved by Langer [2], Phillips [3], and Scheibe [1]. Langer also noticed that a non-degenerate, decomposable space is a Krein space if and only if it is weakly sequentially complete (cf. the assumption of Theorem III.9.3).

That Krein spaces are completions of decomposable non-degenerate spaces was observed in a special case by Iohvidov and Kreĭn [1], and in the general case by Ginzburg and Iohvidov [1]. The uniqueness of the completion seems to be considered in Theorem 2.1 for the first time.

Theorem 3.4 was obtained by Scheibe [1] and Ginzburg [3], independently of each other. For a special case of Theorem 3.5 see Iohvidov and Kreĭn [1].

Theorem 4.2, Theorem 4.4 and Lemma 4.5 belong to Phillips [1], [2]. The first two of these results have also been proved by Ginzburg [3].

Instead of using angular operators, Langer [1], [2] described maximal positive subspaces $\mathfrak{L} \subset \mathfrak{H}$ by means of skew projectors that project $\mathfrak{H}$ onto $\mathfrak{L}$ along $\mathfrak{H}^-$ (cf. also Ginzburg and Iohvidov [1]).

Iohvidov [10] gave a characterization of non-closed maximal definite subspaces of $\mathfrak{H}$.

The concept of plus-defect was introduced by Kreĭn and Šmul'jan [1].

Let $\mathfrak{B}$ be a Banach space, and let P^+, P^- be continuous projectors in $\mathfrak{B}$ such that $P^+ + P^- = I$. The space $\mathfrak{B}$ equipped with the function $J(x) = \|P^+x\|^2 - \|P^-x\|^2$ $(x \in \mathfrak{B})$ is called a *Banach space with a J-metric*. J-semi-definite subspaces of $\mathfrak{B}$ have been studied by Iohvidov [14].

Uniformly definite subspaces of a Kreĭn space were introduced and analysed by Ginzburg [3], [4].

Lemma 6.1 is essentially contained in a paper of Iohvidov and Kreĭn [2]. Theorem 6.3 belongs to Ginzburg [3]. Example 6.5 has its origin in a construction proposed by Iohvidov [6]; for related results cf. also Iohvidov [8].

A special case of Lemma 7.5 is due to Iohvidov and Kreĭn [1]; in the general case it was formulated by Šmul'jan [3]. Lemma 7.6 appears in a paper of Kreĭn and Šmul'jan [3].

Theorem 8.2 was proved by Ginzburg (see Ginzburg and Iohvidov [1]). The present proof is based on his idea, but avoids the spectral theorem. Another method is proposed in the lecture notes of Kreĭn [8].

Alternating pairs were extensively studied by Phillips [1], [2], [3]. The present proof of Theorem 9.1 belongs to Langer (unpublished lecture notes; cf. also Langer [5]). In the original proof Phillips [3] unites the two angular operators $K^+(\mathfrak{L}_1)$, $K^-(\mathfrak{L}_2)$ into a single operator. Both proofs make use of a classical argument of Kreĭn.

The material of Section 10 is borrowed from Phillips [3]. The existence of the maximal extensions in question has independently been obtained by Sz.-Nagy and Foiaş [1; Proposition IV.4.2] without using an indefinite inner product.

Phillips [1], [2] applied his theory of alternating pairs to differential equations; for a short summary cf. also Naĭmark and Ismagilov [1]. Further developments in this direction are to be found in the papers of Cordes [1], Olubummo and Phillips [1], Phillips and Sarason [1], Crandall and Phillips [1], Daleckiĭ and Fadeeva [1]; the latter work also offers an insight into the main idea of these applications.

The survey of Ginzburg and Iohvidov [1] as well as the lecture notes of Kreĭn [8] contain additional facts from the geometry of Krein spaces.

Chapter VI. Unitary and Selfadjoint Operators in Krein Spaces

The study of linear operators in Krein spaces begins with this chapter. Main topics include criteria for the continuity of isometric operators (Section 3) as well as basic properties and the location of spectra of unitary and selfadjoint operators (Sections 4—7). In Section 8 it is proved that every continuous linear operator of a Hilbert space has a unitary dilation in a Krein space. Theorems 3.5, 3.10, 5.1, 5.5, 6.1, 7.5, 8.6 and Lemma 4.7 can be mentioned as representative results.

1. Preliminaries

In Chapter II linear operators of general inner product spaces have been treated. The rest of our book is devoted to linear operators of Krein spaces.

Since every Krein space, when considered with a J-inner product, is a Hilbert space, we shall need some definitions and facts from the theory of Hilbert space operators. A few of them are listed below.

If $\mathfrak{E}_1$, $\mathfrak{E}_2$ are Hilbert spaces, then the class of all continuous linear operators T with $\mathfrak{D}(T) = \mathfrak{E}_1$, $\mathfrak{R}(T) \subset \mathfrak{E}_2$ will be denoted by $\mathfrak{B}(\mathfrak{E}_1, \mathfrak{E}_2)$. In the special case $\mathfrak{E}_1 = \mathfrak{E}_2 = \mathfrak{E}$ we use the symbol $\mathfrak{B}(\mathfrak{E})$ or $\mathfrak{B}$ for short. $\mathfrak{B}(\mathfrak{E}_1, \mathfrak{E}_2)$ is a Banach space with respect to the usual operator norm $\|T\| = \sup_{\|x\|=1} \|Tx\|$.

Besides the Banach topology, we shall have to do with another separated locally convex topology of $\mathfrak{B}(\mathfrak{E}_1, \mathfrak{E}_2)$, the *weak operator topology* $\tau_{\mathrm{wo}}(\mathfrak{E}_1, \mathfrak{E}_2)$ defined by the semi-norms $p_{x,y}$ $(x \in \mathfrak{E}_1,\ y \in \mathfrak{E}_2)$, where

$$p_{x,y}(T) = |(Tx, y)| \qquad (T \in \mathfrak{B}(\mathfrak{E}_1, \mathfrak{E}_2)) .$$

Recall that a subset $\mathfrak{F}$ of a topological space is said to be *compact*, if from any system of open sets whose union contains $\mathfrak{F}$ one can select a finite subsystem whose union also contains $\mathfrak{F}$.

Let $\mathfrak{E}_1, \mathfrak{E}_2$ be Hilbert spaces. We say that the operator $T \in \mathfrak{B}(\mathfrak{E}_1, \mathfrak{E}_2)$ is *compact*, if for every bounded set $\mathfrak{A} \subset \mathfrak{E}_1$ the closure $\overline{T\mathfrak{A}}$ of its image is a compact set in $\mathfrak{E}_2$. Equivalently, the operator T is compact if for every bounded sequence $\{x_n\}_1^\infty \subset \mathfrak{E}_1$ the sequence $\{Tx_n\}_1^\infty \subset \mathfrak{E}_2$ contains a convergent subsequence.

Linear combinations of compact operators are compact. If $\mathfrak{E}_j$ ($j = 1, 2, 3, 4$) are Hilbert spaces, if $T_j \in \mathfrak{B}(\mathfrak{E}_j, \mathfrak{E}_{j+1})$ ($j = 1, 2, 3$), and if T_2 is compact, then T_1T_2 and T_2T_3 are compact.

The linear operator T from the Hilbert space $\mathfrak{E}_1$ to the Hilbert space $\mathfrak{E}_2$ is said to be *closed*, if the relations $\{x_n\}_1^\infty \subset \mathfrak{D}(T)$, $x_n \to x$, $Tx_n \to y$ imply $x \in \mathfrak{D}(T)$, $Tx = y$. This property is equivalent to the closedness of the *graph* $\{\{x, Tx\}: x \in \mathfrak{D}(T)\}$ of T in the Hilbert space $\mathfrak{E}_1 \times \mathfrak{E}_2$.

A linear operator T may or may not have closed extensions. In the first case there exists a least closed extension, denoted by $\overline{T}$ and called the *closure* of T.

Let T be a closed linear operator in the Hilbert space $\mathfrak{E}$. With respect to T, the following classification of complex numbers λ is customary. If $T - \lambda I$ is invertible and its range coincides with $\mathfrak{E}$, then λ is said to belong to the *resolvent set* $\varrho(T)$ of T. If $T - \lambda I$ is invertible and its range is a dense proper subspace of $\mathfrak{E}$, then λ is said to lie in the *continuous spectrum* $\sigma_c(T)$ of T. If $T - \lambda I$ is invertible and its range is not dense in $\mathfrak{E}$, then λ is assigned to the *residual spectrum* $\sigma_r(T)$ of T. Finally, in case $T - \lambda I$ is not invertible, λ is said to belong to the *point spectrum* $\sigma_p(T)$ of T.

The set

$$\sigma(T) = \sigma_c(T) \cup \sigma_r(T) \cup \sigma_p(T),$$

complementary to $\varrho(T)$ in $\mathbf{C}$, is called the *spectrum* of T.

The points belonging to $\varrho(T)$ are called *regular points* of T. By the closed graph principle, $\lambda \in \varrho(T)$ if and only if $(T - \lambda I)^{-1} \in \mathfrak{B}$.

If $\lambda \in \sigma_c(T)$, then $(T - \lambda I)^{-1}$ is discontinuous on its domain.

Obviously, $\sigma_p(T)$ consists of the eigenvalues of T.

An important subset of $\sigma_p(T)$ is the set of "normal" eigenvalues. The number λ is said to be a *normal eigenvalue* of the closed linear operator T in the Hilbert space $\mathfrak{E}$, if 1) $0 < \dim \mathfrak{S}_\lambda(T) < \infty$; 2) $\mathfrak{S}_\lambda(T)$ has a complementary subspace $\mathfrak{L}_\lambda$ in $\mathfrak{E}$ which is closed, invariant for T, and such that λ is a regular point of $T|\mathfrak{L}_\lambda$.

2. The Adjoint of an Operator

Let T be a linear operator with dense domain in the Krein space $\mathfrak{H}$. We define the *adjoint* T^* of T as follows. The vector $y \in \mathfrak{H}$ belongs to $\mathfrak{D}(T^*)$ if and only if there exists a $y^* \in \mathfrak{H}$ such that

$$(Tx, y) = (x, y^*) \qquad (x \in \mathfrak{D}(T)). \tag{2.1}$$

In this case y^* is unique; we set $T^*y = y^*$.

Evidently, 0 always belongs to $\mathfrak{D}(T^*)$, and T^* is a linear operator. Moreover, T^* satisfies the relation

(2.2) $$(Tx, y) = (x, T^*y) \qquad (x \in \mathfrak{D}(T),\ y \in \mathfrak{D}(T^*))$$

and has largest possible domain.

The adjoint operator in $\mathfrak{H}$ can be expressed through the adjoint in a Hilbert space.

Lemma 2.1. *Let T be a densely defined linear operator in the Krein space $\mathfrak{H}$. For some fundamental symmetry $J \in \mathfrak{B}(\mathfrak{H})$, denote the J-adjoint of T by $T^{[*]}$; thus $T^{[*]}$ is the operator satisfying*

(2.3) $$(Tx, z)_J = (x, T^{[*]}z)_J \qquad (x \in \mathfrak{D}(T),\quad z \in \mathfrak{D}(T^{[*]}))$$

and having largest possible domain. Then

(2.4) $$T^* = JT^{[*]}J\,.$$

Proof. Relation (2.3) with the help of Lemma II.11.1 yields:

$$(Tx, y) = (x, JT^{[*]}\,Jy) \qquad (x \in \mathfrak{D}(T),\quad y \in \mathfrak{D}(JT^{[*]}J)).$$

Therefore $JT^{[*]}J \subset T^*$. Similarly, starting from equation (2.2) we derive $JT^*J \subset T^{[*]}$ i.e. $T^* \subset JT^{[*]}J$. □

Numerous properties of adjoint operators in a Krein space formally duplicate those known from Hilbert space theory. Below we list some of them in a more or less logical order, but somewhat arbitrary grouping. The proofs can be obtained either by imitating the Hilbert space argument, or by applying the Hilbert space result itself combined with Lemma 2.1.

Theorem 2.2. a) $0^* = 0$; b) $I^* = I$; c) T^* *is closed (provided it exists)*; d) $T \in \mathfrak{B}(\mathfrak{H})$ *implies* $T^* \in \mathfrak{B}(\mathfrak{H})$. □

Theorem 2.3. *If the adjoints in question exist, we have*: a) $(\alpha T)^* = \bar{\alpha}T^*$ $(\alpha \in \boldsymbol{C})$; b) $(T_1 + T_2)^* \supset T_1^* + T_2^*$; c) $(T_1T_2)^* \supset T_2^*T_1^*$. □

Theorem 2.4. *If T^* and $(T^{-1})^*$ exist, then $(T^*)^{-1}$ exists and is equal to $(T^{-1})^*$.* □

Theorem 2.5. *The linear operator T with dense domain in the Krein space $\mathfrak{H}$ admits closed extensions if and only if T^{**} exists; in this case T^{**} is the closure of T.* □

Theorem 2.6. *If the adjoints in question exist, we have*: a) $T_1 \subset T_2$ *implies* $T_1^* \supset T_2^*$; b) *invariance relative to T of the subspace $\mathfrak{L} \subset \mathfrak{H}$ implies invariance of $\mathfrak{L}^\perp$ relative to T^* provided $\mathfrak{L} \cap \mathfrak{D}(T)$ is dense in $\mathfrak{L}$*; c) $\mathfrak{N}(T^*) = \mathfrak{R}(T)^\perp$; d) $\overline{\mathfrak{R}(T^*)} = \mathfrak{N}(T)^\perp$ *provided T is closed.* □

Theorem 2.7. *Let T be a densely defined, closed operator in the Krein space $\mathfrak{H}$. Then*

a) $\lambda \in \varrho(T)$ *implies* $\bar{\lambda} \in \varrho(T^*)$;
b) $\lambda \in \sigma_c(T)$ *implies* $\bar{\lambda} \in \sigma_c(T^*)$;
c) $\lambda \in \sigma_r(T)$ *implies* $\bar{\lambda} \in \sigma_p(T^*)$;
d) $\lambda \in \sigma_p(T)$ *implies* $\bar{\lambda} \in \sigma_p(T^*) \cup \sigma_r(T^*)$. □

Theorem 2.8. *If T is a linear operator in $\mathfrak{H}$ such that $\mathfrak{D}(T) = \mathfrak{D}(T^*) = \mathfrak{H}$, then T is continuous.* □

The following result is also known in the Hilbert space case. Since, however, it is less standard, we present the proof, adapted to the Krein space situation.

Theorem 2.9. *Let T be a closed linear operator with dense domain in the Krein space $\mathfrak{H}$. If $\mathfrak{R}(T)$ is closed, so is $\mathfrak{R}(T^*)$.*

Proof. Let $z \perp \mathfrak{N}(T)$. Then the relation

$$\varphi(Tx) = (x, z) \qquad (x \in \mathfrak{D}(T))$$

uniquely defines a linear form φ on $\mathfrak{R}(T)$.

Since T is closed, so is $\mathfrak{N}(T)$. Let $\mathfrak{L}$ be a closed complementary subspace to $\mathfrak{N}(T)$ (say, the J-orthogonal companion of $\mathfrak{N}(T)$, where J is a fundamental symmetry on $\mathfrak{H}$). For $u \in \mathfrak{R}(T)$ let $T^{(-1)}u$ denote the unique element $x \in \mathfrak{L}$ satisfying $Tx = u$. By the closed graph principle, $T^{(-1)}$ is continuous. On the other hand, $\varphi(u) = (T^{(-1)}u, z)$ $(u \in \mathfrak{R}(T))$. Thus φ is continuous.

Extending φ to all of $\mathfrak{H}$ as a continuous linear form and applying Theorem V.1.5 we find an element $y \in \mathfrak{H}$ such that $\varphi(Tx) = (Tx, y)$ $(x \in \mathfrak{D}(T))$, i.e.,

$$(Tx, y) = (x, z) \qquad (x \in \mathfrak{D}(T)) .$$

Hence $z = T^*y$, $z \in \mathfrak{R}(T^*)$, $\mathfrak{N}(T)^\perp \subset \mathfrak{R}(T^*)$. Now the assertion follows from Theorem 2.6 d). □

The next lemma exhibits a specific feature of indefinite Krein spaces as compared to Hilbert spaces.

Lemma 2.10. *Let T be a linear operator with dense domain in the Krein space $\mathfrak{H}$. The equation $\mathfrak{N}(T^*T) = \mathfrak{N}(T)$ holds if and only if $\mathfrak{R}(T)$ is non-degenerate.*

Proof. For $x \in \mathfrak{D}(T)$ the relations $Tx \in \mathfrak{D}(T^*)$, $T^*Tx = 0$ are equivalent to $(Tx, Ty) = 0$ $(y \in \mathfrak{D}(T))$, i.e., to Tx being an isotropic vector of $\mathfrak{R}(T)$. □

3. Isometric Operators

Let U be an isometric operator in the Krein space $\mathfrak{H}$. Though U preserves intrinsic norms, it need not preserve natural norms; what is more, U can be discontinuous (with respect to the strong topology). This will turn out from the criteria of continuity given below mainly in terms of $\mathfrak{D}(U)$ and $\mathfrak{R}(U)$.

Remark 3.1. When looking for sufficient conditions of continuity in terms of $\mathfrak{D}(U)$ and $\mathfrak{R}(U)$, we may restrict ourselves to the case where $\mathfrak{D}(U)$ and $\mathfrak{R}(U)$ are non-degenerate. To see this, we distinguish between two possibilities.

a) $\mathfrak{R}(U)$ is degenerate. We neglect the trivial case $\dim\mathfrak{D}(U) < < \infty$, but assume that U is continuous. Choosing a non-zero isotropic vector of $\mathfrak{R}(U)$, say Ue, and a discontinuous linear form φ on $\mathfrak{D}(U)$ with $\varphi(e) \neq -1$, the formula $U_1 x = Ux + \varphi(x)\, Ue$ $(x \in \mathfrak{D}(U))$ defines a discontinuous isometric operator U_1 having the same domain and range as U. Consequently, in this case $\mathfrak{D}(U)$ and $\mathfrak{R}(U)$ do not decide on the continuity of U.

b) $\mathfrak{D}(U)$ is degenerate, but $\mathfrak{R}(U)$ is non-degenerate. Then, on account of Lemma II.2.1, $U\mathfrak{D}^0 = 0$, where $\mathfrak{D}^0$ denotes the isotropic part of $\mathfrak{D}(U)$. Thus the problem of continuity can be split into two pieces (see Theorem I.5.4): one concerning the restriction of U to a maximal non-degenerate subspace $\mathfrak{D}^1$ of $\mathfrak{D}(U)$ (by Lemma II.2.1 the range of this restriction is also non-degenerate), and another concerning the dependence of x^0 and x^1 on the vector $x^0 + x^1$ $(x^0 \in \mathfrak{D}^0,\ x^1 \in \mathfrak{D}^1)$.

The main body of the present section is based on the following simple fact.

Lemma 3.2. *Let U be an isometric operator in the Krein space $\mathfrak{H}$. If U is continuous and $\mathfrak{D}(U)$ is uniformly positive (uniformly negative), then $\mathfrak{R}(U)$ is uniformly positive (uniformly negative). On the other hand, if $\mathfrak{R}(U)$ is uniformly definite, then U is continuous.*

Proof. If $\mathfrak{D}(U)$ is uniformly positive and non-zero, while U is continuous, then for a fundamental symmetry J and suitable positive number α we have

$$(Ux, Ux) = (x, x) \geqq \alpha\, \|x\|_J^2 \geqq \frac{\alpha}{\|U\|_J^2} \|Ux\|_J^2 \qquad (x \in \mathfrak{D}(U));$$

hence $\mathfrak{R}(U)$ is uniformly positive. A similar reasoning applies when $\mathfrak{D}(U)$ is uniformly negative. If, in turn, $\mathfrak{R}(U)$ is uniformly definite, then (cf. Lemma II.11.4)

$$\|Ux\|_J^2 \leqq \beta\, |(Ux, Ux)| = \beta\, |(x, x)| \leqq \beta\, \|x\|_J^2 \qquad (x \in \mathfrak{D}(U))$$

for some $\beta > 0$; so U is continuous. □

Making use of Corollary II.2.2 we obtain:

Corollary 3.3. *If U is an isometric operator with uniformly definite domain in $\mathfrak{H}$, then U^{-1} is continuous.* □

If $\varkappa(\mathfrak{H}) = \infty$, then with the help of Lemma 3.2 we can construct a discontinuous isometric operator that is defined throughout $\mathfrak{H}$.

Example 3.4. Let $\varkappa(\mathfrak{H}) = \infty$. Consider a fundamental decomposition

$$\mathfrak{H} = \mathfrak{H}^+ (\dotplus) \mathfrak{H}^-; \quad \mathfrak{H}^+ \subset \mathfrak{P}^{++}, \quad \mathfrak{H}^- \subset \mathfrak{P}^{--}.$$

Decompose $\mathfrak{H}^+$ and $\mathfrak{H}^-$ as follows:

$$\mathfrak{H}^+ = \mathfrak{H}_1^+ (\dotplus) \mathfrak{H}_2^+, \quad \dim \mathfrak{H}_1^+ \text{ countably infinite},$$
$$\mathfrak{H}^- = \mathfrak{H}_1^- (\dotplus) \mathfrak{H}_2^-, \quad \dim \mathfrak{H}_1^- = 1 .$$

Example V.6.5 shows that the Krein space $\mathfrak{H}_1 = \mathfrak{H}_1^+ (\dotplus) \mathfrak{H}_1^-$ contains a positive definite, intrinsically complete, non-uniformly definite subspace $\mathfrak{L}$. The intrinsic dimension of $\mathfrak{L}$ must be infinite and, by Lemma IV.7.3, not greater than that of $\mathfrak{H}_1^+$. Consequently, there exists an isometric operator U_1^+ with $\mathfrak{D}(U_1^+) = \mathfrak{H}_1^+$, $\mathfrak{R}(U_1^+) = \mathfrak{L}$. Let U^- be an isometric operator that maps $\mathfrak{H}^-$ onto $\mathfrak{H}_2^-$. For

$$x = x_1^+ + x_2^+ + x^-; \quad x_1^+ \in \mathfrak{H}_1^+, \quad x_2^+ \in \mathfrak{H}_2^+, \quad x^- \in \mathfrak{H}^-$$

set

$$Ux = U_1^+ x_1^+ + x_2^+ + U^- x^- .$$

Then U is an isometric operator with $\mathfrak{D}(U) = \mathfrak{H}$. Applying Lemma 3.2 to U_1^+ we see that U is discontinuous.

The next result extends the necessary condition and a weakening of the sufficient condition contained in Lemma 3.2 to the case of indefinite domains.

Theorem 3.5. *Let U be an isometric operator in the Krein space $\mathfrak{H}$. Assume that $\mathfrak{D}(U)$ is the orthogonal direct sum of a uniformly positive subspace $\mathfrak{D}^+$ and a uniformly negative subspace $\mathfrak{D}^-$. Then U will be continuous if and only if $U\mathfrak{D}^+$ and $U\mathfrak{D}^-$ are uniformly definite.*

Proof. Lemma V.5.4, Theorem V.5.3 and Theorem V.3.5 yield a fundamental decomposition

$$\mathfrak{H} = \mathfrak{H}^+ (\dotplus) \mathfrak{H}^-; \quad \mathfrak{H}^+ \subset \mathfrak{P}^{++}, \quad \mathfrak{H}^- \subset \mathfrak{P}^{--} \tag{3.1}$$

such that $\mathfrak{D}^+ \subset \mathfrak{H}^+$, $\mathfrak{D}^- \subset \mathfrak{H}^-$. Denote the corresponding fundamental projectors and fundamental symmetry by P^+, P^- and J, respectively. If $U\mathfrak{D}^+$, $U\mathfrak{D}^-$ are uniformly definite then, according to Lemma 3.2 and Lemma II.11.2,

$$\|Ux\|_J \leqq \|UP^+x\|_J + \|UP^-x\|_J \leqq \beta_1 \|P^+x\|_J + \beta_2 \|P^-x\|_J \leqq$$
$$\leqq (\beta_1 + \beta_2) \|x\|_J$$

for suitable $\beta_1, \beta_2 > 0$ and every $x \in \mathfrak{D}(U)$; hence U is continuous. The reverse implication directly follows from Lemma 3.2. □

Corollary 3.6. *If U is a continuous isometric operator in $\mathfrak{H}$, and $\mathfrak{D}(U)$ is the orthogonal direct sum of a uniformly positive and a uniformly negative subspace, then U^{-1} is continuous.* □

In Theorem 3.5 the uniform definiteness of the components of two fundamental decompositions $\mathfrak{D}(U) = \mathfrak{D}^+ (\dotplus) \mathfrak{D}^-$, $\mathfrak{R}(U) = \mathfrak{R}^+ (\dotplus) \mathfrak{R}^-$ plays a part. It is essential that these decompositions are required to be "coherent": $\mathfrak{R}^\pm = U\mathfrak{D}^\pm$.

Example 3.7. Let $\varkappa(\mathfrak{H}) = \infty$. Consider a fundamental decomposition (3.1) of $\mathfrak{H}$. Let $\{e_j\}_1^\infty \subset \mathfrak{H}^+$, $\{f_j\}_1^\infty \subset \mathfrak{H}^-$ be orthonormal systems. Denote by $\mathfrak{D}^+ (\mathfrak{D}^-)$ the span of all vectors e_j (f_j). Furthermore, let $\mathfrak{D}_1 (\mathfrak{D}_2)$ be the span of the vectors $e_j + \frac{j}{j+1} f_j$ $\left(\text{resp. } f_j + \frac{j}{j+1} e_j\right)$, $j = 1, 2, \ldots$. Then $\mathfrak{D}_1$, $\mathfrak{D}_2$ are orthogonal, non-uniformly definite subspaces satisfying

$$\mathfrak{D}_1 (\dotplus) \mathfrak{D}_2 = \mathfrak{D}^+ (\dotplus) \mathfrak{D}^- . \tag{3.2}$$

As each of $\mathfrak{D}_1$, $\mathfrak{D}_2$, $\mathfrak{D}^+$, $\mathfrak{D}^-$ is the span of a countably infinite set of vectors, there exists an isometric operator U defined on $\mathfrak{D}^+ (\dotplus) \mathfrak{D}^-$ such that

$$U\mathfrak{D}^+ = \mathfrak{D}_1, \qquad U\mathfrak{D}^- = \mathfrak{D}_2 . \tag{3.3}$$

For this operator U both $\mathfrak{D}(U)$ and $\mathfrak{R}(U)$ can be decomposed into the orthogonal direct sum of the uniformly definite subspaces $\mathfrak{D}^+$, $\mathfrak{D}^-$. Nevertheless, from (3.3) and Theorem 3.5 it follows that U is discontinuous. On the other hand, if V is the identity mapping on $\mathfrak{D}^+ (\dotplus) \mathfrak{D}^-$ then V is continuous and isometric; one of the fundamental decompositions of $\mathfrak{R}(V)$ still consists of the non-uniformly definite subspaces $\mathfrak{D}_1$, $\mathfrak{D}_2$.

This dependence on the choice of the fundamental decompositions ceases when $\mathfrak{D}(U)$ is closed.

Theorem 3.8. *An isometric operator U with ortho-complemented domain $\mathfrak{D}(U)$ in the Krein space $\mathfrak{H}$ is continuous if and only if $\mathfrak{R}(U)$ is ortho-complemented.*

Proof. By Theorem V.5.3 $\mathfrak{D}(U)$ can be written in the form

$$\mathfrak{D}(U) = \mathfrak{D}^+ (\dotplus) \mathfrak{D}^-; \quad \mathfrak{D}^+ \subset \mathfrak{P}^{++}, \quad \mathfrak{D}^- \subset \mathfrak{P}^{--}, \tag{3.4}$$

where $\mathfrak{D}^+$ and $\mathfrak{D}^-$ are closed and uniformly definite. Obviously,

$$\mathfrak{R}(U) = U\mathfrak{D}^+ (\dotplus) U\mathfrak{D}^- . \tag{3.5}$$

If U is continuous, then Theorem 3.5 implies that $U\mathfrak{D}^+$ and $U\mathfrak{D}^-$ are uniformly definite. They are closed as well, since a uniformly definite subspace is closed if and only if it is intrinsically complete (cf. Remark V.6.4), and the latter property is preserved by an isometric operator. Applying Theorem V.5.3 once more, we obtain that $\mathfrak{R}(U)$ is ortho-complemented.

Let, conversely, $\mathfrak{R}(U)$ be ortho-complemented. Then, on account of (3.5) and Lemma I.9.2, $U\mathfrak{D}^+$ and $U\mathfrak{D}^-$ are also ortho-complemented. Therefore the continuity of U follows from Theorem V.5.2 and Theorem 3.5. □

If requiring only that the closure of $\mathfrak{D}(U)$ be ortho-complemented, which is a weaker restriction on $\mathfrak{D}(U)$ than the hypothesis of Theorem 3.5, some conclusions can still be drawn.

Lemma 3.9. *If U is a continuous isometric operator in $\mathfrak{H}$ such that the closure of $\mathfrak{D}(U)$ is ortho-complemented, then the closure of $\mathfrak{R}(U)$ is also ortho-complemented and U^{-1} is continuous.*

Proof. U can be extended to a continuous isometric operator U_1 defined on $\overline{\mathfrak{D}(U)}$. Clearly, $\mathfrak{R}(U) \subset \mathfrak{R}(U_1) \subset \overline{\mathfrak{R}(U)}$. By Theorem 3.8 $\mathfrak{R}(U_1)$ is ortho-complemented. Hence, on account of Theorem V.3.4, $\mathfrak{R}(U_1)$ is closed. In particular, $\mathfrak{R}(U_1) = \overline{\mathfrak{R}(U)}$. Furthermore, according to Corollary 3.6 and Theorem V.5.3, U_1^{-1} is continuous. □

It should be noted that the ortho-complementedness of $\overline{\mathfrak{D}(U)}$ and $\overline{\mathfrak{R}(U)}$ do not imply the continuity of the isometric operator U (cf. Example 3.7).

The following rather strong result differs in nature from the preceding ones. It contains one half of Theorem 3.8, but not of Theorem 3.5.

Theorem 3.10. *Let U be an isometric operator in the Krein space $\mathfrak{H}$. If $\mathfrak{D}(U)$ is closed as well as non-degenerate and the closure of $\mathfrak{R}(U)$ is non-degenerate, then U is continuous.*

Proof. Owing to the closed graph principle we only need to prove that U is a closed operator.

Let $\{x_n\}_1^\infty \subset \mathfrak{D}(U)$, $x_n \to x$, $Ux_n \to y$ $(n \to \infty)$. Then

$$(Ux_n, z) \to (y, z) \qquad (z \in \mathfrak{H}) .$$

On the other hand, making use of the invertibility of U (Corollary II.2.2) and the relation $x \in \mathfrak{D}(U)$, we find:

$$(Ux_n, z) = (x_n, U^{-1}z) \to (x, U^{-1}z) = (Ux, z) \qquad (z \in \mathfrak{R}(U)) .$$

Consequently, $(Ux - y, z) = 0$ for every z in $\overline{\mathfrak{R}(U)}$. But $Ux - y$ belongs to $\overline{\mathfrak{R}(U)}$, and $\overline{\mathfrak{R}(U)}$ is non-degenerate. Hence $Ux = y$. □

Theorem 3.10 and Lemma II.2.1 yield:

Corollary 3.11. *Let U be an isometric operator in $\mathfrak{H}$. If $\mathfrak{D}(U)$, $\mathfrak{R}(U)$ are closed and $\mathfrak{D}(U)$ is non-degenerate, then U is continuous.* □

4. Unitary and Rectangular Isometric Operators

An isometric operator U in the Krein space $\mathfrak{H}$ is said to be *unitary* provided $\mathfrak{D}(U) = \mathfrak{R}(U) = \mathfrak{H}$.

Theorem 3.8 yields:

Theorem 4.1. *Every unitary operator in $\mathfrak{H}$ is continuous.* □

Making use of Corollary II.2.2 the following characterization of unitary operators can easily be established.

Theorem 4.2. *The linear operator U in the Krein space $\mathfrak{H}$ is unitary if and only if it is completely invertible and $U^{-1} = U^*$.* □

Let U be an isometric operator in the Krein space $\mathfrak{H}$. If $\mathfrak{D}(U)$ and $\mathfrak{R}(U)$ are ortho-complemented, we say U is *rectangular*. In case U is rectangular and has no rectangular isometric proper extensions we say that U is *maximal rectangular*.

Unitary operators are a special kind of maximal rectangular isometric operators. Every rectangular isometric operator is continuous (Theorem 3.8) and invertible (Corollary II.2.2).

Theorem 4.3. *Let U be a rectangular isometric operator in the Krein space $\mathfrak{H}$. Set $\mathfrak{D} = \mathfrak{D}(U)$, $\mathfrak{R} = \mathfrak{R}(U)$. The operator U is maximal rectangular isometric if and only if*

$$\min\{\varkappa^+(\mathfrak{D}^\perp),\ \varkappa^+(\mathfrak{R}^\perp)\} = 0 \tag{4.1}$$

and

$$\min\{\varkappa^-(\mathfrak{D}^\perp),\ \varkappa^-(\mathfrak{R}^\perp)\} = 0\,. \tag{4.2}$$

Proof. Lemma I.9.1 and Theorem V.3.4 assure that the values $\varkappa$ appearing in the statement make sense.

Let U_1 be a rectangular isometric proper extension of U. Since $\mathfrak{D}$ is ortho-complemented, we have $\mathfrak{D}(U_1) = \mathfrak{D}(\dot{+})\mathfrak{D}_1$, where $\mathfrak{D}_1 = \mathfrak{D}^\perp \cap \mathfrak{D}(U_1)$. Since $\mathfrak{D}(U_1)$ is ortho-complemented, so is $\mathfrak{D}_1$ (cf. Lemma I.9.2). Moreover, $\mathfrak{D}_1 \neq 0$. The relation $\mathfrak{D}_1 \perp \mathfrak{D}$ implies $U_1\mathfrak{D}_1 \perp U_1\mathfrak{D}$; hence $U_1\mathfrak{D}_1 \subset \mathfrak{R}^\perp$. As a result, the isometric operator U_1 maps the ortho-complemented non-zero subspace $\mathfrak{D}_1 \subset \mathfrak{D}^\perp$ into $\mathfrak{R}^\perp$. Owing to Corollaries I.9.5 and I.4.5 $\mathfrak{D}_1$ contains either a positive or a negative vector; $\mathfrak{R}^\perp$ must contain a vector of the same kind.

If, conversely, $\mathfrak{D}^\perp$ contains a positive or negative vector x and $\mathfrak{R}^\perp$ contains a vector y of the same sign, then it is easy to construct an isometric extension U_1 of U such that $\mathfrak{D}(U_1) = \mathfrak{D}(\dotplus)\langle x\rangle$, $\mathfrak{R}(U_1) = \mathfrak{R}(\dotplus)\langle y\rangle$. In view of Lemma I.9.2 this U_1 will be rectangular. □

In contrast to the theorem above, a rectangular isometric U satisfying (4.1) and (4.2) may have isometric proper extensions U_1 which are non-rectangular. This situation occurs when $\mathfrak{D}^\perp$ is indefinite, $\mathfrak{R}^\perp = 0$, and U_1 is defined to be zero on a neutral non-zero subspace of $\mathfrak{D}^\perp$.

Theorem 4.4. *Let U be a rectangular isometric operator in $\mathfrak{H}$. Then U admits maximal rectangular isometric extensions. U admits a unitary extension if and only if*

$$\varkappa^+(\mathfrak{D}^\perp) = \varkappa^+(\mathfrak{R}^\perp), \qquad \varkappa^-(\mathfrak{D}^\perp) = \varkappa^-(\mathfrak{R}^\perp)\,, \tag{4.3}$$

where $\mathfrak{D} = \mathfrak{D}(U)$, $\mathfrak{R} = \mathfrak{R}(U)$.

Proof. According to Lemma I.9.1, Theorem V.3.4 and Theorem V.1.1 the subspaces $\mathfrak{D}^\perp$, $\mathfrak{R}^\perp$ admit fundamental decompositions of the form

$$\mathfrak{D}^\perp = \mathfrak{D}_1^+(\dotplus)\mathfrak{D}_1^-; \quad \mathfrak{D}_1^+ \subset \mathfrak{P}^{++},\ \mathfrak{D}_1^- \subset \mathfrak{P}^{--},$$
$$\mathfrak{R}^\perp = \mathfrak{R}_1^+(\dotplus)\mathfrak{R}_1^-; \quad \mathfrak{R}_1^+ \subset \mathfrak{P}^{++},\ \mathfrak{R}_1^- \subset \mathfrak{P}^{--},$$

where $\mathfrak{D}_1^+$, $\mathfrak{D}_1^-$, $\mathfrak{R}_1^+$, $\mathfrak{R}_1^-$ are intrinsically complete. By the theory of Hilbert spaces there exists an isometric operator U_1^+ from $\mathfrak{D}_1^+$ into $\mathfrak{R}_1^+$ such that

$$\text{either } \mathfrak{D}(U_1^+) = \mathfrak{D}_1^+ \quad \text{or} \quad \mathfrak{R}(U_1^+) = \mathfrak{R}_1^+\,. \tag{4.4}$$

Similarly, there exists an isometric operator U_1^- from $\mathfrak{D}_1^-$ into $\mathfrak{R}_1^-$ such that

$$\text{either } \mathfrak{D}(U_1^-) = \mathfrak{D}_1^- \quad \text{or} \quad \mathfrak{R}(U_1^-) = \mathfrak{R}_1^-\,. \tag{4.5}$$

We define a linear operator U_0 by the relations

$$\mathfrak{D}(U_0) = \mathfrak{D}(U)\ (\dotplus)\ \mathfrak{D}(U_1^+)\ (\dotplus)\ \mathfrak{D}(U_1^-)\,,$$
$$U_0(x + x_1^+ + x_1^-) = Ux + U_1^+x_1^+ + U_1^-x_1^-$$
$$\big(x \in \mathfrak{D}(U),\ x_1^+ \in \mathfrak{D}(U_1^+),\ x_1^- \in \mathfrak{D}(U_1^-)\big).$$

Obviously, U_0 is an isometric extension of U. Moreover, (4.4) implies that $\mathfrak{D}(U_1^+)$ is intrinsically complete, hence ortho-complemented in $\mathfrak{D}_1^+$. Similarly, (4.5) implies that $\mathfrak{D}(U_1^-)$ is ortho-complemented in $\mathfrak{D}_1^-$. Thus $\mathfrak{D}(U_0)$ and in much the same way $\mathfrak{R}(U_0)$ are ortho-complemented in $\mathfrak{H}$. Therefore Theorem 4.3 may be applied: owing to (4.4) and (4.5) U_0 is maximal rectangular isometric.

If (4.3) holds, then both of the equations under (4.4) and both of the equations under (4.5) can be satisfied. In that case U_0 will be unitary. If, conversely, U admits a unitary extension U_0, then $U_0\mathfrak{D}^\perp = \mathfrak{R}^\perp$; therefore (4.3) is a consequence of Theorem V.1.4. □

In what follows we establish a few elementary results concerning invariant subspaces of unitary operators.

Lemma 4.5. *Let U be a unitary operator in the Krein space $\mathfrak{H}$, and let $\mathfrak{L}$ be a subspace of $\mathfrak{H}$. If $U\mathfrak{L} \subset \mathfrak{L}$, then $U^{-1}\mathfrak{L}^\perp \subset \mathfrak{L}^\perp$.*

Proof. Since U is completely invertible (Theorem 4.2), $U\mathfrak{L} \subset \mathfrak{L}$ yields $\mathfrak{L} \subset U^{-1}\mathfrak{L}$. Now Lemma II.2.9 can be applied. □

Taking into account that the relation $U\mathfrak{L} = \mathfrak{L}$ is equivalent to the invariance of $\mathfrak{L}$ under both of the unitary operators U, U^{-1} (and that for $\mathfrak{L}^\perp$ a similar statement holds), we obtain:

Corollary 4.6. *Let U be a unitary operator in $\mathfrak{H}$, and let $\mathfrak{L}$ be a subspace of $\mathfrak{H}$. If $U\mathfrak{L} = \mathfrak{L}$, then $U\mathfrak{L}^\perp = \mathfrak{L}^\perp$.* □

We are going to show that for a maximal positive or maximal negative subspace $\mathfrak{L}$ the condition $U\mathfrak{L} \subset \mathfrak{L}$ already implies $U\mathfrak{L} = \mathfrak{L}$.

Lemma 4.7. *Let U be a unitary operator in the Krein space $\mathfrak{H}$. If $\mathfrak{L}$ is a maximal positive (resp. maximal negative) subspace of $\mathfrak{H}$, so is $U\mathfrak{L}$.*

Proof. By Theorem 4.2 the relation $(Ux, y) = 0$ is equivalent to $(x, U^{-1}y) = 0$. Thus for any subspace $\mathfrak{L}$ we have $(U\mathfrak{L})^\perp = U\mathfrak{L}^\perp$. If $\mathfrak{L}$ is maximal positive then, in particular, $\mathfrak{L}$ is a closed positive subspace (Theorem V.4.1), while $\mathfrak{L}^\perp$ is negative (Lemma I.6.3). Since U is isometric and $U^{-1} \in \mathfrak{B}(\mathfrak{H})$ (Theorems 4.1—4.2), it follows that $U\mathfrak{L}$ is a closed positive subspace with negative orthogonal companion. Therefore, according to Lemma V.4.5, $U\mathfrak{L}$ is maximal positive. □

Corollary 4.8. *Let U be a unitary operator in $\mathfrak{H}$. If the maximal positive or maximal negative subspace $\mathfrak{L} \subset \mathfrak{H}$ is invariant for U, then $U\mathfrak{L} = \mathfrak{L}$.* □

When the invariant positive or negative subspace is not maximal, we can state something less.

Lemma 4.9. *Let U be a unitary operator in $\mathfrak{H}$. If the positive (negative) subspace $\mathfrak{L} \subset \mathfrak{H}$ is invariant for U, then there exists a positive (negative) subspace $\mathfrak{L}_1 \supset \mathfrak{L}$ such that $U\mathfrak{L}_1 = \mathfrak{L}_1$.*

Proof. Denote by $\mathfrak{L}_1$ the span of the subspaces $U^{-j}\mathfrak{L}$ $(j = 0,1,2, \ldots)$. Then the relations $\mathfrak{L}_1 \supset \mathfrak{L}$, $U\mathfrak{L}_1 = \mathfrak{L}_1$ are obvious. Furthermore, if $x = \sum_{j=0}^{n} U^{-j}x_j$, where $x_j \in \mathfrak{L}$ for every j, then $U^n x \in \mathfrak{L}$. Since $(x, x) = (U^n x, U^n x)$, this proves the assertion concerning the sign of $\mathfrak{L}_1$. □

5. Spectral Properties of Unitary Operators

In this section we shall repeatedly use the notation

$$\nu^* = \frac{1}{\bar{\nu}} \qquad (\nu \in C\,, \quad \nu \neq 0)$$

introduced in Section II.2.

Theorem 5.1. *If U is a unitary operator in the Krein space $\mathfrak{H}$ and ν is a non-zero number, then the following implications are valid:*

a) $\nu \in \varrho(U)$ *implies* $\nu^* \in \varrho(U)$;
b) $\nu \in \sigma_c(U)$ *implies* $\nu^* \in \sigma_c(U)$;
c) $\nu \in \sigma_r(U)$ *implies* $\nu^* \in \sigma_p(U)$;
d) $\nu \in \sigma_p(U)$ *implies* $\nu^* \in \sigma_p(U) \cup \sigma_r(U)$.

Proof. In view of Theorem 4.1 we may apply Theorem 2.7. It remains to observe that, owing to Theorem 4.2,

$$U^* - \bar{\nu} I = -\bar{\nu} U^*(U - \nu^* I) \qquad (\nu \neq 0)\,, \tag{5.1}$$

where the operator $U^* = U^{-1}$ is unitary, hence a homeomorphism. □

Corollary 5.2. *The residual spectrum of a unitary operator in $\mathfrak{H}$ does not intersect the unit circle $|\nu| = 1$.* □

For a unitary operator the point $\nu = 0$ is always regular (cf. Theorem 4.2). Consequently, Theorem 5.1 a) can be reformulated as follows.

Corollary 5.3. *The spectrum of a unitary operator in $\mathfrak{H}$ lies symmetrically with respect to the unit circle.* □

Remark 5.4. A unitary operator in $\mathfrak{H}$ may really have non-unimodular numbers in its spectrum. For the continuous and residual spectra this will be seen from Example 6.4 and Theorems 7.1, 7.3. For the point spectrum cf. Example II.2.4.

If a unitary operator has a non-empty residual spectrum then, by Theorem 5.1 c), its point spectrum cannot be symmetric relative to the unit circle. The set of normal eigenvalues, however, is always symmetric, as shown by the following generalization of Theorems II.2.7—II.2.8.

Theorem 5.5. *Let U be a unitary operator in the Krein space $\mathfrak{H}$. If ν is a normal eigenvalue of U, so is ν^*, and the principal subspaces $\mathfrak{S}_\nu(U)$, $\mathfrak{S}_{\nu^*}(U)$ form a dual pair; moreover, the matrices of $U|\mathfrak{S}_\nu(U)$ and $U|\mathfrak{S}_{\nu^*}(U)$ with respect to any dual pair of bases are the inverses of the transposed conjugates of each other.*

Proof. According to the assumption we have

$$\mathfrak{H} = \mathfrak{S} \dotplus \mathfrak{L}\,, \tag{5.2}$$

where $\mathfrak{S} = \mathfrak{S}_\nu(U)$, $0 < \dim \mathfrak{S} = m < \infty$, $\mathfrak{L}$ is closed, $U\mathfrak{L} \subset \mathfrak{L}$, and $\nu \in \varrho(U|\mathfrak{L})$.

Set

$$\mathfrak{S}^* = \mathfrak{L}^\perp, \quad \mathfrak{L}^* = \mathfrak{S}^\perp\,. \tag{5.3}$$

Then $\mathfrak{S}^* \cap \mathfrak{L}^* = (\mathfrak{L} + \mathfrak{S})^\perp = \mathfrak{H}^\perp = 0$. Moreover, by Theorem I.10.9

$$\operatorname{codim} \mathfrak{L}^* = \dim \mathfrak{S} = m\,.$$

On the other hand, relation (5.3), Theorems I.10.9, III.6.1 and V.1.5 along with relation (5.2) yield

$$\dim \mathfrak{S}^* = \operatorname{codim} \mathfrak{L}^{\perp\perp} = \operatorname{codim} \mathfrak{L} = \dim \mathfrak{S} = m\,.$$

Therefore, owing to Lemma I.1.1,

$$\mathfrak{H} = \mathfrak{S}^* \dotplus \mathfrak{L}^*\,. \tag{5.4}$$

On account of (5.3) and Lemma III.2.4 the subspace $\mathfrak{L}^*$ is closed. That $\mathfrak{S}$ and $\mathfrak{S}^*$ form a dual pair is also easy to verify: $\mathfrak{S} \cap (\mathfrak{S}^*)^\perp = \mathfrak{S} \cap \mathfrak{L}^{\perp\perp} = \mathfrak{S} \cap \mathfrak{L} = 0$, $\mathfrak{S}^* \cap \mathfrak{S}^\perp = \mathfrak{S}^* \cap \mathfrak{L}^* = 0$.

From the definitions of $\mathfrak{S}$ and m, recalling basic facts of linear algebra, we obtain:

$$(U - \nu I)^n \mathfrak{S} = 0 \qquad (n \geqq m)\,. \tag{5.5}$$

At the same time, $\nu \in \varrho(U|\mathfrak{L})$ implies

$$(U - \nu I)^n \mathfrak{L} = \mathfrak{L} \qquad (n = 1, 2, \ldots)\,. \tag{5.6}$$

Comparing (5.5) and (5.6) we find that

$$\mathfrak{L} = \mathfrak{R}\big((U - \nu I)^n\big) \qquad (n \geqq m)\,. \tag{5.7}$$

Hence, on account of relation (5.3) and Theorems 2.6 c), 2.3, 4.1,

$$\mathfrak{S}^* = \mathfrak{N}\big((U^* - \bar{\nu} I)^n\big) \qquad (n \geqq m)\,,$$

or, by (5.1) and Theorem 4.2,

$$\mathfrak{S}^* = \mathfrak{N}\big((U - \nu^* I)^n\big) \qquad (n \geqq m)\,. \tag{5.8}$$

Thus $\mathfrak{S}^* = \mathfrak{S}_{\nu^*}(U)$.

Since $\mathfrak{S}$ is a finite-dimensional principal subspace of U and the operator U is invertible, it follows that $U\mathfrak{S} = \mathfrak{S}$. Therefore, in view of (5.3) and Corollary 4.6,

$$U\mathfrak{L}^* = \mathfrak{L}^*\,. \tag{5.9}$$

Consider the restriction $(U - \nu^* I)|\mathfrak{L}^*$. According to (5.8) and (5.4) it is invertible. Moreover, its range $\mathfrak{L}_*$ is contained in $\mathfrak{L}^*$ (see (5.9)). Let us prove that $\mathfrak{L}_*$ fills $\mathfrak{L}^*$.

Obviously, $(U - \nu^* I)^m \mathfrak{L}^* \subset \mathfrak{L}_*$. By (5.8) and (5.4) this is equivalent to $\mathfrak{R}((U - \nu^* I)^m) \subset \mathfrak{L}_*$ or, by (5.1), (5.9) and Theorem 4.2, to $\mathfrak{R}((U^* - \bar{\nu} I)^m) \subset \mathfrak{L}_*$. But, as a consequence of (5.7) and Theorem 2.9, $\mathfrak{R}((U^* - \bar{\nu} I)^m)$ is closed. Therefore, owing to Theorem 2.6 d), $\mathfrak{N}((U - \nu I)^m)^\perp \subset \mathfrak{L}_*$. Recalling the definition of $\mathfrak{S}$ and making use of relations (5.5), (5.3) we obtain $\mathfrak{L}^* \subset \mathfrak{L}_*$.

Consequently, $\nu^* \in \varrho(U|\mathfrak{L}^*)$.

The assertion concerning the matrices of $U|\mathfrak{S}$ and $U|\mathfrak{S}^*$ follows from Theorem II.2.8 applied to the space $\mathfrak{E} = \mathfrak{S} + \mathfrak{S}^*$, which is non-degenerate by Corollary II.2.6 and Lemma I.10.1 in case $|\nu| \neq 1$, and by $\mathfrak{E} = \mathfrak{S}$, $\mathfrak{S} \# \mathfrak{S}$ *in* the opposite case. □

Corollary 5.6. *If ν is a unimodular normal eigenvalue of the unitary operator U in $\mathfrak{H}$, then the principal subspace $\mathfrak{S}_\nu(U)$ is non-degenerate.* □

6. Selfadjoint Operators

The linear operator A in the Krein space $\mathfrak{H}$ is said to be *selfadjoint* if $\mathfrak{D}(A)$ is dense in $\mathfrak{H}$ and $A^* = A$.

In particular, every selfadjoint operator is symmetric and closed (cf. Theorem 2.2 c). On the other hand, if A is symmetric and $\mathfrak{D}(A) = \mathfrak{H}$, then A is selfadjoint.

From Theorem 2.7 we obtain the following analogue of Theorem 5.1.

Theorem 6.1. *If A is a selfadjoint operator in the Krein space $\mathfrak{H}$ and $\lambda \in \mathbf{C}$, then*

a) $\lambda \in \varrho(A)$ *implies* $\bar{\lambda} \in \varrho(A)$;

b) $\lambda \in \sigma_c(A)$ *implies* $\bar{\lambda} \in \sigma_c(A)$;

c) $\lambda \in \sigma_r(A)$ *implies* $\bar{\lambda} \in \sigma_p(A)$;

d) $\lambda \in \sigma_p(A)$ *implies* $\bar{\lambda} \in \sigma_p(A) \cup \sigma_r(A)$. □

Corollary 6.2. *The residual spectrum of a selfadjoint operator in $\mathfrak{H}$ does not intersect the real axis.* □

Corollary 6.3. *The spectrum of a selfadjoint operator in $\mathfrak{H}$ is symmetric with respect to the real axis.* □

That the point spectrum of a selfadjoint operator in a Krein space may contain non-real points is shown by Example II.3.2. The corresponding fact for the residual and continuous spectra can be established as follows.

Example 6.4. Let $\mathfrak{H}_0$ be a Hilbert space of countably infinite dimension. Denote the inner product by $[.\,,.]_0$. Consider the cartesian product $\mathfrak{H}=\mathfrak{H}_0\times\mathfrak{H}_0$, i.e., the Hilbert space of all ordered pairs $\{x,y\}$ $(x,y \in \mathfrak{H}_0)$ with inner product

$$[\{x_1, y_1\}, \{x_2, y_2\}] = [x_1, x_2]_0 + [y_1, y_2]_0 .$$

The new inner product

$$(\{x_1, y_1\}, \{x_2, y_2\}) = [x_1, y_2]_0 + [y_1, x_2]_0$$

turns $\mathfrak{H}$ into an indefinite Krein space (see Section V.10). For any $T_0 \in \mathfrak{B}(\mathfrak{H}_0)$ with adjoint $T_0^{[*]}$ the formula $A\{x, y\} = \{T_0x, T_0^{[*]}y\}$ defines a selfadjoint operator on the Krein space $\mathfrak{H}$.

Let $\lambda \in \boldsymbol{C}$, $\lambda \neq \bar{\lambda}$. If we fix T_0 by the equations $T_0e_j = \lambda e_j + e_{j+1}$ $(j = 1,2, \ldots)$, where $\{e_j\}_1^\infty$ is a complete orthonormal system in the Hilbert space $\mathfrak{H}_0$, then $\lambda \in \sigma_r(T_0)$, $\bar{\lambda} \in \varrho(T_0)$. Owing to Theorem 2.7 a) applied to the Hilbert space $\mathfrak{H}_0$, the latter inclusion is equivalent to $\lambda \in \varrho(T_0^{[*]})$. Thus $\lambda \in \sigma_r(A)$. On the other hand, if $T_0e_j = \lambda e_j + \frac{1}{j}e_j$ $(j = 1,2, \ldots)$, then $\lambda \in \sigma_c(T_0)$ and $\lambda \in \varrho(T_0^{[*]})$, so that $\lambda \in \sigma_c(A)$.

The next result is an analogue of Theorem 5.5 and, at the same time, an extension of Theorems II.3.5—II.3.6. A further extension, relating to certain discontinuous selfadjoint operators, will be given in Section 7.

Theorem 6.5. *Let A be a continuous selfadjoint operator in the Krein space $\mathfrak{H}$. If λ is a normal eigenvalue of A, so is $\bar{\lambda}$, and the principal subspaces $\mathfrak{S}_\lambda(A)$, $\mathfrak{S}_{\bar{\lambda}}(A)$ form a dual pair; moreover, the matrices of $A|\mathfrak{S}_\lambda(A)$ and $A|\mathfrak{S}_{\bar{\lambda}}(A)$ with respect to any dual pair of bases are the transposed conjugates of each other.*

The *proof* essentially duplicates that of Theorem 5.5. One has to replace U, ν^* and $|\nu| = 1$ by A, $\bar{\lambda}$ and $\operatorname{Im}\lambda = 0$, respectively. As a counterpart of $U\mathfrak{S} = \mathfrak{S}$ the inclusion $A\mathfrak{S} \subset \mathfrak{S}$ can only be stated; then Theorem II.3.7 yields $A\mathfrak{L}^* \subset \mathfrak{L}^*$, which suffices in the rest of the argument. At the end of the proof the references Theorem II.2.8 and Corollary II.2.6 are to be replaced by Theorem II.3.6 and Corollary II.3.4. □

Corollary 6.6. *If λ is a real, normal eigenvalue of the selfadjoint operator $A \in \mathfrak{B}(\mathfrak{H})$, then $\mathfrak{S}_\lambda(A)$ is non-degenerate.* □

The following lemma, actually a Hilbert space result, serves as a point of departure when analysing selfadjoint operators with the help of fundamental decompositions (see Chapters VIII—IX).

Lemma 6.7. *Let A be a symmetric operator in the Krein space $\mathfrak{H}$. Consider a fundamental decomposition*

$$\mathfrak{H} = \mathfrak{H}^+ (\dot{+}) \mathfrak{H}^-; \quad \mathfrak{H}^+ \subset \mathfrak{P}^{++}, \ \mathfrak{H}^- \subset \mathfrak{P}^{--} \tag{6.1}$$

and the corresponding fundamental projectors P^+, P^-. *Suppose that* $\mathfrak{H}^+ \subset \mathfrak{D}(A)$. *Suppose also that either* A *is closed or* $\varkappa^+(\mathfrak{H})$ *is finite. Set*

$$(6.2)\qquad \begin{aligned} A_{11} &= P^+A|\mathfrak{H}^+, & A_{12} &= P^+A|\mathfrak{H}^-, \\ A_{21} &= P^-A|\mathfrak{H}^+, & A_{22} &= P^-A|\mathfrak{H}^-. \end{aligned}$$

Then:

a) $\mathfrak{D}(A_{11})=\mathfrak{D}(A_{21})=\mathfrak{H}^+$, $\quad \mathfrak{D}(A_{12})=\mathfrak{D}(A_{22})=\mathfrak{D}(A)\cap\mathfrak{H}^-$;

b) A_{11} *and* A_{22} *are symmetric;*

c) A_{11}, A_{12} *and* A_{21} *are continuous;*

d) *in order that* A_{22} *be densely defined, closed, or selfadjoint in* $\mathfrak{H}^-$ *it is necessary and sufficient that* A *have the respective property in* $\mathfrak{H}$.

Proof. a) and b) are clear.

Let us prove c). Owing to a) and b) the operator A_{11} is selfadjoint on $\mathfrak{H}^+$; therefore, by the closed graph principle, it is continuous. Further, if A is closed, then A_{21}, as difference of the closed operator $A|\mathfrak{H}^+$ and the continuous operator A_{11}, is closed; being defined throughout $\mathfrak{H}^+$, it is continuous. On the other hand, if $\varkappa^+(\mathfrak{H}) < \infty$ then A_{21} is defined on a finite-dimensional space, hence continuous again. Finally, for $x^- \in \mathfrak{D}(A_{12})$, $y^+ \in \mathfrak{H}^+$, $J = P^+ - P^-$ the symmetry of A, Lemma II.11.4 and the continuity of A_{21} yield

$$|(A_{12}x^-, y^+)| = |(x^-, A_{21}y^+)| \leqq \|x^-\|_J \, \|A_{21}\|_J \, \|y^+\|_J ;$$

setting $y^+ = A_{12}x^-$ we obtain

$$(6.3)\qquad \|A_{12}x^-\|_J \leqq \|A_{21}\|_J \, \|x^-\|_J \qquad (x^- \in \mathfrak{D}(A_{12})) .$$

As to the proof of d), the assertion concerning dense domains immediately follows from a). That A_{22} is closed if and only if A is so can be established with the aid of a) and c). It remains to consider the property of selfadjointness.

Let A be selfadjoint. Then, in particular, $\mathfrak{D}(A_{12})$ is dense in $\mathfrak{H}^-$. This circumstance, the symmetry of A and the continuity of A_{12} (see c)) imply

$$(6.4)\qquad (A_{21}x^+, y^-) = (x^+, \bar{A}_{12}y^-) \qquad (x^+ \in \mathfrak{H}^+, y^- \in \mathfrak{H}^-) .$$

Suppose now that for a certain $y^- \in \mathfrak{H}^-$ there exists a $y^-_* \in \mathfrak{H}^-$ with

$$(6.5)\qquad (A_{22}x^-, y^-) = (x^-, y^-_*) \qquad (x^- \in \mathfrak{D}(A_{22})) .$$

Leaning on (6.2), (6.4) and (6.5) we derive

$$\begin{aligned}(Ax, y^-) = (A_{21}x^+, y^-) + (A_{22}x^-, y^-) = (x^+, \bar{A}_{12}y^-) + \\ + (x^-, y^-_*) = (x, \bar{A}_{12}y^- + y^-_*) \qquad (x \in \mathfrak{D}(A)) ,\end{aligned}$$

where $x^\pm = P^\pm x$. Hence, A being selfadjoint, $y^- \in \mathfrak{D}(A)$ i.e. $y^- \in \mathfrak{D}(A_{22})$.

Conversely, let A_{22} be selfadjoint in $\mathfrak{H}^-$. Suppose that for some fixed $y \in \mathfrak{H}$ there exists a $y_* \in \mathfrak{H}$ with

$$(Ax, y) = (x, y_*) \qquad (x \in \mathfrak{D}(A)) .$$

Setting $y^{\pm} = P^{\pm}y$ we find:

$$(A_{22}x^-, y^-) = (A_{22}x^-, y) = (Ax^-, y) - (A_{12}x^-, y) =$$
$$= (x^-, y_* - A_{21}y^+) \qquad (x^- \in \mathfrak{D}(A_{22})) .$$

Therefore $y^- \in \mathfrak{D}(A_{22})$, i.e., $y \in \mathfrak{D}(A)$. □

7. Cayley Transformations

Applied to operators in a Krein space $\mathfrak{H}$, the Cayley transformations defined in Section II.4 prove to be a source of the analogies between unitary and selfadjoint operators of $\mathfrak{H}$.

Theorem 7.1. *Let A be a selfadjoint operator in the Krein space $\mathfrak{H}$. Let ε, ζ be complex numbers such that $|\varepsilon|=1$, $\bar{\zeta} \neq \zeta$, $\zeta \in \varrho(A)$. Then the Cayley transform*

$$U = \varepsilon(A - \bar{\zeta}I)(A - \zeta I)^{-1} \tag{7.1}$$

is unitary and we have $\varepsilon \notin \sigma_p(U)$.

Proof. It was shown in Lemma II.4.1 (for the more general case $\zeta \notin \sigma_p(A)$) that U is isometric and ε is not an eigenvalue of U. Since, according to (7.1),

$$\mathfrak{D}(U) = \mathfrak{R}(A - \zeta I) , \qquad \mathfrak{R}(U) = \mathfrak{R}(A - \bar{\zeta}I) , \tag{7.2}$$

from the assumption $\zeta \in \varrho(A)$ and its consequence $\bar{\zeta} \in \varrho(A)$ (see Theorem 6.1 a)) we obtain $\mathfrak{D}(U) = \mathfrak{R}(U) = \mathfrak{H}$. □

Theorem 7.2. *Let U be a unitary operator in the Krein space $\mathfrak{H}$. Let ε, ζ be complex numbers such that $|\varepsilon|=1$, $\bar{\zeta} \neq \zeta$, $\varepsilon \notin \sigma_p(U)$. Then the Cayley transform*

$$A = (\zeta U - \varepsilon\bar{\zeta}I)(U - \varepsilon I)^{-1} \tag{7.3}$$

is selfadjoint and we have $\zeta \in \varrho(A)$.

Proof. By Lemma II.4.1 the operator A is symmetric and ζ is not an eigenvalue of A; moreover, (7.1) and (7.3) define mutually inverse transformations. The latter circumstance implies that the relations (7.2) still hold. In particular, $\mathfrak{R}(A - \zeta I) = \mathfrak{H}$. Hence $\zeta \in \varrho(A)$. It remains to show that A^* exists and $\mathfrak{D}(A^*) \subset \mathfrak{D}(A)$.

From (7.3) we obtain:

$$\mathfrak{D}(A) = \mathfrak{R}(U - \varepsilon I) . \tag{7.4}$$

Since $\varepsilon \notin \sigma_p(U)$ by assumption and $\varepsilon \notin \sigma_r(U)$ by Corollary 5.2, it follows that $\mathfrak{D}(A)$ is dense in $\mathfrak{H}$. Thus A^* is defined.

Let $g \in \mathfrak{D}(A^*)$. Denoting A^*g by g^* we have

$$(Af, g) = (f, g^*) \qquad (f \in \mathfrak{D}(A)) .$$

In view of (7.4) and (7.3) this can be written in the form

$$((\zeta U - \varepsilon\bar{\zeta} I)x, g) = ((U - \varepsilon I)x, g^*) \qquad (x \in \mathfrak{H}) .$$

Taking into account that U is unitary we find:

$$(Ux , \bar{\zeta} g - \bar{\varepsilon}\zeta U g - g^* + \bar{\varepsilon} U g^*) = 0 \qquad (x \in \mathfrak{H}) .$$

Hence

$$\bar{\zeta} g - \bar{\varepsilon}\zeta U g - g^* + \bar{\varepsilon} U g^* = 0 ,$$

i.e.,

$$g = \frac{1}{\varepsilon(\zeta - \bar{\zeta})} (U - \varepsilon I)(g^* - \zeta g) .$$

Therefore, in view of (7.4), $g \in \mathfrak{D}(A)$. □

Theorem 7.3. *Let A be a closed linear operator in $\mathfrak{H}$. Let ε, ζ be complex numbers such that $|\varepsilon| = 1$, $\bar{\zeta} \neq \zeta$, $\zeta \notin \sigma_p(A)$. Then the Cayley transform*

$$U = \varepsilon(A - \bar{\zeta} I)(A - \zeta I)^{-1} \tag{7.5}$$

is closed. Moreover, if $\lambda \neq \zeta$ is a number belonging to $\sigma_p(A)$, $\sigma_r(A)$, $\sigma_c(A)$ or $\varrho(A)$, then the number

$$\nu = \varepsilon \frac{\lambda - \bar{\zeta}}{\lambda - \zeta} \tag{7.6}$$

belongs to $\sigma_p(U)$, $\sigma_r(U)$, $\sigma_c(U)$ or $\varrho(U)$, respectively. If, in particular, λ is a normal eigenvalue of A, then ν is a normal eigenvalue of U.

Proof. We have

$$U = \varepsilon I + \varepsilon(\zeta - \bar{\zeta})(A - \zeta I)^{-1} . \tag{7.7}$$

As inverses, multiples and translates of closed operators are closed, U is closed.

Let $\lambda \in \boldsymbol{C}$, $\lambda \neq \zeta$, and let ν be the number furnished by (7.6). If $\lambda \in \sigma_p(A)$ then, owing to Lemma II.5.1, $\nu \in \sigma_p(U)$. If, on the other hand, $\lambda \notin \sigma_p(A)$, then Lemma II.5.2 applied to the inverses of the transformations (7.5)—(7.6) (see Lemma II.4.1 and Remark II.5.3) yields $\nu \notin \sigma_p(U)$. Moreover, from (7.5) and (7.6) we derive

$$U - \nu I = \left[\varepsilon(A - \bar{\zeta} I) - \varepsilon \frac{\lambda - \bar{\zeta}}{\lambda - \zeta}(A - \zeta I)\right](A - \zeta I)^{-1} =$$

$$= \varepsilon \frac{\bar{\zeta} - \zeta}{\lambda - \zeta}(A - \lambda I)(A - \zeta I)^{-1} .$$

Hence $\mathfrak{R}(U - \nu I) = \mathfrak{R}(A - \lambda I)$.

Finally, let λ be a normal eigenvalue of A. Then $\mathfrak{H} = \mathfrak{S}_\lambda(A) \dotplus \mathfrak{L}$, where $\dim \mathfrak{S}_\lambda(A) < \infty$, while $\mathfrak{L}$ is closed, invariant for A and such that $\lambda \in \varrho(A|\mathfrak{L})$. This decomposition reduces A, since $\mathfrak{S}_\lambda(A) \subset \mathfrak{D}(A)$; therefore, on account of (7.5), it reduces U as well. Applying one of the assertions just established to $A|\mathfrak{L}$, a legitimate procedure because the proof only involved the Banach (and not the Krein) space structure of $\mathfrak{H}$, we obtain that $\nu \in \varrho(U|\mathfrak{L})$. On the other hand, Lemma II.5.5 yields $\mathfrak{S}_\lambda(A) = \mathfrak{S}_\nu(U)$. Thus ν is a normal eigenvalue of U. □

Theorem 7.4. *Let U be a closed linear operator in $\mathfrak{H}$. Let ε, ζ be complex numbers such that $|\varepsilon| = 1$, $\bar{\zeta} \neq \zeta$, $\varepsilon \notin \sigma_p(U)$. Then the Cayley transform*

(7.8) $$A = (\zeta U - \varepsilon \bar{\zeta} I)(U - \varepsilon I)^{-1}$$

is closed. Moreover, if $\nu \neq \varepsilon$ is a number belonging to $\sigma_p(U)$, $\sigma_r(U)$, $\sigma_c(U)$ or $\varrho(U)$, then the number

(7.9) $$\lambda = \frac{\zeta \nu - \varepsilon \bar{\zeta}}{\nu - \varepsilon}$$

belongs to $\sigma_p(A)$, $\sigma_r(A)$, $\sigma_c(A)$ or $\varrho(A)$, respectively. If, in particular, ν is a normal eigenvalue of U, then λ is a normal eigenvalue of A.

Proof. We have

(7.10) $$A = \zeta I + \varepsilon(\zeta - \bar{\zeta})(U - \varepsilon I)^{-1} .$$

Thus A is closed.

The assertion concerning $\sigma_p(U)$, $\sigma_r(U)$, $\sigma_c(U)$ and $\varrho(U)$ follows from the similar statement of Theorem 7.3, since (7.5) and (7.8) as well as (7.6) and (7.9) are mutually inverse transformations, while $\varepsilon \notin \sigma_p(U)$ and $\nu \neq \varepsilon$ are equivalent to $\zeta \notin \sigma_p(A)$ and $\lambda \neq \zeta$, respectively (cf. Lemma II.4.1 and Remark II.5.3).

For a normal eigenvalue ν of U we argue just as in the proof of Theorem 7.3, replacing A, λ, U, ν by U, ν, A, λ, respectively, and referring to (7.8) instead of (7.5). □

Now we are able to extend Theorem 6.5 to a large class of discontinuous selfadjoint operators.

Theorem 7.5. *Let A be a selfadjoint operator in $\mathfrak{H}$. Suppose that $\sigma(A) \neq C$. If λ is a normal eigenvalue of A, so is $\bar{\lambda}$, and we have $\mathfrak{S}_\lambda(A) \#\# \mathfrak{S}_{\bar{\lambda}}(A)$; moreover, the matrices of $A|\mathfrak{S}_\lambda(A)$ and $A|\mathfrak{S}_{\bar{\lambda}}(A)$ with respect to any dual pair of bases are the transposed conjugates of each other.*

Proof. Since $\varrho(A)$ is open and non-empty, it contains a non-real point ζ. By Theorem 7.1 the operator $U = (A - \bar{\zeta} I)(A - \zeta I)^{-1}$ is unitary. According to Theorems 2.2 c) and 7.3 the number $\nu = \dfrac{\lambda - \bar{\zeta}}{\lambda - \zeta}$ is a normal eigenvalue of U. On account of Theorem 5.5 also ν^* is a normal eigenvalue of U, whereas $\mathfrak{S}_\nu(U) \#\# \mathfrak{S}_{\nu^*}(U)$. But $\nu^* = \dfrac{\bar{\lambda} - \bar{\zeta}}{\bar{\lambda} - \zeta}$. Expressing A through U and $\bar{\lambda}$ through ν^* (cf. Lemma II.4.1 and Remark II.5.3), with the help of Theorem 7.4 we obtain that $\bar{\lambda}$ is a normal eigenvalue of A. Moreover, $\mathfrak{S}_\lambda(A)$ and $\mathfrak{S}_{\bar{\lambda}}(A)$ form a dual pair, since they coincide with $\mathfrak{S}_\nu(U)$ and $\mathfrak{S}_{\nu^*}(U)$, respectively (Lemma II.5.5).

To get the desired conclusion for the matrices of $A|\mathfrak{S}_\lambda(A)$ and $A|\mathfrak{S}_{\bar{\lambda}}(A)$, we apply Theorem II.3.6 to the space $\mathfrak{E} = \mathfrak{S}_\lambda(A) + \mathfrak{S}_{\bar{\lambda}}(A)$, whose non-degeneracy is a consequence of Corollary II.3.4 and Lemma I.10.1 in the case $\lambda \neq \bar{\lambda}$, and of the relations $\mathfrak{E} = \mathfrak{S}_\lambda(A)$, $\mathfrak{S}_\lambda(A) \#\# \mathfrak{S}_\lambda(A)$ in the case $\lambda = \bar{\lambda}$. □

8. Unitary Dilations

Consider a Hilbert space $\mathfrak{H}_0$ and a linear operator $T \in \mathfrak{B}(\mathfrak{H}_0)$. Suppose there exists a Krein space $\mathfrak{H}$ containing an ortho-complemented subspace $\mathfrak{H}_{(0)}$ that is isometrically isomorphic to $\mathfrak{H}_0$, and a unitary operator $U \in \mathfrak{B}(\mathfrak{H})$ satisfying the relations

$$(8.1) \qquad T^n x = P U^n x\,, \qquad (T^*)^n x = P U^{-n} x \qquad (x \in \mathfrak{H}_{(0)};\quad n = 1, 2, \ldots)\,,$$

where P is the orthogonal projector from $\mathfrak{D}(P) = \mathfrak{H}$ onto $\mathfrak{R}(P) = \mathfrak{H}_{(0)}$. This situation will be expressed by saying that U is a *unitary dilation* of T on the Krein space $\mathfrak{H} \supset \mathfrak{H}_0$. If the span of the subspaces $U^n \mathfrak{H}_{(0)}$ ($n = 0, \pm 1, \pm 2, \ldots$) is dense in $\mathfrak{H}$, we say the unitary dilation U is *minimal*.

One sees immediately that in the case where indefinite inner products are not admitted T can have a unitary dilation only if T is a contraction (i.e., its norm does not exceed 1). This condition is known to be sufficient as well. It turns out that in the general case every $T \in \mathfrak{B}(\mathfrak{H}_0)$ has a unitary dilation. Before proving this, however, we need some preparation.

Let $\mathfrak{H}_0$ be a Hilbert space. Recall that a positive operator $B \in \mathfrak{B}(\mathfrak{H}_0)$ has a unique positive square root $B^{1/2} \in \mathfrak{B}(\mathfrak{H}_0)$; moreover, $B^{1/2}$ is the strong limit of a sequence of polynomials of B (see e.g. Riesz and Sz.-Nagy [1; Section 104]). In particular, we have:

Lemma 8.1. *Let $\mathfrak{H}_0$ be a Hilbert space. If two positive operators $B_1, B_2 \in \mathfrak{B}(\mathfrak{H}_0)$ and an operator $S \in \mathfrak{B}(\mathfrak{H}_0)$ fulfil the condition $SB_1 = B_2S$, then $SB_1^{1/2} = B_2^{1/2}S$.* □

The following special case of Lemma 2.10 is also related to square roots.

Lemma 8.2. *Let $\mathfrak{H}_0$ be a Hilbert space. If $A \in \mathfrak{B}(\mathfrak{H}_0)$ is selfadjoint, then $\mathfrak{R}(A^2) = \mathfrak{R}(A)$.* □

For a selfadjoint operator $A \in \mathfrak{B}(\mathfrak{H}_0)$ ($\mathfrak{H}_0$ a Hilbert space) it is usual to set

(8.2) $$|A| = (A^2)^{1/2}, \qquad \operatorname{sgn} A = E_A^+ - E_A^-,$$

where E_A^+ and E_A^- are the orthogonal projectors onto the subspaces

(8.3) $$\mathfrak{R}(E_A^+) = \mathfrak{H}_A^+ = \mathfrak{N}(A)^\perp \cap \mathfrak{N}(A - |A|)$$

and

(8.4) $$\mathfrak{R}(E_A^-) = \mathfrak{H}_A^- = \mathfrak{N}(A)^\perp \cap \mathfrak{N}(A + |A|),$$

respectively.

The following two results, though seldom mentioned explicitly, are known. We state and prove them in order to make the discussions involving $\operatorname{sgn} A$ possibly elementary and self-contained.

Lemma 8.3. *The subspaces defined by* (8.3) *and* (8.4) *satisfy the relations $\mathfrak{H}_A^+ \perp \mathfrak{H}_A^-$, $\mathfrak{H}_A^+ \dotplus \mathfrak{H}_A^- = \mathfrak{N}(A)^\perp$.*

Proof. As a consequence of Lemma 8.1 the operator A commutes with $|A|$. Therefore, in view of Theorem 2.6 d), $A\mathfrak{H}_A^\pm \subset \mathfrak{H}_A^\pm$. Moreover, again by Theorem 2.6 d), $A\mathfrak{H}_A^\pm$ is dense in $\mathfrak{H}_A^\pm$, since $\mathfrak{N}(A|\mathfrak{H}_A^\pm) = 0$. But for $x \in \mathfrak{H}_A^+$, $y \in \mathfrak{H}_A^-$ (denoting the inner product on $\mathfrak{H}_0$ by $(\cdot\,,\cdot)_0$) we have:

$$(Ax, Ay)_0 = \big(1/2\,(A + |A|)x,\ 1/2\,(A - |A|)y\big)_0 = \\ = 1/4\,\big((A^2 - |A|^2)x,\ y\big)_0 = 0\,.$$

Thus $A\mathfrak{H}_A^+ \perp A\mathfrak{H}_A^-$ and, taking closures, $\mathfrak{H}_A^+ \perp \mathfrak{H}_A^-$.

In order to establish the second half of the statement consider an element $z \in \mathfrak{H}_0$ and put

$$x = 1/2\,(A + |A|)z\,, \qquad y = 1/2\,(A - |A|)z\,.$$

Then $Az = x + y$, where $(A - |A|)x = 0$ and $(A + |A|)y = 0$. Making use of the relations $\Re(A) \subset \mathfrak{N}(A)^\perp$, $\Re(|A|) \subset \mathfrak{N}(|A|)^\perp$, $\mathfrak{N}(|A|) = \mathfrak{N}(|A|^2) = \mathfrak{N}(A^2) = \mathfrak{N}(A)$ (cf. Theorem 2.6 d) and Lemma 8.2) we obtain that $x \in \mathfrak{H}_A^+$, $y \in \mathfrak{H}_A^-$. Consequently, $\Re(A) \subset \mathfrak{H}_A^+ \dotplus \mathfrak{H}_A^-$. It remains to observe that $\mathfrak{H}_A^+ \dotplus \mathfrak{H}_A^-$ is an orthogonal direct sum of closed subspaces in a Hilbert space, hence closed, while $\Re(A)$ is dense in $\mathfrak{N}(A)^\perp$. □

Lemma 8.4. *Let $\mathfrak{H}_0$ be a Hilbert space. Let $S, A_1, A_2 \in \mathfrak{B}(\mathfrak{H}_0)$, $A_1^* = A_1$, $A_2^* = A_2$. Denote by P_j $(j = 1,2)$ the orthogonal projector with range $\mathfrak{N}(A_j)$. If $SA_1 = A_2S$, then $SP_1 = P_2S$.*

Proof. $SA_1 = A_2S$ implies that $S\mathfrak{N}(A_1) \subset \mathfrak{N}(A_2)$. Hence

$$SP_1 = P_2SP_1\,. \tag{8.5}$$

Similarly, $A_1S^* = S^*A_2$ yields $S^*\mathfrak{N}(A_2) \subset \mathfrak{N}(A_1)$ i.e.

$$S^*P_2 = P_1S^*P_2\,. \tag{8.6}$$

Taking adjoints in (8.6) and comparing with (8.5) we obtain $SP_1 = P_2S$. □

It is via the next assertion that Lemmas 8.1—8.4 contribute to the proof of the main result of this section.

Lemma 8.5. *Let $\mathfrak{H}_0$ be a Hilbert space. For $T \in \mathfrak{B}(\mathfrak{H}_0)$ put*

$$Q_T = |I - T^*T|^{1/2}\,, \qquad Q_{T^*} = |I - TT^*|^{1/2}\,, \tag{8.7}$$

$$J_T = \operatorname{sgn}(I - T^*T)\,, \qquad J_{T^*} = \operatorname{sgn}(I - TT^*) \tag{8.8}$$

(cf. the definitions (8.2)). Then the following relations hold true:

$$TQ_T = Q_{T^*}T\,, \qquad T^*Q_{T^*} = Q_TT^*\,, \tag{8.9}$$

$$TJ_T = J_{T^*}T\,, \qquad T^*J_{T^*} = J_TT^*\,, \tag{8.10}$$

$$Q_TJ_T = J_TQ_T\,, \qquad Q_{T^*}J_{T^*} = J_{T^*}Q_{T^*}\,, \tag{8.11}$$

$$Q_T^2J_T = I - T^*T\,, \qquad Q_{T^*}^2J_{T^*} = I - TT^*\,, \tag{8.12}$$

$$Q_TJ_T^2 = Q_T\,, \qquad Q_{T^*}J_{T^*}^2 = Q_{T^*}\,. \tag{8.13}$$

Proof. Set

$$A_T = I - T^*T\,, \qquad A_{T^*} = I - TT^*\,. \tag{8.14}$$

Denote by F_T^+, F_T^-, F_T^0, F_{T*}^+, F_{T*}^-, F_{T*}^0 the orthogonal projectors satisfying the conditions

$$\mathfrak{R}(F_T^\pm) = \mathfrak{N}(A_T)^\perp \cap \mathfrak{N}(A_T \mp |A_T|)\,, \tag{8.15}$$

$$\mathfrak{R}(F_T^0) = \mathfrak{N}(A_T)\,, \tag{8.16}$$

$$\mathfrak{R}(F_{T*}^\pm) = \mathfrak{N}(A_{T*})^\perp \cap \mathfrak{N}(A_{T*} \mp |A_{T*}|)\,, \tag{8.17}$$

$$\mathfrak{R}(F_{T*}^0) = \mathfrak{N}(A_{T*})\,. \tag{8.18}$$

Now, in view of (8.2)—(8.4), the definitions (8.7)—(8.8) can be written as

$$Q_T = |A_T|^{1/2}\,, \qquad Q_{T*} = |A_{T*}|^{1/2}\,, \tag{8.19}$$

$$J_T = F_T^+ - F_T^-\,, \qquad J_{T*} = F_{T*}^+ - F_{T*}^-\,. \tag{8.20}$$

Lemma 8.2 yields $\mathfrak{N}(|A_T|)=\mathfrak{N}(A_T)$, so that $\mathfrak{N}(A_T)\subset\mathfrak{N}(A_T\mp|A_T|)$. Thus the orthogonal projectors $F_T^+ + F_T^0$, $F_T^- + F_T^0$ have the ranges

$$\mathfrak{R}(F_T^\pm + F_T^0) = \mathfrak{N}(A_T \mp |A_T|)\,, \tag{8.21}$$

respectively. In the same way, for the orthogonal projectors $F_{T*}^\pm + F_{T*}^0$ we have

$$\mathfrak{R}(F_{T*}^\pm + F_{T*}^0) = \mathfrak{N}(A_{T*} \mp |A_{T*}|)\,. \tag{8.22}$$

Observe that in each of the items (8.9)—(8.13) it is sufficient to verify the first equation, since applying it to the operator T^* we obtain the second one.

After these preliminaries, let us prove (8.9). The definitions (8.14) yield $TA_T = A_{T*}T$. Hence we deduce the relation $TQ_T = Q_{T*}T$ with the help of (8.19), (8.2) and a repeated application of Lemma 8.1.

Again, owing to (8.14) and Lemma 8.1,

$$T(A_T \mp |A_T|) = (A_{T*} \mp |A_{T*}|)T\,.$$

Therefore, in view of (8.21), (8.22) and Lemma 8.4,

$$T(F_T^\pm + F_T^0) = (F_{T*}^\pm + F_{T*}^0)T\,.$$

Hence, by the definition (8.20), $TJ_T = J_{T*}T$.

Further, making use of the formulas (8.19), (8.2) and (twice) of Lemma 8.1, we obtain that $Q_TA_T = A_TQ_T$. Applying Lemma 8.1 once more, we find:

$$Q_T(A_T \mp |A_T|) = (A_T \mp |A_T|)Q_T\,.$$

The same steps as in the preceding paragraph lead to the conclusion $Q_TJ_T = J_TQ_T$.

To prove (8.12) we start from the relations

$$|A_T|\,F_T^+ = A_TF_T^+\,, \qquad |A_T|\,F_T^- = -A_TF_T^-$$

(see (8.15)). Along with (8.16) and Lemma 8.3 they imply:

$$|A_T|\,(F_T^+ - F_T^-) = A_T(F_T^+ + F_T^-) = A_T(I - F_T^0) = A_T\,.$$

Hence, in view of (8.19), (8.20) and (8.14), $Q_T^2 J_T = I - T^*T$.

Finally, from (8.20), (8.15), (8.16) and Lemma 8.3 we infer

$$J_T^2 = (F_T^+ - F_T^-)^2 = F_T^+ + F_T^- = I - F_T^0$$

so that, according to (8.19) and Lemma 8.2,

$$Q_T J_T^2 = |A_T|^{1/2}(I - F_T^0) = |A_T|^{1/2} = Q_T\,. \quad \square$$

We are now coming back to the dilation problem.

Theorem 8.6. *Let $\mathfrak{H}_0$ be a Hilbert space, and let $T \in \mathfrak{B}(\mathfrak{H}_0)$. Then T has at least one minimal unitary dilation on a suitable Krein space $\mathfrak{H} \supset \mathfrak{H}_0$.*

Proof. We shall use the notations (8.7)—(8.8). The inner product and the norm on $\mathfrak{H}_0$ will be denoted by $(.\,,.)_0$ and $\|\cdot\|_0$, respectively.

Consider the space

$$\mathfrak{H} = \mathop{\times}_{j=-\infty}^{\infty} \mathfrak{H}_j\,,$$

the cartesian product of the Hilbert spaces $\mathfrak{H}_j$ $(j = 0, \pm 1, \pm 2, \ldots)$, where

$$(8.23)\qquad \mathfrak{H}_j = \begin{cases} \mathfrak{H}_0 & \text{if} \quad j = 0\,, \\ \overline{\mathfrak{R}(Q_T)} & \text{if} \quad j > 0\,, \\ \overline{\mathfrak{R}(Q_{T^*})} & \text{if} \quad j < 0\,. \end{cases}$$

In other words, $\mathfrak{H}$ is the Hilbert space of all sequences $x = \{x_j\}_{j=-\infty}^{\infty}$ with $x_j \in \mathfrak{H}_j$ $(j = 0, \pm 1, \pm 2, \ldots)$ and $\sum_{j=-\infty}^{\infty} \|x_j\|_0^2 < \infty$; the (positive definite) inner product on $\mathfrak{H}$ is given by the formula

$$(8.24)\qquad [x, y] = \sum_{j=-\infty}^{\infty} (x_j, y_j)_0 \qquad (x, y \in \mathfrak{H})\,.$$

Define a linear operator J on $\mathfrak{H}$ setting

$$(8.25)\qquad (Jx)_j = \begin{cases} x_0 & \text{if} \quad j = 0\,, \\ J_T x_j & \text{if} \quad j > 0\,, \\ J_{T^*} x_j & \text{if} \quad j < 0\,. \end{cases}$$

By the aid of Lemma 8.5 it is easy to see that $\mathfrak{R}(J) \subset \mathfrak{H}$ and $J^{-1} = J$. Moreover, J is continuous and selfadjoint. Thus, according to

Theorem V.1.3, the inner product

$$(8.26)\qquad (x, y) = [Jx, y] \qquad (x,y \in \mathfrak{H})$$

turns $\mathfrak{H}$ into a Krein space.

The subspace

$$(8.27)\qquad \mathfrak{H}_{(0)} = \{x \in \mathfrak{H}: \quad x_j = 0 \quad \text{for} \quad j \neq 0\}$$

is isometrically isomorphic to $\mathfrak{H}_0$, and the complementary subspace $\{x \in \mathfrak{H}: x_0 = 0\}$ is orthogonal to $\mathfrak{H}_{(0)}$ with respect to both of the inner products (8.24), (8.26).

Next define a linear operator U on $\mathfrak{H}$ by the formula

$$(8.28)\qquad (Ux)_j = \begin{cases} Q_{T^*}x_{-1} + Tx_0 & \text{if} \quad j = 0\,, \\ -J_T T^* x_{-1} + Q_T x_0 & \text{if} \quad j = 1\,, \\ x_{j-1} & \text{otherwise.} \end{cases}$$

Leaning on Lemma 8.5 one verifies the relations $\mathfrak{R}(U) \subset \mathfrak{H}$, $(Ux, Uy) = (x, y)$ $(x,y \in \mathfrak{H})$. Moreover, with the help of the same lemma it can be checked that U^{-1} is defined on all of $\mathfrak{H}$; namely

$$(8.29)\qquad (U^{-1}x)_j = \begin{cases} T^*x_0 + J_T Q_T x_1 & \text{if} \quad j = 0\,, \\ J_{T^*}Q_{T^*}x_0 - J_{T^*}Tx_1 & \text{if} \quad j = -1\,, \\ x_{j+1} & \text{otherwise} \end{cases}$$

for every $x \in \mathfrak{H}$. Hence U is unitary with respect to $(.\,,.)$.

Let $x \in \mathfrak{H}_{(0)}$ (see (8.27)). It follows by induction relative to n that in this case

$$(8.30)\qquad (U^n x)_j = \begin{cases} T^n x_0 & \text{if} \quad j = 0\,, \\ Q_T T^{n-j} x_0 & \text{if} \quad 1 \leqq j \leqq n\,, \\ 0 & \text{otherwise} \end{cases}$$

and

$$(8.31)\qquad (U^{-n}x)_j = \begin{cases} (T^*)^n x_0 & \text{if} \quad j = 0\,, \\ J_{T^*}Q_{T^*}(T^*)^{n+j}x_0 & \text{if} \quad -n \leqq j \leqq -1\,, \\ 0 & \text{otherwise} \end{cases}$$

$(n = 1,2,\ldots)$. Considering $j = 0$ we obtain the dilation property (8.1) of U.

Let $x,y,z \in \mathfrak{H}_{(0)}$, $y_0 = Tx_0$, $z_0 = T^*x_0$. Then (8.30) and (8.31) yield:

$$(U^n x - U^{n-1} y)_j = \begin{cases} Q_T x_0 & \text{if} \quad j = n\,, \\ 0 & \text{otherwise,} \end{cases}$$

$$(U^{-n} x - U^{-(n-1)} z)_j = \begin{cases} J_{T^*} Q_{T^*} x_0 & \text{if} \quad j = -n\,, \\ 0 & \text{otherwise.} \end{cases}$$

But, according to (8.23), $\mathfrak{R}(Q_T)$ is dense in $\mathfrak{H}_n$, and the subspace

$$\mathfrak{R}(J_{T^*} Q_{T^*}) \supset \mathfrak{R}(J_{T^*} Q_{T^*} J_{T^*}) = \mathfrak{R}(Q_{T^*})$$

(cf. Lemma 8.5) is dense in $\mathfrak{H}_{-n}$ $(n = 1, 2, \ldots)$. Thus the span of the sets $U^m \mathfrak{H}_{(0)}$ $(m = 0, \pm 1, \pm 2, \ldots)$ is dense in each of the subspaces

$$\mathfrak{H}_{(r)} = \{x \in \mathfrak{H}\colon \quad x_j = 0 \text{ for } j \neq r\} \qquad (r = 0, \pm 1, \pm 2, \ldots)$$

and, consequently, in $\mathfrak{H}$. □

Notes to Chapter VI

Linear operators in quasi-positive Krein spaces have first been studied by Pontrjagin [1]. The investigation of operators in arbitrary Krein spaces was begun by Ginzburg [1], [2].

Properties of the adjoint in inner product spaces with a Hilbert majorant are discussed in the expository paper of Azizov and Iohvidov [2]. Lemma 2.10 (for $T \in \mathfrak{B}(\mathfrak{H})$) was formulated by Kreĭn and Šmul'jan [2].

The propositions of Section 3, some of them even for inner product spaces with a Hilbert majorant, belong to Iohvidov [4], [11]; cf. also Azizov and Iohvidov [2]. For analogous questions concerning Banach spaces with a J-metric see Iohvidov [14], Iohvidov and Senderov [1].

Theorems 4.3 and 4.4 are due to Kreĭn and Šmul'jan [2]. Lemma 4.9 appears in a paper of Langer [5], and more explicitly in unpublished lecture notes by the same author.

A class of "almost unitary" operators in $\mathfrak{H}$ has been investigated by Kužel' [1].

The results of Section 5 are natural counterparts for those of Section 6, obtained by Langer [1], [2]. In contrast to our geometrical treatment of normal eigenvalues, Langer established Theorem 6.5, and its extension to spectral sets of a possibly discontinuous selfadjoint operator, with the help of general spectral theory (integration of the resolvent).

Let us note that our method of proving Theorem 6.5 applies to discontinuous selfadjoint operators A as well, but in this case the auxiliary relation $((A - \lambda I)^n)^* = (A - \bar{\lambda} I)^n$ is far from being trivial. If we take the latter equation for granted, Theorem 7.5 becomes superfluous.

Examples of selfadjoint operators in $\mathfrak{H}$ whose residual or continuous spectrum contains non-real points have first been published by Azizov [1], while the construction presented here as Example 6.4 is due to Langer (personal communication).

Azizov and Iohvidov [2] found that symmetry properties of spectra do not as a rule remain valid in non-degenerate inner product spaces with a Hilbert majorant. Noël [2] generalized Langer's results to operators $T \in \mathfrak{B}(\mathfrak{H})$ such that T^* is a function of T. Senderov [1] examined the spectral radius and principal subspaces of J-isometric operators in Banach spaces with a J-metric.

Hess [2] studied polynomials of certain selfadjoint operators in $\mathfrak{H}$. Berezin [1] defined selfadjoint operators in arbitrary non-degenerate inner product spaces and described selfadjoint extensions of symmetric operators.

The assertions of Theorems 7.1—7.4 in the case of a quasi-negative $\mathfrak{H}$ largely belong to Iohvidov and Kreĭn [1]. For an exposition in the case of non-degenerate inner product spaces with a Hilbert majorant see Azizov and Iohvidov [2]. In the second half of the proof of Theorem 7.2 we follow Riesz and Sz.-Nagy [1].

Theorem 8.6 is due to Davis [1], who admitted discontinuous operators and discussed unicity of the dilation too. For the definite case see e.g. Sz.-Nagy and Foiaș [1]. Davis [2] extended his result to semigroups. An application to characteristic operator functions was given by Davis and Foiaș [1].

Jalava [2] carried over several results of the present chapter to Banach spaces with a continuous, but possibly non-symmetric bilinear form.

Chapter VII. Positive Operators and Plus-operators in Krein Spaces

Positive operators in a Krein space (Sections 1 and 3) have turned out (see Chapter VIII, especially Section 6 and the Notes) to be the most fruitful generalizations of positive operators in Hilbert space. In Section 2 an alternative extension of the latter concept is considered. Sections 4—6 resume the study of plus-operators begun in Section II.8. A unifying feature of the chapter is the repeated application of the Schwarz inequality. Theorems 1.3, 3.1, 5.2, 5.4, 5.5 and 6.1 deserve particular attention.

1. Positive Operators

A symmetric operator A in the Krein space $\mathfrak{H}$ is said to be *positive* if $(Ax, x) \geqq 0$ for every $x \in \mathfrak{D}(A)$.

We begin with two elementary but substantial facts.

Lemma 1.1. *Let A be a densely defined positive operator in $\mathfrak{H}$. If an element $x_0 \in \mathfrak{D}(A)$ satisfies $(Ax_0, x_0) = 0$, then $Ax_0 = 0$.*

Proof. Applying Lemma I.4.4 to the space $\mathfrak{E} = \mathfrak{D}(A)$ and the inner product $(x, y)_A = (Ax, y)$ $(x, y \in \mathfrak{E})$ we obtain that $(Ax_0, y) = 0$ $(y \in \mathfrak{D}(A))$. Thus $Ax_0 \perp \mathfrak{H}$, $Ax_0 = 0$. □

Theorem 1.2. *Let A be a densely defined positive operator in the Krein space $\mathfrak{H}$. If λ is a positive (negative) eigenvalue of A, then the corresponding eigenspace is positive definite (negative definite).*

Proof. Let $Ax_0 = \lambda x_0$, where $\lambda > 0$ and $x_0 \neq 0$. Then $Ax_0 \neq 0$, so that by Lemma 1.1 $(Ax_0, x_0) \neq 0$. Hence $(x_0, x_0) = 1/\lambda\,(Ax_0, x_0) > 0$. For $\lambda < 0$ the reasoning is similar. □

The spectrum of a selfadjoint operator $A \in \mathfrak{B}(\mathfrak{H})$ need not be contained in the real line (cf. Examples II.3.2 and VI.6.4). For a positive A, however, the situation is different.

Theorem 1.3. *The spectrum of a positive operator $A \in \mathfrak{B}(\mathfrak{H})$ is real.*

Proof. Suppose that $\lambda \in \sigma_c(A) \cup \sigma_p(A)$. Let J be a fundamental symmetry on $\mathfrak{H}$. Then there exists a sequence $\{x_n\}_1^\infty \subset \mathfrak{H}$ with the properties

$$(1.1) \qquad \|x_n\|_J = 1 \qquad (n = 1, 2, \ldots),$$

$$(1.2) \qquad \|Ax_n - \lambda x_n\|_J \to 0 \qquad (n \to \infty).$$

Making use of Lemma II.11.4 we find:

$$(1.3) \qquad (Ax_n - \lambda x_n, \ x_n) \to 0 \qquad (n \to \infty).$$

Since $|(x_n, x_n)| \leqq \|x_n\|_J^2 = 1$ for every n, we may assume that

$$(1.4) \qquad (x_n, x_n) \to \alpha,$$

a real number. Then (1.3) yields:

$$(1.5) \qquad (Ax_n, x_n) \to \lambda\alpha \qquad (n \to \infty).$$

According to (1.5), $\alpha > 0$ implies $\lambda \geqq 0$, whereas $\alpha < 0$ implies $\lambda \leqq 0$. Next let $\alpha = 0$, i.e.,

$$(1.6) \qquad (Ax_n, x_n) \to 0 \qquad (n \to \infty).$$

Applying Lemma I.2.2 to the A-inner product, making use of Lemma II.11.4 and recalling (1.1) we obtain:

$$\|Ax_n\|_J^2 = (Ax_n, JAx_n) = (x_n, JAx_n)_A \leqq$$
$$\leqq (x_n, x_n)_A^{1/2} \, (JAx_n, JAx_n)_A^{1/2} =$$
$$= (Ax_n, x_n)^{1/2} \, (AJAx_n, JAx_n)^{1/2} \leqq (Ax_n, x_n)^{1/2} \|A\|_J^{3/2}.$$

Thus, in view of (1.6), $\|Ax_n\|_J \to 0$ $(n \to \infty)$. Hence, on account of (1.2) and (1.1), $\lambda = 0$.

Consequently, $\sigma_c(A)$ and $\sigma_p(A)$ are real. From Theorem VI.6.1 c) we conclude that $\sigma_r(A)$ is empty. □

Under the assumption and with the notations of Theorem II.9.4 $A - \mu_A I$ is a positive operator. Therefore, by Theorem 1.3 and Theorem II.9.1, the following supplement to Corollary II.9.9 holds:

Corollary 1.4. *The spectrum of a Pesonen operator $A \in \mathfrak{B}(\mathfrak{H})$ is real.* □

The spectrum of a discontinuous positive selfadjoint operator need not be real.

Example 1.5. Let the space $\mathfrak{H} = \mathfrak{H}_0 \times \mathfrak{H}_0$ and the inner products $[.\,,.]$, $(.\,,.)$ be defined as in Example VI.6.4. Let B_0 be a discontinuous positive selfadjoint operator in the Hilbert space $\mathfrak{H}_0$. For $x \in \mathfrak{D}(B_0)$, $y \in \mathfrak{H}_0$ set $A\{x, y\} = \{0, B_0 x\}$. The operator A is positive and selfadjoint with respect to $(.\,,.)$. On the other hand, choosing an $x_0 \notin \mathfrak{D}(B_0)$ the element $\{x_0, 0\}$ does not belong to $\mathfrak{R}(A - \lambda I)$ for any λ. Hence $\sigma(A) = \boldsymbol{C}$.

If $\mathfrak{H}$ is a Hilbert space $(\varkappa^-(\mathfrak{H}) = 0)$, then each of the following conditions is known to be necessary and sufficient in order that a selfadjoint operator $A \in \mathfrak{B}(\mathfrak{H})$ be positive: a) the spectrum of A is nonnegative; b) there exists a selfadjoint operator $B \in \mathfrak{B}(\mathfrak{H})$ such that $A = B^2$; c) there exists a linear operator $T \in \mathfrak{B}(\mathfrak{H})$ such that $A = T^*T$. On the other hand, if $\mathfrak{H}$ is an indefinite Krein space then I is not a positive operator on $\mathfrak{H}$, though it satisfies the above conditions. Thus in the general case a), b) and c) are not sufficient for the positivity of A. It will turn out from Examples 1.6 and 2.3 below that they are not necessary either.

Example 1.6. Let $\dim \mathfrak{H} < \infty$, $\varkappa(\mathfrak{H}) > 0$. Any fundamental symmetry J on $\mathfrak{H}$ is positive. Suppose that $J = B^2$ for a selfadjoint operator B. Since -1 is an eigenvalue of J, making use of the Jordan normal form it follows that i or $-i$ is an eigenvalue of B, i.e., $Bx = ix$ or $Bx = -ix$ for some $x \neq 0$. Corollary II.3.4 yields $(x, x) = 0$. But $Jx = B^2x = -x$ and the definition of J imply $(x, x) < 0$. Thus J has no selfadjoint square root.

2. Operators of the Form T^*T

Let $\mathfrak{H}$ be a Krein space and $A \in \mathfrak{B}(\mathfrak{H})$ a selfadjoint operator. The vector space $\mathfrak{H}$ equipped with the A-inner product

$$(x, y)_A = (Ax, y) \qquad (x, y \in \mathfrak{H}) \tag{2.1}$$

can be regarded as an inner product space $\mathfrak{H}_A$. Since $\mathfrak{H}_A$ admits the strong topology of $\mathfrak{H}$ as a majorant, $\mathfrak{H}_A$ is decomposable (cf. Theorem IV.5.2). The rank of positivity (negativity) of $\mathfrak{H}_A$ will be called the *rank of A-positivity (A-negativity)* of $\mathfrak{H}$ and denoted by $\varkappa_A^+(\mathfrak{H})$ $(\varkappa_A^-(\mathfrak{H}))$. Thus $\varkappa_A^+(\mathfrak{H}) = \varkappa^+(\mathfrak{H}_A)$, $\varkappa_A^-(\mathfrak{H}) = \varkappa^-(\mathfrak{H}_A)$.

Theorem 2.1. *Consider a Krein space $\mathfrak{H}$ and a selfadjoint operator $A \in \mathfrak{B}(\mathfrak{H})$. There exists a linear operator $T \in \mathfrak{B}(\mathfrak{H})$ satisfying the equation*

$$A = T^*T \tag{2.2}$$

if and only if

$$\varkappa_A^+(\mathfrak{H}) \leqq \varkappa^+(\mathfrak{H}) \tag{2.3}$$

and

$$\varkappa_A^-(\mathfrak{H}) \leqq \varkappa^-(\mathfrak{H}) \,. \tag{2.4}$$

Proof. We first note that for $T \in \mathfrak{B}(\mathfrak{H})$ (2.2) is equivalent to the relation

$$(x, y)_A = (Tx, \; Ty) \qquad (x, y \in \mathfrak{H}) \,. \tag{2.5}$$

In other words, T must be an isometric linear mapping of $\mathfrak{H}_A$ into $\mathfrak{H}$.

Let us fix a fundamental and an A-fundamental decomposition of $\mathfrak{H}$:

$$\mathfrak{H} = \mathfrak{H}^+ (\dotplus) \mathfrak{H}^- , \tag{2.6}$$

$$\mathfrak{H}_A = \mathfrak{H}_A^0 \, (\dotplus)_A \, \mathfrak{H}_A^+ \, (\dotplus)_A \, \mathfrak{H}_A^- , \tag{2.7}$$

the superscript indicating, as usual, the sign of the respective inner square on the subspace in question.

Suppose that (2.5) is satisfied by some T. Then the range of $T|\mathfrak{H}_A^+$ is a positive definite subspace $\mathfrak{N}^+ \subset \mathfrak{H}$ with $\dim_{\text{int}} \mathfrak{N}^+ = \varkappa_A^+(\mathfrak{H})$. But, according to Lemma IV.7.3, $\dim_{\text{int}} \mathfrak{N}^+ \leqq \varkappa^+(\mathfrak{H})$. Thus (2.3) holds. The inequality (2.4) can be proved in a similar way.

Suppose, conversely, that (2.3) and (2.4) are valid. Owing to (2.3) and the intrinsic completeness of $\mathfrak{H}^+$ there exists an isometric isomorphism T_+ of $\mathfrak{H}_A^+ \subset \mathfrak{H}_A$ onto a subspace of $\mathfrak{H}^+ \subset \mathfrak{H}$. Similarly, (2.4) implies the existence of an isometric isomorphism T_- carrying $\mathfrak{H}_A^-$ into a subspace of $\mathfrak{H}^-$. If for

$$x = x_A^0 + x_A^+ + x_A^- ; \qquad x_A^0 \in \mathfrak{H}_A^0 , \quad x_A^+ \in \mathfrak{H}_A^+ , \quad x_A^- \in \mathfrak{H}_A^-$$

we set

$$Tx = T_+ x_A^+ + T_- x_A^- ,$$

then (2.5) will be satisfied. Moreover, for the fundamental symmetry J that corresponds to (2.6) we have

$$\begin{aligned} \|Tx\|_J^2 &= (T_+ x_A^+ , T_+ x_A^+) - (T_- x_A^- , T_- x_A^-) = \\ &= (A x_A^+ , x_A^+) - (A x_A^- , x_A^-) \leqq \|A\|_J \left(\|x_A^+\|_J^2 + \|x_A^-\|_J^2 \right) . \end{aligned}$$

Therefore if we additionally require from the decomposition (2.7) that x_A^+ , x_A^- continuously depend on x (by Theorem IV.5.2 and the closed graph principle this can be achieved), then $T \in \mathfrak{B}(\mathfrak{H})$. $\square$

The following result is a special case of Theorem 2.1.

Theorem 2.2. *Let $\mathfrak{H}$ be a Krein space with* $\dim \mathfrak{H} = \infty$, $\varkappa^+(\mathfrak{H}) = \varkappa^-(\mathfrak{H})$. *Then for every selfadjoint operator $A \in \mathfrak{B}(\mathfrak{H})$ there exists a linear operator $T \in \mathfrak{B}(\mathfrak{H})$ such that* $A = T^*T$.

Proof. In view of the assumption, $\varkappa^+(\mathfrak{H}) = \varkappa^-(\mathfrak{H}) = \varkappa^+(\mathfrak{H}) + \varkappa^-(\mathfrak{H}) = \dim \mathfrak{H}$. Therefore, by Theorem 2.1, it is sufficient to prove that $\varkappa_A^+(\mathfrak{H}) \leqq \dim \mathfrak{H}$ and $\varkappa_A^-(\mathfrak{H}) \leqq \dim \mathfrak{H}$ for every selfadjoint $A \in \mathfrak{B}(\mathfrak{H})$. But this follows from the (strong) continuity of $(.\,,.)_A$; namely the intrinsic A-dimension of any A-definite subspace $\mathfrak{L} \subset \mathfrak{H}$ is less than or equal to the (strong) dimension of $\mathfrak{L}$. $\square$

In contrast to Theorem 2.2, there are numerous cases where (2.3) or (2.4) is not satisfied.

Example 2.3. Let $\mathfrak{H}$ be a Krein space with $\varkappa^+(\mathfrak{H}) < \dim \mathfrak{H}$, and let J denote a fundamental symmetry on $\mathfrak{H}$. Then $\varkappa_J^+(\mathfrak{H}) > \varkappa^+(\mathfrak{H})$, so that by Theorem 2.1 J cannot be written in the form T^*T.

3. Uniformly Positive Operators

Consider a Krein space $\mathfrak{H}$ and a selfadjoint operator $A \in \mathfrak{B}(\mathfrak{H})$. We shall say that A is *uniformly positive*, if for a fundamental symmetry J and a suitable positive number α it satisfies the inequality

$$(3.1) \qquad (Ax, x) \geqq \alpha\|x\|_J^2 \qquad (x \in \mathfrak{H}).$$

In other words, A is uniformly positive if the function

$$(3.2) \qquad p_A(x) = (Ax, x)^{1/2} \qquad (x \in \mathfrak{H})$$

is a norm defining the strong topology $\tau_M(\mathfrak{H})$.

Every uniformly positive operator on an indefinite $\mathfrak{H}$ is a Pesonen operator (cf. Section II.9).

The next result should be compared with the following special case of Theorem II.9.4:

Let $\mathfrak{H}$ be indefinite, $A \in \mathfrak{B}(\mathfrak{H})$, $A^* = A$, and $(Ax, x) > 0$ for $(x, x) \geqq 0$ $(x \neq 0)$. Then $A - \mu_A I$, where

$$(3.3) \qquad \mu_A = \inf_{(x,x)=1} (Ax, x),$$

is a positive operator.

Theorem 3.1. *Let $\mathfrak{H}$ be an indefinite Krein space, $A \in \mathfrak{B}(\mathfrak{H})$, and $A^* = A$. Suppose that for a fundamental symmetry J and suitable positive number α the inequality*

$$(3.4) \qquad (Ax, x) \geqq \alpha\|x\|_J^2 \qquad (x \in \mathfrak{P}^+)$$

holds. Then there exists an $\varepsilon_0 > 0$ such that for every $0 < \varepsilon < \varepsilon_0$ the operator $A - (\mu_A - \varepsilon)I$ (cf. (3.3)) is uniformly positive.

Proof. From (3.3) and (3.4), writing $\mu_A = \mu$ and applying Lemma II.11.4, we obtain:

$$\mu = \inf_{(x,x)>0} \frac{(Ax, x)}{(x, x)} \geqq \inf_{(x,x)>0} \frac{(Ax, x)}{\|x\|_J^2} \geqq \alpha > 0.$$

Further, by Theorem II.9.4,

$$(3.5) \qquad (Ax, x) \geqq \mu(x, x) \qquad (x \in \mathfrak{H}).$$

For $x \in \mathfrak{P}^+$, $0 < \varepsilon < \mu$ relations (3.5) and (3.4) yield:

$$(Ax, x) - (\mu - \varepsilon)(x, x) \geqq (Ax, x)\left(1 - \frac{\mu - \varepsilon}{\mu}\right) \geqq \frac{\alpha\varepsilon}{\mu}\|x\|_J^2.$$

Consequently, if we prove the existence of a number $\delta > 0$ satisfying

$$(3.6)\qquad (Ax, x) - \mu(x, x) \geqq \delta\|x\|_J^2 \qquad (x \in \mathfrak{P}^{--}),$$

our theorem will be established with $\varepsilon_0 = \min\{\mu, \delta\}$.

Assume that (3.6) is not fulfilled by any positive δ. Then there exists a sequence $\{x_n\}_1^\infty \subset \mathfrak{H}$ satisfying the relations

$$(3.7)\qquad (x_n, x_n) < 0, \quad \|x_n\|_J = 1 \qquad (n = 1, 2, \ldots),$$

$$(3.8)\qquad (Ax_n, x_n) - \mu(x_n, x_n) \to 0 \qquad (n \to \infty).$$

On the other hand, in view of (3.3), (3.7) and the estimate $|(x_n, x_n)| \leqq \|x_n\|_J^2$, we can find a sequence $\{y_n\}_1^\infty \subset \mathfrak{H}$ such that

$$(3.9)\qquad (y_n, y_n) = -(x_n, x_n), \quad \mathrm{Re}(x_n, y_n) = 0 \qquad (n = 1, 2, \ldots),$$

$$(3.10)\qquad (Ay_n, y_n) - \mu(y_n, y_n) \to 0 \qquad (n \to \infty).$$

Consider the neutral vectors $z_n = x_n - y_n$ $(n = 1, 2, \ldots)$. According to condition (3.4) we have:

$$\|x_n - y_n\|_J = \|z_n\|_J \leqq \alpha^{-1/2}(Az_n, z_n)^{1/2} = \alpha^{-1/2}((A - \mu I)z_n, z_n)^{1/2}.$$

Hence, applying the triangle inequality to the positive semi-definite inner product $(.\,,.)_{A-\mu I}$ (see (3.5)) and making use of relations (3.8), (3.10) we obtain:

$$(3.11)\qquad \|x_n - y_n\|_J \to 0 \qquad (n \to \infty).$$

But (3.4) is valid for $x = y_n$ too:

$$(3.12)\qquad \frac{(Ay_n, y_n)}{\|y_n\|_J^2} \geqq \alpha \qquad (n = 1, 2, \ldots).$$

From (3.12), (3.11) and (3.7) we conclude that

$$(Ax_n, x_n) > \frac{\alpha}{2} \qquad (n > n_0),$$

contrary to the assumption (3.7)—(3.8). □

In Chapter VIII we shall make use of the following simple fact.

Lemma 3.2. *Every uniformly positive operator $A \in \mathfrak{B}(\mathfrak{H})$ is completely invertible.*

Proof. Relation (3.1) and Lemma II.11.4 yield $\|Ax\|_J\|x\|_J \geqq \alpha\|x\|_J^2$ i.e. $\|Ax\|_J \geqq \alpha\|x\|_J$ with an $\alpha > 0$ for every $x \in \mathfrak{H}$. As in the Hilbert space case, Theorem VI.2.6 d) leads to the desired conclusion. □

4. Plus-operators

In this section we assume that $\varkappa(\mathfrak{H}) > 0$.

We are going to study plus-operators $T \in \mathfrak{B}(\mathfrak{H})$, i.e., continuous linear operators T defined on the indefinite Krein space $\mathfrak{H}$, the value

$$(4.1) \qquad \mu(T) = \inf_{(x,x)=1} (Tx, Tx)$$

being non-negative (cf. Section II.8). Recall that for every operator of this kind

$$(4.2) \qquad (Tx, Tx) \geqq \mu(T)\,(x, x) \qquad (x \in \mathfrak{H})$$

(Theorem II.8.1).

According to relation (4.2) the operator

$$(4.3) \qquad A = T^*T - \mu(T)I$$

is positive. Thus from Theorem 1.3 we obtain:

Lemma 4.1. *If $T \in \mathfrak{B}(\mathfrak{H})$ is a plus-operator, then $\sigma(T^*T)$ is real.* □

In the case where both T and T^* are plus-operators we can prove more.

Theorem 4.2. *If the adjoint of the plus-operator $T \in \mathfrak{B}(\mathfrak{H})$ is also a plus-operator, then $\sigma(T^*T)$ is non-negative.*

Proof. We use the notations (4.1), (4.3). By relation (4.2) the operator A is positive. Let $\lambda \in \sigma_p(A) \cup \sigma_c(A)$. Repeating the proof of Theorem 1.3 we find a sequence $\{x_n\}_1^\infty \subset \mathfrak{H}$ such that

$$(4.4) \qquad \|x_n\|_J = 1 \quad (n = 1, 2, \ldots), \qquad \|Ax_n - \lambda x_n\|_J \to 0 \quad (n \to \infty),$$

$$(4.5) \qquad (x_n, x_n) \to \alpha, \quad (Ax_n, x_n) \to \lambda\alpha \quad (n \to \infty),$$

where J is a fixed fundamental symmetry and $\alpha \in \boldsymbol{R}$. According to the same proof $\alpha \geqq 0$ implies $\lambda \geqq 0$, while $\alpha < 0$ yields $\lambda \leqq 0$. In particular, $\sigma_r(A) = \emptyset$ (cf. Theorem VI.6.1 c)). It remains to show that $\lambda \geqq -\mu(T)$ even in the case $\alpha < 0$.

Let

$$(4.6) \qquad \alpha < 0, \quad \lambda < -\mu(T).$$

From (4.3)—(4.5) we obtain:

$$(T^*Tx_n, T^*Tx_n) = (Ax_n, Ax_n - \lambda x_n) + (Ax_n, \lambda x_n) + \\ + 2\mu(T)(Ax_n, x_n) + \mu^2(T)(x_n, x_n) \to \alpha(\lambda + \mu(T))^2.$$

Therefore, in view of (4.6),

$$(4.7) \qquad \lim_{n\to\infty} (T^*Tx_n, T^*Tx_n) < 0.$$

On the other hand, from (4.3) and (4.5)

$$(Tx_n, Tx_n) = (Ax_n, x_n) + \mu(T)(x_n, x_n) \to \alpha(\lambda + \mu(T)),$$

so that (cf. (4.6))

$$\lim_{n\to\infty} (Tx_n, Tx_n) > 0. \tag{4.8}$$

Furthermore, Theorem II.8.1 applied to T^* (which is a plus-operator by assumption) yields

$$(T^*Tx_n, T^*Tx_n) \geqq \mu(T^*)(Tx_n, Tx_n) \qquad (n = 1,2, \ldots),$$

where

$$\mu(T^*) = \inf_{(x,x)=1} (T^*x, T^*x) \geqq 0.$$

Hence, on account of (4.8),

$$(T^*Tx_n, T^*Tx_n) \geqq 0 \qquad (n \geqq n_0).$$

This, however, contradicts (4.7). □

In connection with Theorem 4.2 it should be noted that the adjoint of a plus-operator need not be a plus-operator.

Example 4.3. Let $\mathfrak{H} = \mathfrak{H}^+(\dot{+})\mathfrak{H}^-$, $\mathfrak{H}^+ \subset \mathfrak{P}^{++}$, $\mathfrak{H}^- \subset \mathfrak{P}^{--}$, $\dim \mathfrak{H}^+$ countably infinite, $\dim \mathfrak{H}^- = 1$. Let $\{e_j\}_1^\infty$ and $\{f\}$ be complete orthonormal systems in $\mathfrak{H}^+$ and $\mathfrak{H}^-$, respectively. Define a linear operator $T \in \mathfrak{B}(\mathfrak{H})$ by the relations $Te_j = e_{j+1}$ $(j = 1,2, \ldots)$, $Tf = e_1$. Then T is a plus-operator, but T^* carries the positive e_1 into the negative $-f$.

5. Strict Plus-operators

We still assume that $\varkappa(\mathfrak{H}) > 0$.

According to the definition given in Section II.8, the plus-operator T is said to be strict if the value $\mu(T)$ (see (4.1)) is positive.

Lemma 5.1. *Let $T \in \mathfrak{B}(\mathfrak{H})$ be a strict plus-operator. If $\mathfrak{L}$ is a positive (positive definite, uniformly positive) subspace of $\mathfrak{H}$, so is $T\mathfrak{L}$.*

Proof. For a positive or positive definite $\mathfrak{L}$ the assertion directly follows from (4.2). Next let $\mathfrak{L}$ be uniformly positive:

$$(x, x) \geqq \alpha \|x\|_J^2 \qquad (x \in \mathfrak{L})$$

for a fundamental symmetry J and positive number α. Then, again by (4.2),

$$(Tx, Tx) \geqq \mu(T)(x, x) \geqq \mu(T)\alpha\|x\|_J^2 \geqq \frac{\mu(T)\alpha}{\|T\|_J^2}\|Tx\|_J^2$$

for every $x \in \mathfrak{L}$. □

Theorem 5.2. *If $T \in \mathfrak{B}(\mathfrak{H})$ is a strict plus-operator and J is a fundamental symmetry on $\mathfrak{H}$, then there exists a number $\delta > 0$ such that*

$$\|Tx\|_J \geqq \delta \|x\|_J \qquad (x \in \mathfrak{P}^+) . \tag{5.1}$$

Proof. The operator $A = T^*T - \mu I$, where $\mu = \mu(T) > 0$, is positive (cf. (4.1) and (4.2)). Let $e \in \mathfrak{P}^+$, $\|e\|_J = 1$, $x \in \mathfrak{H}$. Lemma I.2.2 applied to the A-inner product $(.\,,.)_A$ yields:

$$|(Ax, e)|^2 \leqq (Ax, x)(Ae, e) \leqq [(Tx, Tx) - \mu(x, x)](Te, Te) \leqq$$
$$\leqq (\|T\|_J^2 + \mu)\, \|x\|_J^2\, \|Te\|_J^2 .$$

Setting $x = JAe$ we obtain:

$$\|Ae\|_J^2 \leqq (\|T\|_J^2 + \mu)\, \|Te\|_J^2 .$$

On the other hand,

$$\|Ae\|_J \geqq \mu\|e\|_J - \|T^*Te\|_J \geqq \mu - \|T^*\|_J\, \|Te\|_J .$$

As a result,

$$\mu - \|T^*\|_J\, \|Te\|_J \leqq (\|T\|_J^2 + \mu)^{1/2}\, \|Te\|_J ,$$

i.e.,

$$\|Te\|_J \geqq \frac{\mu}{(\|T\|_J^2 + \mu)^{1/2} + \|T^*\|_J} . \qquad \square$$

Corollary 5.3. *If $T \in \mathfrak{B}(\mathfrak{H})$ is a strict plus-operator and $\mathfrak{L}$ is a closed positive subspace of $\mathfrak{H}$, then $T\mathfrak{L}$ is closed.* $\square$

Theorem 5.4. *If $T \in \mathfrak{B}(\mathfrak{H})$ is a strict plus-operator, then the null space $\mathfrak{N}(T)$ is uniformly negative.*

Proof. The assumption ensures the existence of a number $\mu > 0$ such that the operator $A = T^*T - \mu I$ is positive (cf. (4.1) and (4.2)). Let us fix a fundamental symmetry J and an element $x \in \mathfrak{H}$. Setting $JAx = y$, with the help of Lemma I.2.2 (applied to the A-inner product) and Lemma II.11.4 we obtain:

$$\|Ax\|_J^4 = (Ax, Ax)_J^2 = (Ax, y)^2 \leqq (Ax, x)(Ay, y) \leqq$$
$$\leqq (Ax, x)\, \|A\|_J\, \|y\|_J^2 = (Ax, x)\, \|A\|_J\, \|Ax\|_J^2 .$$

Consequently,

$$\|Ax\|_J^2 \leqq \|A\|_J\, (Ax, x) \qquad (x \in \mathfrak{H}) . \tag{5.2}$$

If $x \in \mathfrak{N}(T)$, $x \neq 0$, then $Ax = -\mu x$ (in particular, $A \neq 0$), so that from (5.2)

$$(x, x) \leqq -\frac{\mu}{\|A\|_J}\, \|x\|_J^2 . \qquad \square$$

The following theorem yields a classification of strict plus-operators $T \in \mathfrak{B}(\mathfrak{H})$.

Theorem 5.5. *Let $T \in \mathfrak{B}(\mathfrak{H})$ be a strict plus-operator. Then for every maximal positive subspace $\mathfrak{L} \subset \mathfrak{H}$ the plus-defect $d^+(T\mathfrak{L})$ is the same.*

Proof. By Theorem V.4.1 and Corollary 5.3 $d^+(T\mathfrak{L})$ exists.
Consider a fundamental decomposition

$$(5.3) \qquad \mathfrak{H} = \mathfrak{H}^+ (\dot{+}) \mathfrak{H}^-; \qquad \mathfrak{H}^+ \subset \mathfrak{P}^{++}, \qquad \mathfrak{H}^- \subset \mathfrak{P}^{--},$$

the corresponding fundamental projectors P^+, P^-, and the fundamental symmetry $J = P^+ - P^-$. Put $x^+ = P^+x$, $x^- = P^-x$ for every $x \in \mathfrak{H}$. Let

$$\begin{pmatrix} T_{11} & T_{12} \\ T_{21} & T_{22} \end{pmatrix}$$

be the operator matrix of T with respect to (5.3), the indices 1, 2 referring to the components $\mathfrak{H}^+$ and $\mathfrak{H}^-$, respectively. Then

$$(5.4) \qquad P^+Tx = T_{11}x^+ + T_{12}x^- \qquad (x \in \mathfrak{H}).$$

Define

$$(5.5) \qquad \mathfrak{K} = \mathfrak{K}(\mathfrak{H}^+, \mathfrak{H}^-) = \{K \in \mathfrak{B}(\mathfrak{H}^+, \mathfrak{H}^-): \quad \|K\|_J \leqq 1\}.$$

In view of Theorems II.11.7 and V.4.2 a subspace $\mathfrak{L} \subset \mathfrak{H}$ is maximal positive if and only if it can be written in the form

$$(5.6) \qquad \mathfrak{L} = \{x^+ + Kx^+: \quad x^+ \in \mathfrak{H}^+\},$$

where $K \in \mathfrak{K}$. From (5.4) and (5.6) we obtain:

$$(5.7) \qquad P^+T\mathfrak{L} = \{(T_{11} + T_{12}K)x^+: \quad x^+ \in \mathfrak{H}^+\}.$$

Thus we have to prove (cf. Lemma V.4.6) that

$$(5.8) \qquad \operatorname{codim}_{\mathfrak{H}^+}(T_{11} + T_{12}K)\mathfrak{H}^+ = \operatorname{codim}_{\mathfrak{H}^+}(T_{11} + T_{12}K')\mathfrak{H}^+ \qquad (K, K' \in \mathfrak{K}).$$

Since, however, $\mathfrak{K}$ is convex, it is sufficient to prove (5.8) for $\|K' - K\|_J$ smaller than some $\varepsilon > 0$. We also may assume that $T_{12} \neq 0$.

Theorem 5.2 and the relation $\mathfrak{L} \subset \mathfrak{P}^+$ yield $\|Tx\|_J \geqq \delta \|x\|_J$ $(x \in \mathfrak{L})$, where the number $\delta > 0$ is independent of $\mathfrak{L}$. Making use of the inequality $(x, x) \geqq 0$ once more, we derive:

$$\|P^+Tx\|_J \geqq \frac{\delta}{\sqrt{2}} \|x\|_J \geqq \frac{\delta}{\sqrt{2}} \|x^+\|_J \qquad (x \in \mathfrak{L}).$$

Therefore, on account of (5.7),

$$(5.9) \qquad \|(T_{11} + T_{12}K)x^+\|_J \geqq \frac{\delta}{\sqrt{2}} \|x^+\|_J \qquad (x^+ \in \mathfrak{H}^+).$$

Set $B = T_{11} + T_{12}K$ and define a linear operator $B^{(-1)}$ on $\mathfrak{H}^+$ by the relations

$$B^{(-1)}y = B^{-1}y \qquad (y \in \mathfrak{R}(B)),$$
$$B^{(-1)}y = 0 \qquad (y \in \mathfrak{H}^+, \ y \perp_J \mathfrak{R}(B)).$$

For every $K' \in \mathfrak{K}$ we have

$$T_{11} + T_{12}K' = [I + T_{12}(K' - K)B^{(-1)}]B. \tag{5.10}$$

But (5.9) implies $\|B^{(-1)}\|_J \leqq \frac{\sqrt{2}}{\delta}$. Consequently, if

$$\|K' - K\|_J < \frac{\delta}{\sqrt{2}\,\|T_{12}\|_J} \tag{5.11}$$

then $\|T_{12}(K' - K)B^{(-1)}\|_J < 1$, so that in (5.10) B is multiplied by a completely invertible operator of the space $\mathfrak{H}^+$. In other words, under the assumption (5.11) there exists a homeomorphic isomorphism between $\mathfrak{H}^+$ and $\mathfrak{H}^+$ which maps $(T_{11} + T_{12}K)\mathfrak{H}^+$ onto $(T_{11} + T_{12}K')\mathfrak{H}^+$. This proves (5.8) for K, K' satisfying (5.11). □

Theorem 5.5 justifies the following definition.

Let $T \in \mathfrak{B}(\mathfrak{H})$ be a strict plus-operator. The cardinal number $d^+(T\mathfrak{L})$, where $\mathfrak{L}$ is any maximal positive subspace of the Krein space $\mathfrak{H}$, will be called the *plus-defect* of the operator T and denoted by $d^+(T)$.

Below we present two simple summation rules for plus-defects.

Theorem 5.6. *If $T \in \mathfrak{B}(\mathfrak{H})$ is a strict plus-operator and $\mathfrak{L}$ is a closed positive subspace of $\mathfrak{H}$, then $d^+(T\mathfrak{L}) = d^+(T) + d^+(\mathfrak{L})$.*

Proof. Choosing two maximal positive subspaces $\mathfrak{L}_1, \mathfrak{M} \subset \mathfrak{H}$ such that $\mathfrak{L} \subset \mathfrak{L}_1$, $T\mathfrak{L} \subset \mathfrak{M}$ we find:

$$\operatorname{codim}_{\mathfrak{M}} T\mathfrak{L} = \operatorname{codim}_{\mathfrak{M}} T\mathfrak{L}_1 + \operatorname{codim}_{T\mathfrak{L}_1} T\mathfrak{L}.$$

But the last term is equal to $\operatorname{codim}_{\mathfrak{L}_1}\mathfrak{L}$, since by Theorem 5.2 T is a homeomorphic isomorphism between $\mathfrak{L}_1$ and $T\mathfrak{L}_1$. □

Theorem 5.7. *If $T_1, T_2 \in \mathfrak{B}(\mathfrak{H})$ are strict plus-operators, so is T_1T_2. Moreover, $d^+(T_1T_2) = d^+(T_1) + d^+(T_2)$.*

Proof. From the relations

$$(T_2x, T_2x) \geqq \mu_2 > 0 \quad \text{if} \quad (x, x) = 1,$$
$$(T_1T_2x, T_1T_2x) \geqq \mu_1 > 0 \quad \text{if} \quad (T_2x, T_2x) = 1$$

it follows that

$$(T_1T_2x, T_1T_2x) \geqq \mu_1\mu_2 > 0 \quad \text{if } (x, x) = 1.$$

Hence T_1T_2 is a strict plus-operator. Let $\mathfrak{L}$ be a maximal positive subspace of $\mathfrak{H}$. Then, owing to Theorem V.4.1 and Corollary 5.3, $T_2\mathfrak{L}$ is closed. Consequently, we may apply Theorem 5.6:

$$d^+(T_1T_2) = d^+(T_1T_2\mathfrak{L}) = d^+(T_1) + d^+(T_2\mathfrak{L}) = \\ = d^+(T_1) + d^+(T_2). \quad \square$$

6. Doubly Strict Plus-operators

A linear operator $T \in \mathfrak{B}(\mathfrak{H})$ is called a *doubly strict plus-operator* provided both T and T^* are strict plus-operators. In particular, the underlying Krein space is required to be indefinite (cf. Sections 4—5 and II.8).

Example 4.3 exhibits a strict plus-operator whose adjoint is not even a plus-operator.

The following result characterizes doubly strict plus-operators among strict plus-operators.

Theorem 6.1. *The strict plus-operator $T \in \mathfrak{B}(\mathfrak{H})$ is doubly strict if and only if $d^+(T) = 0$.*

Proof. Assume that the value

$$\mu(T^*) = \inf_{(x,x)=1} (T^*x, T^*x)$$

is positive. Let $\mathfrak{L} \subset \mathfrak{H}$ be a maximal positive subspace. Then $T\mathfrak{L}$ is positive and closed (cf. Theorem V.4.1 and Corollary 5.3). For $y \perp T\mathfrak{L}$ we have $T^*y \perp \mathfrak{L}$; hence by Lemma I.6.3 $(T^*y, T^*y) \leqq 0$, so that in view of Theorem II.8.1

$$(y, y) \leqq \frac{1}{\mu(T^*)} (T^*y, T^*y) \leqq 0 .$$

Thus, on account of Lemma V.4.5, $T\mathfrak{L}$ is maximal positive, i.e., $d^+(T) = 0$.

Let, conversely, $d^+(T) = 0$. Suppose that $(T^*x, T^*x) < 0$ for some $x \in \mathfrak{H}$. Then, owing to Theorems V.5.6 and V.7.1, there exists a maximal uniformly positive subspace $\mathfrak{L}$ which is orthogonal to T^*x. The relation $T^*x \perp \mathfrak{L}$ implies $x \perp T\mathfrak{L}$. On the other hand, in view of Corollary V.7.2, the assumption $d^+(T) = 0$ and Lemma 5.1, the subspace $T\mathfrak{L}$ is maximal uniformly positive. Thus Lemma I.6.6 yields $(x, x) < 0$. Hence T^* is a plus-operator.

Let us prove that the plus-operator T^* is strict. According to Corollary II.8.3 we may confine ourselves to the case where $\mathfrak{R}(T^*) \subset \mathfrak{P}^+$ i.e. (see Lemma II.11.4)

$$\overline{\mathfrak{R}(T^*)} \subset \mathfrak{P}^+ . \tag{6.1}$$

Since $\mathfrak{N}(T)$ is closed and, by Theorem 5.4, uniformly negative, from Theorems VI.2.6 d) and V.5.2 we obtain:

$$\mathfrak{N}(T)\ (\dotplus)\ \overline{\mathfrak{R}(T^*)} = \mathfrak{H}\,. \tag{6.2}$$

Thus, in view of (6.1) and Lemma I.6.2, the subspace $\overline{\mathfrak{R}(T^*)}$ is maximal positive; being ortho-complemented (see (6.2)), Theorem V.5.2 assures that it is uniformly positive as well.

On account of (6.2) we have:

$$\mathfrak{R}(T) = T\mathfrak{H} = T\overline{\mathfrak{R}(T^*)}\,. \tag{6.3}$$

Therefore, owing to the assumption $d^+(T) = 0$ and Lemma 5.1, $\mathfrak{R}(T)$ is maximal positive and uniformly positive. In particular, $\mathfrak{R}(T)$ is closed (Theorem V.4.1) and, by Theorem VI.2.9, so is $\mathfrak{R}(T^*)$.

Since $\mathfrak{R}(T)$ is maximal uniformly positive, Theorems V.7.1 and VI.2.6 c) yield the decomposition

$$\mathfrak{R}(T)\ (\dotplus)\ \mathfrak{N}(T^*) = \mathfrak{H}\,, \tag{6.4}$$

where $\mathfrak{N}(T^*)$ is maximal uniformly negative. Thus every $x \in \mathfrak{H}$ can be written in the form

$$x = x^+ + x^-;\quad x^+ \in \mathfrak{R}(T),\ x^- \in \mathfrak{N}(T^*)\,.$$

Relying on the uniform positiveness of $\mathfrak{R}(T^*)$ (see (6.1)—(6.2)) we find

$$(T^*x, T^*x) = (T^*x^+, T^*x^+) \geqq \alpha\|T^*x^+\|_J^2\,,$$

where J is a fundamental symmetry on $\mathfrak{H}$ and $\alpha > 0$ is independent of x. On the other hand, the continuous operator $T^*|\mathfrak{R}(T)$ is invertible (see (6.4)) and has closed range $\mathfrak{R}(T^*)$; hence, according to the closed graph principle and Lemma II.11.4,

$$\|T^*x^+\|_J^2 \geqq \beta\|x^+\|_J^2 \geqq \beta(x^+, x^+) \geqq \beta(x, x)$$

with $\beta > 0$ independent of x. As a result,

$$(T^*x, T^*x) \geqq \alpha\beta(x, x) \qquad (x \in \mathfrak{H})\,.$$

So T^* is a strict plus-operator. □

From Lemma II.10.2 and Theorems 5.2, 6.1 we derive:

Corollary 6.2. *In a quasi-negative Krein space $\mathfrak{H}$ every strict plus-operator $T \in \mathfrak{B}(\mathfrak{H})$ is doubly strict.* □

The next result should be compared with Theorem II.8.9.

Theorem 6.3. *Let* $\varkappa(\mathfrak{H}) > 0$. *For an operator* $T \in \mathfrak{B}(\mathfrak{H})$ *the following assertions* a)—c) *are equivalent:*

a) $T\mathfrak{P}^+ = \mathfrak{P}^+$;

b) $T\mathfrak{P}^{++} = \mathfrak{P}^{++}$;

c) $\mathfrak{N}(T)$ *is uniformly negative, and* $T|\mathfrak{N}(T)^\perp$ *is a positive number times an invertible isometric operator with range* $\mathfrak{H}$.

If a), b) *or* c) *holds, then* T *is a doubly strict plus-operator.*

Proof. Suppose that a) is valid. Then, according to Theorem II.8.6, T is a strict plus-operator. Hence, by Theorem 5.4, $\mathfrak{N}(T)$ is uniformly negative. In particular, $\mathfrak{N}(T)$ is ortho-complemented (cf. Theorem V.5.2).

Introduce the notations

$$\mathfrak{N}(T)^\perp = \mathfrak{H}_1, \qquad \mathfrak{H}_1 \cap \mathfrak{P}^+ = \mathfrak{P}_1^+, \qquad T|\mathfrak{H}_1 = T_1 .$$

Obviously, T_1 is invertible. We claim that $T_1\mathfrak{P}_1^+ = \mathfrak{P}^+$. The inclusion $T_1\mathfrak{P}_1^+ \subset \mathfrak{P}^+$ is clear. On the other hand, $y \in \mathfrak{P}^+$ implies $y = Tx$ for some $x \in \mathfrak{P}^+$; decomposing x as

$$x = x_0 + x_1; \quad x_0 \in \mathfrak{N}(T), \quad x_1 \in \mathfrak{H}_1 , \tag{6.5}$$

we find the relations

$$Tx = T_1 x_1 , \tag{6.6}$$

$$(x_1, x_1) = (x, x) - (x_0, x_0) \geqq (x, x) \geqq 0 . \tag{6.7}$$

Furthermore, $\mathfrak{H}_1$ is indefinite. Namely $\mathfrak{H}_1 \subset \mathfrak{P}^-$ along with Lemma I.3.1 yields $\mathfrak{H} = \mathfrak{H}_1(\dot{+})\mathfrak{N}(T) \subset \mathfrak{P}^-$, contrary to the assumption $\varkappa(\mathfrak{H}) > 0$. If, in turn, $\mathfrak{H}_1 \subset \mathfrak{P}^+$, then $\mathfrak{R}(T) = T\mathfrak{H}_1 \subset \mathfrak{P}^+$ or, making use of a) once more, $\mathfrak{R}(T) = \mathfrak{P}^+$, which is impossible by Corollary I.2.7. Applying Theorem II.8.10 to T_1 we obtain c).

That b) implies c) can be established similarly.

Let c) be satisfied. Then $T_1\mathfrak{P}_1^+ = \mathfrak{P}^+$, so that $T\mathfrak{P}^+ \supset \mathfrak{P}^+$. On the other hand, for $x \in \mathfrak{P}^+$ we have (cf. (6.5) — (6.7)) $Tx \in T_1\mathfrak{P}_1^+ = \mathfrak{P}^+$. Thus a) holds. The assertion b) can be deduced from c) in just the same manner.

As already mentioned, every operator T with property a) is a strict plus-operator. Moreover, from c) it follows that T carries every fundamental decomposition of $\mathfrak{N}(T)^\perp$ into a fundamental decomposition of $\mathfrak{H}$. Hence $d^+(T) = 0$ and, consequently, T is doubly strict (cf. Theorem 6.1). □

Notes to Chapter VII

Weighted integral operators which are positive in an indefinite space $\mathfrak{H}$ have been examined, without explicit use of the indefinite inner product, by Kreĭn [1] in 1940. In this special case Theorems 1.2—1.3 are due to him. The general concept of positive operator in a finite-dimensional $\mathfrak{H}$ belongs to Potapov [1], and in an arbitrary Krein space to Ginzburg [2], who proved Theorem 1.3. Using more efficient tools, Kreĭn and Šmul'jan [2] showed that the eigenspaces appearing in Theorem 1.2 are uniformly definite. Example 1.5 was proposed by Langer (personal communication).

Other possible extensions of the concept of positive operator in Hilbert space, for the special case $\varkappa^+(\mathfrak{H}) < \infty$, have been investigated by Bognár [1], [3] — [6]. Theorem 2.1 for a finite-dimensional $\mathfrak{H}$ belongs to Potapov [1], and for arbitrary $\mathfrak{H}$ to Bognár and Krámli [1].

Ginzburg [2] proved that every selfadjoint $A \in \mathfrak{B}(\mathfrak{H})$ with positive spectrum admits one and only one square root of the same type. Another sufficient condition for the existence of a selfadjoint square root was given by Jalava [3].

Theorem 3.1 is due to Kühne [1].

Sections 4—6 are mainly based on two papers of Kreĭn and Šmul'jan [1], [2]. Theorems 5.6—5.7 are borrowed from Šmul'jan [2]. Special cases of Theorem 4.2 have been obtained by Potapov [1] and Ginzburg [2]; our argument follows the latter author. The present proof of Theorem 5.4 is due to Iohvidov [14]. Our reasoning in the second half of the proof of Theorem 6.1 (cf. Bognár [9]) seems to have methodical advantage over the original version; however, we miss the corollary $\mu(T^*) = \mu(T)$. For $\dim\mathfrak{H} < \infty$ Corollary 6.2 belongs to Potapov [1].

According to Theorem 6.1, a doubly strict plus-operator T carries every maximal positive subspace $\mathfrak{L}$ into a maximal positive subspace $T\mathfrak{L}$. Introducing a fundamental decomposition (5.3) of $\mathfrak{H}$, the corresponding matrix $(T_{jl})_{j,l=1,2}$ of T, and the angular operators $K = K^+(\mathfrak{L})$, $\Phi_T(K) = K^+(T\mathfrak{L})$, one easily verifies that $T_{11} + T_{12}K$ is completely invertible on $\mathfrak{H}^+$ and

$$\Phi_T(K) = (T_{21} + T_{22}K)(T_{11} + T_{12}K)^{-1} \qquad (K \in \mathfrak{K}(\mathfrak{H}^+, \mathfrak{H}^-)) .$$

Thus Φ_T is a linear fractional transformation of the "operator unit ball" $\mathfrak{K}(\mathfrak{H}^+, \mathfrak{H}^-)$ to itself. The geometric properties of Φ_T for doubly strict plus-operators as well as for another class of operators T have been studied in a paper of Kreĭn and Šmul'jan [3] and in subsequent works of Šmul'jan [5]—[7]. In the same context Šmul'jan [4] considered operators of the form T^*T, where T is a doubly strict plus-operator. For a unitary T Kreĭn and Šmul'jan [3] found the general form of the matrix (T_{jl})

(in the special case $\varkappa(\mathfrak{H}) < \infty$ see Iohvidov and Kreĭn [1]). Helton [2] examined the iterates of Φ_T.

Conditions for the continuity or closedness of plus-operators in Banach spaces with a J-metric have been given by Iohvidov [14].

Operators $T \in \mathfrak{B}(\mathfrak{H})$ with the property that $(Tx, Tx) \geqq (x, x)$ (resp. $(Tx, Tx) \leqq (x, x)$) for every x may be called *augmenting* (*diminishing*) operators.

Evidently, an operator is augmenting in $\mathfrak{H}$ if and only if it is diminishing in the anti-space of $\mathfrak{H}$. Furthermore, an operator $T \in \mathfrak{B}(\mathfrak{H})$ is a strict plus-operator if and only if it is a non-zero multiple of an augmenting operator. Hence many results can equally well be established for augmenting (or diminishing) operators and strict plus-operators, as displayed by the history of the subject. This is, however, not always the case; for instance, the divisibility conditions obtained by Šmul'jan [1], [2] are better suited to augmenting operators.

It can be shown that the formula

$$C = \Gamma(D) = (P^- - P^+D)(P^+ - P^-D)^{-1}$$

defines a one-to-one correspondence between a subclass of diminishing operators $D \in \mathfrak{B}(\mathfrak{H})$ and a subclass of operators $C \in \mathfrak{B}(\mathfrak{H})$ having the property $\|C\|_J \leqq 1$, where $J = P^+ - P^-$. The linear fractional transformation Γ (which has nothing to do with Φ_T above) has been investigated by Potapov [1], Ginzburg [2], and Iohvidov [12], [15].

Analytic functions whose values are diminishing operators in $\mathfrak{H}$ have first been studied by Potapov [1] (finite-dimensional $\mathfrak{H}$) and Ginzburg [1] (arbitrary $\mathfrak{H}$). For an idea of how these functions intervene in the theory of characteristic operator functions see e.g. V. M. Brodskiĭ [1]. Concerning operator valued analytic functions cf. also Šmul'jan [8].

Chapter VIII. Invariant Semi-definite Subspaces of Linear Operators in Krein Spaces

The existence of invariant maximal uniformly definite (Section 1) resp. maximal semi-definite (Sections 2—3) subspaces is treated. In Sections 4—5 an application to quadratic pencils of Hilbert space operators is given. Section 6 contains the statement of some results concerning the spectral function of a positive or positizable selfadjoint operator in a Krein space. Theorem 2.1 is the culminating point of the book. Theorems 1.2, 1.4, 3.1, 3.2 and 5.5 should also be mentioned.

1. Fundamentally Reducible Operators

Let T be a linear operator in the Krein space $\mathfrak{H}$. If some fundamental decomposition of $\mathfrak{H}$ reduces T, we say T is *fundamentally reducible.*

The study of a fundamentally reducible operator is equivalent to the treatment of two Hilbert space operators.

For simplicity, we shall consider continuous operators only.

Lemma 1.1. *If a fundamental decomposition of* $\mathfrak{H}$ *reduces the operator* $T \in \mathfrak{B}(\mathfrak{H})$, *then the corresponding fundamental projectors* P^+, P^- *and fundamental symmetry* $J = P^+ - P^-$ *commute with* T.

Proof. The assumption yields $TP^+ = P^+TP^+$, $TP^- = P^-TP^-$. From the latter equation, substituting $P^- = I - P^+$, we obtain $P^+T = P^+TP^+$. Thus $P^+T = TP^+$. It follows that P^- and J also commute with T. □

Theorem 1.2. *The operator* $T \in \mathfrak{B}(\mathfrak{H})$ *is fundamentally reducible if and only if there exists a uniformly positive operator commuting with* T.

Proof. If T is fundamentally reducible then, according to Lemma 1.1, there exists a fundamental symmetry J that commutes with T. But J is uniformly positive. Thus the condition is necessary.

Suppose, conversely, that $AT = TA$, where the operator $A \in \mathfrak{B}(\mathfrak{H})$ is uniformly positive. Replacing the inner product $(.\,,.)$ by the A-inner product $(.\,,.)_A$, the Krein space $\mathfrak{H}$ turns into a Hilbert space $\mathfrak{H}_A$ whose strong topology coincides with $\tau_M(\mathfrak{H})$. Obviously, A is a selfadjoint operator on $\mathfrak{H}_A$. In addition, by Lemma VII.3.2, A is completely invertible: $\mathfrak{N}(A) = 0$, $\mathfrak{R}(A) = \mathfrak{H} = \mathfrak{H}_A$.

Working in the space $\mathfrak{H}_A$, set

(1.1) $$\mathfrak{H}^+ = \mathfrak{N}(A - |A|_A), \quad \mathfrak{H}^- = \mathfrak{N}(A + |A|_A),$$

where $|A|_A$ denotes the unique A-positive square root of A^2. In view of Lemma VI.8.3 we have:

(1.2) $$\mathfrak{H}^+ \dotplus \mathfrak{H}^- = \mathfrak{H}_A = \mathfrak{H},$$

(1.3) $$(x, y)_A = 0 \quad (x \in \mathfrak{H}^+, \ y \in \mathfrak{H}^-).$$

Moreover, since the subspaces (1.1) are invariant under A (cf. Lemma VI.8.1), the relations $A|\mathfrak{H}^\pm = \pm|A|_A|\mathfrak{H}^\pm$ and $\mathfrak{N}(A) = 0$ with the aid of Lemma VII.1.1 yield:

(1.4) $$(Ax^+, x^+)_A > 0 \quad (x^+ \in \mathfrak{H}^+, \ x^+ \neq 0),$$

(1.5) $$(Ax^-, x^-)_A < 0 \quad (x^- \in \mathfrak{H}^-, \ x^- \neq 0).$$

Returning to the original inner product $(.\,,.)$ and taking into account that $A\mathfrak{H} = \mathfrak{H}$ implies $A\mathfrak{H}^\pm = \mathfrak{H}^\pm$, from (1.3) — (1.5) we obtain:

$$\mathfrak{H}^+ \perp \mathfrak{H}^-, \quad \mathfrak{H}^+ \subset \mathfrak{P}^{++}, \quad \mathfrak{H}^- \subset \mathfrak{P}^{--}.$$

In other words, (1.2) is a fundamental decomposition of $\mathfrak{H}$. Finally, by the assumption and Lemma VI.8.1, this decomposition reduces T. □

Theorem 1.2 and Theorem VII.3.1 yield:

Corollary 1.3. *Let $\varkappa(\mathfrak{H}) > 0$ and $T \in \mathfrak{B}(\mathfrak{H})$. If there exists a selfadjoint operator $A \in \mathfrak{B}(\mathfrak{H})$ commuting with T and satisfying the inequality $(Ax, x) \geqq \alpha\|x\|_J^2$ with a fundamental symmetry J and positive number α for every $x \in \mathfrak{P}^+$, then T is fundamentally reducible.* □

Next we apply Theorem 1.2 to unitary operators of $\mathfrak{H}$. Loosely speaking, our result says that a unitary operator similar to a Hilbert space unitary operator is fundamentally reducible.

Theorem 1.4. *The unitary operator U on the Krein space $\mathfrak{H}$ is fundamentally reducible if and only if there exist a completely invertible operator T and a fundamental symmetry J on $\mathfrak{H}$ such that TUT^{-1} is J-unitary (i.e., unitary relative to the J-inner product).*

Proof. If U is fundamentally reducible then, by Lemma 1.1, a certain fundamental symmetry J commutes with U. For this J we have:

$$(JUx, Uy) = (UJx, Uy) = (Jx, y) \qquad (x, y \in \mathfrak{H}).$$

Thus, choosing $T = I$, the operator TUT^{-1} is J-unitary.

Assume, conversely, that the operator

$$V = TUT^{-1}$$

is J-unitary, where J denotes a fundamental symmetry and T is completely invertible. Set

$$A = T^*JT\,.$$

Then $A \in \mathfrak{B}(\mathfrak{H})$ and, in view of Theorem VI.2.3 c), $A^* = A$. Moreover, A is uniformly positive, since T^{-1} is continuous and $(Ax, x) = \|Tx\|_J^2$. Finally, applying Theorem VI.4.2 and Lemma VI.2.1 we obtain:

$$AU = T^*JTU = UU^*T^*JTU = UT^*V^*JVT =$$
$$= UT^*JJV^*JVT = UT^*JT = UA\,.$$

Therefore (see Theorem 1.2) U is fundamentally reducible. □

2. Invariant Positive Subspaces of Plus-operators

Fundamental reducibility, at least for operators defined throughout $\mathfrak{H}$, means the existence of two, mutually orthogonal, invariant subspaces: one maximal uniformly positive and one maximal uniformly negative (cf. Theorem V.7.1).

A weaker, yet useful property can be obtained by omitting the words "uniformly". In the next section we shall verify this weaker property for an important subclass of unitary and selfadjoint operators. The proof will be based on a related result concerning plus-operators that we are going to treat just now.

Theorem 2.1. *Consider a Krein space $\mathfrak{H}$ and a plus-operator $T \in \mathfrak{B}(\mathfrak{H})$. Suppose that for some fundamental decomposition*

(2.1) $$\mathfrak{H} = \mathfrak{H}^+(\dot{+})\mathfrak{H}^-; \quad \mathfrak{H}^+ \subset \mathfrak{P}^{++}, \quad \mathfrak{H}^- \subset \mathfrak{P}^{--}$$

and corresponding fundamental projectors P^+, P^- the operator P^+TP^- is compact. Then every closed positive subspace $\mathfrak{L} \subset \mathfrak{H}$ with the property $\overline{T\mathfrak{L}} = \mathfrak{L}$ can be extended to a maximal positive subspace $\mathfrak{L}_1 \subset \mathfrak{H}$ such that $T\mathfrak{L}_1 \subset \mathfrak{L}_1$.

Proof. Let $\mathfrak{L}_1$ be any maximal positive subspace containing $\mathfrak{L}$, and let $\mathfrak{M}_1$ be any maximal positive subspace containing $T\mathfrak{L}_1$ (cf. Section I.6). Then, in view of Theorem V.4.1,

$$\mathfrak{M}_1 = \overline{\mathfrak{M}_1} \supset \overline{T\mathfrak{L}_1} \supset \overline{T\mathfrak{L}} = \mathfrak{L}\,.$$

Denoting the angular operators relative to $\mathfrak{H}^+$ of $\mathfrak{L}$, $\mathfrak{L}_1$ and $\mathfrak{M}_1$ (cf. Theorems II.11.7 and V.4.2) by K, K_1 and M_1, respectively, the identity operator of the space $\mathfrak{H}^+$ by I_1, the orthogonal projector of the space $\mathfrak{H}^+$ onto $P^+\mathfrak{L}$ (cf. Lemma IV.7.1) by P, and setting

(2.2) $$K_1|(I_1 - P)\mathfrak{H}^+ = X\,, \qquad M_1|(I_1 - P)\mathfrak{H}^+ = Y\,,$$

we have:

$$K_1 = KP + X(I_1 - P), \qquad M_1 = KP + Y(I_1 - P). \tag{2.3}$$

In particular,

$$X, Y \in \mathfrak{B}((I_1 - P)\mathfrak{H}^+, \mathfrak{H}^-) \tag{2.4}$$

and

$$\|KP + X(I_1 - P)\|_J \leqq 1, \qquad \|KP + Y(I_1 - P)\|_J \leqq 1, \tag{2.5}$$

where $J = P^+ - P^-$ is the fundamental symmetry belonging to the decomposition (2.1), and $\mathfrak{H}^+$, $\mathfrak{H}^-$ are considered as Hilbert spaces relative to the J-inner product.

Let $x^+ \in \mathfrak{H}^+$. Then, according to Theorems II.11.6 and V.4.2, $x^+ + K_1x^+ \in \mathfrak{L}_1$. On the other hand, if we denote the operator matrix of T with respect to (2.1) by $(T_{jl})_{j,l=1,2}$, where $T_{jl} \in \mathfrak{B}(\mathfrak{H}_l, \mathfrak{H}_j)$, $\mathfrak{H}_1 = \mathfrak{H}^+$, $\mathfrak{H}_2 = \mathfrak{H}^-$, then $T(x^+ + K_1x^+)$ can be written in the form

$$(T_{11}x^+ + T_{12}K_1x^+) + (T_{21}x^+ + T_{22}K_1x^+).$$

Now the assumption $T\mathfrak{L}_1 \subset \mathfrak{M}_1$ yields

$$T_{21}x^+ + T_{22}K_1x^+ = M_1(T_{11}x^+ + T_{12}K_1x^+)$$

or, x^+ being arbitrary,

$$T_{21} + T_{22}K_1 = M_1(T_{11} + T_{12}K_1).$$

Replacing K_1 and M_1 by their expressions (2.3) we obtain:

$$\begin{aligned} &T_{21} + T_{22}(KP + X(I_1 - P)) - \\ &- (KP + Y(I_1 - P))\{T_{11} + T_{12}(KP + X(I_1 - P))\} = 0. \end{aligned} \tag{2.6}$$

Consequently, in order that the maximal positive subspaces $\mathfrak{L}_1$, $\mathfrak{M}_1$ meet the requirements

$$\mathfrak{L} \subset \mathfrak{L}_1, \quad T\mathfrak{L}_1 \subset \mathfrak{M}_1 \tag{2.7}$$

it is necessary that the operators X, Y defined by (2.2) satisfy the relations (2.4)—(2.6).

It is easy to see, performing the steps in reverse order, that this condition is sufficient as well. More exactly, if X and Y obey the relations (2.4)—(2.6), then they define, via (2.3), maximal positive subspaces $\mathfrak{L}_1$, $\mathfrak{M}_1$ satisfying (2.7).

In addition, $\mathfrak{L}_1$ is invariant under T if and only if $\mathfrak{M}_1$ appearing in (2.7) can be chosen equal to $\mathfrak{L}_1$, i.e., if in (2.4) — (2.6) the choice $Y = X$ is possible.

Consider the space

$$\mathfrak{B}_1 = \mathfrak{B}((I_1 - P)\mathfrak{H}^+, \mathfrak{H}^-)$$

with the weak operator topology

$$\tau_{\text{wo}} = \tau_{\text{wo}}\big((I_1 - P)\mathfrak{H}^+, \mathfrak{H}^-\big).$$

Put

$$\mathfrak{A} = \{X \in \mathfrak{B}_1\colon \ \|KP + X(I_1 - P)\|_J \leqq 1\},$$

and define a set-valued mapping Φ on $\mathfrak{A}$ by the formula

$$\Phi(X) = \{Y \in \mathfrak{A}\colon \ (2.6) \text{ holds for } X, Y\} \qquad (X \in \mathfrak{A}).$$

The set $\mathfrak{A}$ is τ_{wo}-closed. In fact, if

$$\|KP + X(I_1 - P)\|_J > 1,$$

then for suitable vectors $f \in \mathfrak{H}^+$, $g \in \mathfrak{H}^-$ and a number $\varepsilon > 0$ we have

$$|(\{KP + X(I_1 - P)\}f, g)| > \|f\|_J \|g\|_J + \varepsilon;$$

thus every operator $X' \in \mathfrak{B}_1$ in the τ_{wo}-neighbourhood of X characterized by the inequality

$$|((X' - X)(I_1 - P)f, g)| < \varepsilon$$

also lies outside $\mathfrak{A}$. On the other hand, $\mathfrak{A}$ is bounded:

(2.8) $$\|X\|_J = \|X(I_1 - P)\|_J \leqq 1 + \|KP\|_J \leqq 2 \qquad (X \in \mathfrak{A}).$$

Therefore (see e.g. Dunford and Schwartz [1; Exercise VI.9.6]) $\mathfrak{A}$ is τ_{wo}-compact. Besides, $\mathfrak{A}$ is easily seen to be convex.

For fixed X the set $\Phi(X)$ is obviously convex and, by the argument leading to equation (2.6), non-empty.

With the aim of proving the τ_{wo}-closedness of the set

(2.9) $$\{\{X, Y\}\colon \ X \in \mathfrak{A}, \quad Y \in \Phi(X)\} \subset \mathfrak{B}_1 \times \mathfrak{B}_1,$$

we will show that the left-hand side of (2.6) is a jointly continuous function of the variables $X, Y \in \mathfrak{A}$ in the sense of the weak operator topologies of the domain space $\mathfrak{B}_1$ and the range space $\mathfrak{B}(\mathfrak{H}^+, \mathfrak{H}^-)$, respectively (a τ_{wo}-continuous function for short).

The only term demanding a closer attention is

$$\Psi(X, Y; T_{12}) = Y(I_1 - P)T_{12}X(I_1 - P).$$

Since, by assumption, $T_{12} = P^+TP^-|\mathfrak{H}^-$ is a compact operator, for every $\delta > 0$ there exists an operator $R_\delta \in \mathfrak{B}(\mathfrak{H}^-, \mathfrak{H}^+)$ of finite rank such that $\|T_{12} - R_\delta\|_J < \delta$ (see e.g. Riesz and Sz.-Nagy [1; Section 85]). Making use of (2.8) we find:

$$\|\Psi(X, Y; T_{12}) - \Psi(X, Y; R_\delta)\|_J < 4\delta \qquad (X, Y \in \mathfrak{A}).$$

Hence it is sufficient to prove the τ_{wo}-continuity of $\Psi(X, Y; R)$, where $R \in \mathfrak{B}(\mathfrak{H}^-, \mathfrak{H}^+)$ is a fixed operator of finite rank. In order to establish the latter property of Ψ, write

$$Rx = \sum_{j=1}^{n}(x, v_j)z_j \qquad (x \in \mathfrak{H}^-);$$

then

$$(\Psi(X, Y; R)f, g) = \sum_{j=1}^{n} (X(I_1 - P)f, v_j)(Y(I_1 - P)z_j, g)$$

$$(f \in \mathfrak{H}^+, \ g \in \mathfrak{H}^-),$$

a polynomial of manifestly τ_{wo}-continuous numerical expressions.

To sum up, $\mathfrak{A}$ is a non-empty compact convex set in the separated locally convex $\mathfrak{B}_1$, the mapping Φ is defined on $\mathfrak{A}$, its values are non-empty convex subsets of $\mathfrak{A}$, and the "graph" (2.9) of Φ is closed. In these circumstances a theorem independently obtained by Fan [1] and Glicksberg [1] (cf. also Berge [1; Sections IX.5 and VI.1]) ensures the existence of a point $X_0 \in \mathfrak{A}$ such that $X_0 \in \Phi(X_0)$. Thus for $Y = X = X_0$ the conditions (2.4)—(2.6) are fulfilled. □

Remark 2.2. In Theorem 2.1 the choice $\mathfrak{L} = 0$ is always possible.

3. Invariant Semi-definite Subspaces of Unitary and Selfadjoint Operators

The invariant subspace theorems, presented below in a weakened form, rank among the most celebrated achievements of the theory of Krein spaces.

Theorem 3.1. *Let U be a unitary operator in the Krein space $\mathfrak{H}$. Suppose that for a fundamental decomposition*

(3.1) $$\mathfrak{H} = \mathfrak{H}^+(\dot{+})\mathfrak{H}^-; \quad \mathfrak{H}^+ \subset \mathfrak{P}^{++}, \quad \mathfrak{H}^- \subset \mathfrak{P}^{--}$$

and corresponding fundamental projectors P^+, P^- the operator P^+UP^- is compact. Let σ_0 denote any set of non-unimodular eigenvalues of U such that $\nu \in \sigma_0$ implies $\nu^* \notin \sigma_0$.

Then U has an invariant maximal positive subspace $\mathfrak{L}_1$ and an invariant maximal negative subspace $\mathfrak{L}_2$, each containing all principal subspaces $\mathfrak{S}_\nu(U)$ with $\nu \in \sigma_0$. For every $\mathfrak{L}_1$ and $\mathfrak{L}_2$ of this kind and every normal eigenvalue $\nu \in \sigma_0$ we have the relations $\mathfrak{S}_{\nu^*}(U) \cap \mathfrak{L}_1 = \mathfrak{S}_{\nu^*}(U) \cap \mathfrak{L}_2 = 0$. One can choose the subspaces $\mathfrak{L}_1$, $\mathfrak{L}_2$ to be the orthogonal companions of each other.

Proof. Denote the span of the subspaces $\mathfrak{S}_\nu(U)$ $(\nu \in \sigma_0)$ by $\mathfrak{L}_0$. According to Corollary II.2.6, Theorem II.2.5 and Lemma I.3.1, $\mathfrak{L}_0$ is a neutral subspace. On the other hand, $\mathfrak{L}_0$ is invariant for U. Moreover, since the relations $\nu \in \sigma_p(U)$, $(U - \nu I)^r x = 0$ imply $\nu \neq 0$, $x = 1/\nu\,(Ux - x_1)$, where $x_1 = (U - \nu I)x$, $(U - \nu I)^{r-1} x_1 = 0$, by recursion we obtain $U\mathfrak{S}_\nu(U) = \mathfrak{S}_\nu(U)$ $(\nu \in \sigma_0)$ i.e. $U\mathfrak{L}_0 = \mathfrak{L}_0$. Applying now Theorem 2.1 to the operator $T = U$ and the subspace $\mathfrak{L} = \overline{\mathfrak{L}_0}$, we find a maximal positive subspace $\mathfrak{L}_1$ which is invariant for U and contains $\mathfrak{L}_0$.

The subspace $\mathfrak{L}_2 = \mathfrak{L}_1^\perp$ is maximal negative (Theorem V.4.4) and invariant for U (Corollaries VI.4.8 and VI.4.6). Since $\mathfrak{L}_0$ is a neutral subspace of the positive subspace $\mathfrak{L}_1$, Lemma I.4.4 assures that $\mathfrak{L}_0 \subset \mathfrak{L}_2$. Theorems V.3.6 and V.4.1 yield the relation $\mathfrak{L}_2^\perp = \mathfrak{L}_1$.

Let $\nu \in \sigma_0$ be a normal eigenvalue of U. As we have just mentioned, $\mathfrak{L}_0 \perp \mathfrak{L}_1$. In particular, $x \in \mathfrak{L}_1$ implies $x \perp \mathfrak{S}_\nu(U)$. Thus, by Theorem VI.5.5, x cannot belong to $\mathfrak{S}_{\nu^*}(U)$ unless $x = 0$. For $\mathfrak{L}_2$ the reasoning is similar. □

Theorem 3.2. *Let A be a selfadjoint operator in the Krein space $\mathfrak{H}$. Suppose that $\mathfrak{H}^+ \subset \mathfrak{D}(A)$, where*

$$\mathfrak{H} = \mathfrak{H}^+ (\dotplus) \mathfrak{H}^-; \quad \mathfrak{H}^+ \subset \mathfrak{P}^{++}, \quad \mathfrak{H}^- \subset \mathfrak{P}^{--} \tag{3.2}$$

is a suitable fundamental decomposition of $\mathfrak{H}$.

Then A has at least one non-real regular point.

Further, denoting the fundamental projectors onto $\mathfrak{H}^+$ resp. $\mathfrak{H}^-$ by P^+ and P^-, setting $I_1 = I|\mathfrak{H}^+$, $I_2 = I|\mathfrak{H}^-$ and introducing the operator matrix $A = (A_{jl})_{j,l=1,2}$ by the formulas

$$A_{11} = P^+A|\mathfrak{H}^+, \qquad A_{12} = P^+A|\mathfrak{H}^-,$$
$$A_{21} = P^-A|\mathfrak{H}^+, \qquad A_{22} = P^-A|\mathfrak{H}^-,$$

suppose also that for some $\zeta_0 \in \varrho(A_{22})$ (cf. Theorem VI.2.2 c) *and* Lemma VI.6.7 d)) *the operator $A_{12}(A_{22} - \zeta_0 I_2)^{-1}$ is compact. Finally, let σ_0 be any set of non-real eigenvalues of A such that $\lambda \in \sigma_0$ implies $\bar{\lambda} \notin \sigma_0$.*

Then A has an invariant maximal positive subspace $\mathfrak{L}_1 \subset \mathfrak{D}(A)$ and an invariant maximal negative subspace $\mathfrak{L}_2$, each containing all principal subspaces $\mathfrak{S}_\lambda(A)$ with $\lambda \in \sigma_0$. For every $\mathfrak{L}_1$ and $\mathfrak{L}_2$ of this kind and every normal eigenvalue $\lambda \in \sigma_0$ we have the relations $\mathfrak{S}_{\bar{\lambda}}(A) \cap \mathfrak{L}_1 = \mathfrak{S}_{\bar{\lambda}}(A) \cap \mathfrak{L}_2 = 0$. One can choose the subspaces $\mathfrak{L}_1$, $\mathfrak{L}_2$ to be the orthogonal companions of each other.

Proof. Once a non-real $\zeta \in \varrho(A)$ is found, we shall try, with the help of the Cayley transformation, to reduce the rest of the statement to Theorem 3.1.

Reasoning for the moment heuristically, we have

$$A - \zeta I = \begin{pmatrix} A_{11} - \zeta I_1 & 0 \\ 0 & A_{22} - \zeta I_2 \end{pmatrix} \begin{pmatrix} I_1 & B_{12} \\ B_{21} & I_2 \end{pmatrix},$$

where $\zeta \in \boldsymbol{C}$ and

$$(3.3)\qquad B_{12} = (A_{11} - \zeta I_1)^{-1} A_{12}\,, \qquad B_{21} = (A_{22} - \zeta I_2)^{-1} A_{21}\,.$$

Thus, still formally,

$$(3.4)\qquad (A - \zeta I)^{-1} =$$

$$= \begin{pmatrix} (I_1 - T_1)^{-1} & 0 \\ 0 & (I_2 - T_2)^{-1} \end{pmatrix} \begin{pmatrix} I_1 & -B_{12} \\ -B_{21} & I_2 \end{pmatrix} \begin{pmatrix} (A_{11} - \zeta I_1)^{-1} & 0 \\ 0 & (A_{22} - \zeta I_2)^{-1} \end{pmatrix},$$

where

$$(3.5)\qquad T_1 = B_{12} B_{21}\,, \qquad T_2 = B_{21} B_{12}\,.$$

Let $\zeta \neq \bar{\zeta}$. The spectrum of a selfadjoint operator in Hilbert space being real, from (3.3) and Lemma VI.6.7 we conclude that $B_{21} \in \mathfrak{B}(\mathfrak{H}^+, \mathfrak{H}^-)$, whereas B_{12} admits a continuous closure $\bar{B}_{12} \in \mathfrak{B}(\mathfrak{H}^-, \mathfrak{H}^+)$. Since $\mathfrak{R}(B_{21}) \subset \mathfrak{D}(A_{22}) = \mathfrak{D}(A_{12}) = \mathfrak{D}(B_{12})$, with the aid of (3.5) it follows that $T_1 \in \mathfrak{B}(\mathfrak{H}^+)$, while T_2 admits a continuous closure $\bar{T}_2 \in \mathfrak{B}(\mathfrak{H}^-)$.

But, in order to ascribe an exact meaning to equation (3.4), we need more. The elementary identity

$$\|A_{11}x - \zeta x\|_J^2 = \|A_{11}x - \operatorname{Re}\zeta \cdot x\|_J^2 + (\operatorname{Im}\zeta)^2 \|x\|_J^2 \qquad (x \in \mathfrak{H}^+)\,,$$

where $J = P^+ - P^-$, yields:

$$(3.6)\qquad \|(A_{11} - \zeta I_1)^{-1}\|_J \leqq \frac{1}{|\operatorname{Im}\zeta|} \qquad (\zeta \neq \bar{\zeta})\,.$$

The estimate

$$(3.7)\qquad \|(A_{22} - \zeta I_2)^{-1}\|_J \leqq \frac{1}{|\operatorname{Im}\zeta|} \qquad (\zeta \neq \bar{\zeta})$$

can be verified similarly. Now if

$$(3.8)\qquad |\operatorname{Im}\zeta| > \max\{\|A_{12}\|_J\,,\ \|A_{21}\|_J\}\,,$$

then by (3.5), (3.3), (3.6) and (3.7) we have $\|T_1\|_J < 1$, $\|\bar{T}_2\|_J < 1$, so that

$$(I_1 - T_1)^{-1} \in \mathfrak{B}(\mathfrak{H}^+)\,, \qquad (I_2 - \bar{T}_2)^{-1} \in \mathfrak{B}(\mathfrak{H}^-)\,.$$

As a result, for every ζ satisfying condition (3.8) the right-hand side of (3.4) is a well-defined linear operator $S(\zeta)$, which is continuous on its domain. Moreover, an easy calculation shows that

$$S(\zeta)(A - \zeta I)x = x \qquad (x \in \mathfrak{D}(A))\,.$$

Consequently, ζ either belongs to $\varrho(A)$ or to $\sigma_r(A)$. A similar statement holds for $\bar{\zeta}$. Thus, by Theorem VI.6.1 c), $\zeta \in \varrho(A)$. In particular, equation (3.4) is correct.

We fix ζ in compliance with (3.8) and build the Cayley transform

$$U = (A - \bar{\zeta} I)(A - \zeta I)^{-1} = I + (\zeta - \bar{\zeta})(A - \zeta I)^{-1}. \tag{3.9}$$

According to Theorem VI.7.1 U is unitary. Further, from (3.4) and (3.3) we obtain

$$U_{11} = P^+U|\mathfrak{H}^+ = I_1 + (\zeta - \bar{\zeta})(I_1 - T_1)^{-1}(A_{11} - \zeta I_1)^{-1}, \tag{3.10}$$

$$U_{12} = P^+U|\mathfrak{H}^- = \tag{3.11}$$

$$= -(\zeta - \bar{\zeta})(I_1 - T_1)^{-1}(A_{11} - \zeta I_1)^{-1} A_{12}(A_{22} - \zeta I_2)^{-1};$$

the matrix elements U_{21}, U_{22} will not be used in the sequel. The assumption concerning the compactness of $A_{12}(A_{22} - \zeta_0 I_2)^{-1}$ for a certain ζ_0, along with the well-known equation

$$R_{\zeta_0} - R_\zeta = (\zeta_0 - \zeta) R_{\zeta_0} R_\zeta$$

where $R_\mu = (A_{22} - \mu I_2)^{-1}$ $(\mu \in \varrho(A_{22}))$, imply the compactness of $A_{12}(A_{22} - \zeta I_2)^{-1}$. Hence U_{12} is compact.

Owing to Theorem 3.1 there exists a maximal positive subspace $\mathfrak{L}_1 \subset \mathfrak{H}$ such that $U\mathfrak{L}_1 \subset \mathfrak{L}_1$ and $\mathfrak{S}_\nu(U) \subset \mathfrak{L}_1$ $(\nu \in \sigma_0')$, where

$$\sigma_0' = \left\{ \frac{\lambda - \bar{\zeta}}{\lambda - \zeta} : \quad \lambda \in \sigma_0 \right\}.$$

Obviously, $(U - I)\mathfrak{L}_1 \subset \mathfrak{L}_1$. Let us prove the stronger fact

$$(U - I)\mathfrak{L}_1 = \mathfrak{L}_1. \tag{3.12}$$

In view of Theorem V.4.2 the equation (3.12) is equivalent to $P^+(U - I)\mathfrak{L}_1 = \mathfrak{H}^+$. Expressing here the left-hand side by the aid of the angular operator $K = K^+(\mathfrak{L}_1) \in \mathfrak{K}(\mathfrak{H}^+, \mathfrak{H}^-)$ (cf. Theorems II.11.7, V.4.2 and II.11.6), we find

$$P^+(U - I)\mathfrak{L}_1 = (U_{11} - I_1 + U_{12}K)\mathfrak{H}^+$$

or, making use of (3.10) and (3.11),

$$P^+(U - I)\mathfrak{L}_1 = \tag{3.13}$$

$$= (\zeta - \bar{\zeta})(I_1 - T_1)^{-1}(A_{11} - \zeta I_1)^{-1}[I_1 - A_{12}(A_{22} - \zeta I_2)^{-1} K]\mathfrak{H}^+.$$

But $\mathfrak{D}(T_1) = \mathfrak{D}(A_{11}) = \mathfrak{H}^+$ and, according to (3.7)—(3.8),

$$\|A_{12}(A_{22} - \zeta I_2)^{-1} K\|_J < 1,$$

so that the range of each multiplier of $\mathfrak{H}^+$ in (3.13) coincides with $\mathfrak{H}^+$. This proves (3.12).

Solving equation (3.9) for A, we obtain (cf. Lemma II.4.1):

$$A = (\zeta U - \bar{\zeta} I)(U - I)^{-1}.$$

Hence, leaning on the construction of $\mathfrak{L}_1$ as well as on relation (3.12) and Lemma II.5.5, we derive the properties $\mathfrak{L}_1 \subset \mathfrak{D}(A)$, $A\mathfrak{L}_1 \subset \mathfrak{L}_1$, $\mathfrak{S}_\lambda(A) \subset \mathfrak{L}_1$ $(\lambda \in \sigma_0)$.

Set $\mathfrak{L}_1^\perp = \mathfrak{L}_2$. Then $\mathfrak{L}_2$ is maximal negative (Theorem V.4.4), invariant for A (Theorem II.3.7), and its orthogonal companion is $\mathfrak{L}_1$ (Theorems V.3.6 and V.4.1). Moreover, by Corollary II.3.4 each $\mathfrak{S}_\lambda(A)$ $(\lambda \in \sigma_0)$ is a neutral subspace of the positive subspace $\mathfrak{L}_1$; thus, in view of Lemma I.4.4,

$$\mathfrak{S}_\lambda(A) \perp \mathfrak{L}_1 \qquad (\lambda \in \sigma_0) \tag{3.14}$$

i.e. $\mathfrak{S}_\lambda(A) \subset \mathfrak{L}_2$ $(\lambda \in \sigma_0)$.

Finally, let $\lambda \in \sigma_0$ be a normal eigenvalue of A. Then, owing to Theorem VI.7.5, $\mathfrak{S}_\lambda(A) \# \mathfrak{S}_{\bar\lambda}(A)$. Comparing with (3.14) yields $\mathfrak{S}_{\bar\lambda}(A) \cap \mathfrak{L}_1 = 0$. The proof of the relation $\mathfrak{S}_{\bar\lambda}(A) \cap \mathfrak{L}_2 = 0$ is similar. □

4. Quadratic Pencils of Operators in Hilbert Space

Consider the differential equation

$$\frac{d^2v}{dt^2} + B\frac{dv}{dt} + Cv = 0\,, \tag{4.1}$$

where t is a real variable, $v = v(t)$ is a function with values in a Hilbert space $\mathfrak{H}_0$, the operator $B \in \mathfrak{B}(\mathfrak{H}_0)$ is selfadjoint, and the operator $C \in \mathfrak{B}(\mathfrak{H}_0)$ is positive and invertible. The inner product on $\mathfrak{H}_0$ will be denoted by $(.\,,.)_0$.

A solution of the equation (4.1) is called an *elementary solution* if it can be written in the form

$$v(t) = e^{\lambda_0 t} \sum_{j=0}^{p-1} \frac{t^j}{j!} x_{p-1-j}\,, \tag{4.2}$$

where $\lambda_0 \in \boldsymbol{C}$, $p \geqq 1$, $\{x_j\}_0^{p-1} \subset \mathfrak{H}_0$, $x_0 \neq 0$.

A tool in the study of equation (4.1) is the family of operators

$$L(\lambda) = \lambda^2 I + \lambda B + C \qquad (\lambda \in \boldsymbol{C})\,. \tag{4.3}$$

A family of this type is called a *quadratic pencil*.

The number $\lambda_0 \in \boldsymbol{C}$ is said to be an *eigenvalue* of the pencil L provided $\mathfrak{N}(L(\lambda_0)) \neq 0$. The set $\sigma_p(L)$ of all eigenvalues of L is called the *point spectrum* of L.

We say that the system $\{x_j\}_0^{p-1} \subset \mathfrak{H}_0$, where $p \geqq 1$ and $x_0 \neq 0$, is a *Jordan chain* of the pencil (4.3) to the value $\lambda_0 \in \boldsymbol{C}$, if it satisfies the conditions

$$(4.4)\qquad (\lambda_0^2 I + \lambda_0 B + C)x_j + (2\lambda_0 I + B)x_{j-1} + x_{j-2} = 0$$
$$(j = 0,1,\ldots,p-1;\quad x_{-1} = x_{-2} = 0)\,.$$

Setting $j = 0$ we see that necessarily $\lambda_0 \in \sigma_p(L)$.

The usefulness of the pencil L in treating the differential equation (4.1) is partly due to the following circumstance.

Lemma 4.1. *A function $v(t)$ of the form* (4.2) *is an elementary solution of the equation* (4.1) *if and only if $x_0, x_1, \ldots, x_{p-1}$ is a Jordan chain to the eigenvalue λ_0 of the quadratic pencil* (4.3).

The *proof* is straightforward. □

Having indicated the connection between the equation (4.1) and the pencil (4.3), we are going to show that, in turn, the pencil (4.3) is related to a selfadjoint operator of a suitable Krein space.

Let

$$(4.5)\qquad \mathfrak{H} = \mathfrak{H}_0 \times \mathfrak{H}_0',$$

where $\mathfrak{H}_0'$ stands for the anti-space of $\mathfrak{H}_0$. In other words, $\mathfrak{H}$ is the vector space of ordered pairs of the form $\boldsymbol{x} = \{x^+, x^-\}$, where x^+, x^- run through the elements of $\mathfrak{H}_0$; the inner product of $\boldsymbol{x}$ and $\boldsymbol{y} = \{y^+, y^-\}$ is given by the formula

$$(\boldsymbol{x}, \boldsymbol{y}) = (x^+, y^+)_0 - (x^-, y^-)_0\,.$$

Obviously, the subspaces

$$(4.6)\qquad \begin{aligned} \mathfrak{H}^+ &= \mathfrak{H}_0 \times 0 = \{\{x^+, 0\}:\ x^+ \in \mathfrak{H}_0\}\,,\\ \mathfrak{H}^- &= 0 \times \mathfrak{H}_0' = \{\{0, x^-\}:\ x^- \in \mathfrak{H}_0\} \end{aligned}$$

induce a fundamental decomposition of $\mathfrak{H}$:

$$(4.7)\qquad \mathfrak{H} = \mathfrak{H}^+ (\dot{+}) \mathfrak{H}^-;\qquad \mathfrak{H}^+ \subset \mathfrak{P}^{++},\qquad \mathfrak{H}^- \subset \mathfrak{P}^{--}\,.$$

In particular, $\mathfrak{H}$ is a Krein space.

Consider the operator $\boldsymbol{A} \in \mathfrak{B}(\mathfrak{H})$ whose matrix with respect to (4.7) looks as follows:

$$(4.8)\qquad \boldsymbol{A} = \begin{pmatrix} 0 & C^{1/2} \\ -C^{1/2} & -B \end{pmatrix}.$$

One immediately verifies that $\boldsymbol{A}$ is selfadjoint in $\mathfrak{H}$.

Lemma 4.2. *If $x_0, x_1, \ldots, x_{p-1}$ is a Jordan chain of the quadratic pencil L to the eigenvalue λ_0, then the vectors*

$$(4.9)\qquad \boldsymbol{y}_j = \{C^{1/2}x_j\,,\ \lambda_0 x_j + x_{j-1}\}\qquad (j = 0,1,\ldots,p-1;\quad x_{-1} = 0)$$

constitute a Jordan chain of the operator $\boldsymbol{A}$ *to the same* λ_0. *Conversely, if the vectors* $\boldsymbol{y}_j = \{y_j^+, y_j^-\}$ $(j = 0,1, \ldots, p-1)$ *form a Jordan chain of* $\boldsymbol{A}$ *to the eigenvalue* λ_0, *then* λ_0 *is non-zero, and the vectors* $x_0, x_1, \ldots, x_{p-1}$ *successively obtainable from the formula*

$$(4.10) \qquad x_j = \frac{1}{\lambda_0}(y_j^- - x_{j-1}) \qquad (j = 0,1, \ldots, p-1; \quad x_{-1} = 0)$$

constitute a Jordan chain of L *to the same* λ_0. *The transformations* (4.9) *and* (4.10) *are mutually inverse.*

Proof. From (4.4), (4.9) and (4.8) one easily derives the equations

$$(4.11) \qquad \boldsymbol{A}\boldsymbol{y}_j = \lambda_0 \boldsymbol{y}_j + \boldsymbol{y}_{j-1} \qquad (j = 0,1, \ldots, p-1; \quad \boldsymbol{y}_{-1} = 0).$$

Since, by assumption, $C^{1/2}$ is invertible, $x_0 \neq 0$ implies $\boldsymbol{y}_0 \neq 0$. Moreover, if $\{\boldsymbol{y}_j\}_0^{p-1}$ is defined by (4.9), then $\{x_j\}_0^{p-1}$ can be recovered with the aid of (4.10).

Suppose, conversely, that (4.11) holds and $\boldsymbol{y}_0 \neq 0$. Setting $j = 0$ and making use of the invertibility of $C^{1/2}$ we obtain that $\lambda_0 \neq 0$. If $\{x_j\}_0^{p-1}$ denotes the system furnished by (4.10), then the relations (4.9) and (4.4) can be verified by simultaneous recursion for j. Finally, (4.9) applied to the case $j = 0$ yields $x_0 \neq 0$. □

Corollary 4.3. $\sigma_p(L) = \sigma_p(\boldsymbol{A})$. □

We say that the number $\lambda_0 \in \boldsymbol{C}$ belongs to the *resolvent set* $\varrho(L)$ or to the *spectrum* $\sigma(L)$ of the quadratic pencil L according as $L(\lambda_0)$ is completely invertible or not. Obviously, $\sigma_p(L) \subset \sigma(L)$.

In addition to Corollary 4.3 we have:

Lemma 4.4. $\sigma(L) = \sigma(\boldsymbol{A})$.

Proof. If λ does not belong to the set $\sigma_p(L) = \sigma_p(\boldsymbol{A})$ (cf. Corollary 4.3) and $\lambda \neq 0$, then solving the equation $(\boldsymbol{A} - \lambda \boldsymbol{I})\boldsymbol{y} = \boldsymbol{g}$ we find:

$$(\boldsymbol{A} - \lambda \boldsymbol{I})^{-1} = \begin{pmatrix} -\frac{1}{\lambda} I + \frac{1}{\lambda} C^{1/2} L(\lambda)^{-1} C^{1/2} & -C^{1/2} L(\lambda)^{-1} \\ L(\lambda)^{-1} C^{1/2} & -\lambda L(\lambda)^{-1} \end{pmatrix}.$$

Therefore $(\boldsymbol{A} - \lambda \boldsymbol{I})^{-1}$ is defined throughout $\mathfrak{H}$ if and only if $L(\lambda)^{-1}$ is defined throughout $\mathfrak{H}_0$.

In the case $\lambda = 0$ we obtain:

$$\boldsymbol{A}^{-1} = \begin{pmatrix} -C^{-1/2} B C^{-1/2} & -C^{-1/2} \\ C^{-1/2} & 0 \end{pmatrix}$$

Hence $\mathfrak{D}(A^{-1})=\mathfrak{H}$ if and only if $\mathfrak{D}(C^{-1/2})=\mathfrak{H}_0$. But the latter equation is easily seen to be equivalent to $\mathfrak{D}(C^{-1})=\mathfrak{H}_0$, i.e., to the complete invertibility of $L(0)$. □

Lemmas 4.2 and 4.4 serve as a point of departure for utilizing Krein space theory in the study of the quadratic pencil (4.3).

5. Quadratic Operator Equations in Hilbert Space

We pursue our investigations concerning the quadratic pencil

$$L(\lambda) = \lambda^2 I + \lambda B + C \qquad (\lambda \in \boldsymbol{C}) \tag{5.1}$$

of operators in the Hilbert space $\mathfrak{H}_0$ with inner product $(.\,,.)_0$, still assuming that $B \in \mathfrak{B}(\mathfrak{H}_0)$ is selfadjoint, while $C \in \mathfrak{B}(\mathfrak{H}_0)$ is positive and invertible.

Writing

$$L(Z) = Z^2 + BZ + C \qquad (Z \in \mathfrak{B}(\mathfrak{H}_0)), \tag{5.2}$$

we shall exhibit a relationship between the pencil (5.1) and the roots of the operator equation $L(Z) = 0$.

Let us see first a few general properties of the equation $L(Z) = 0$.

Lemma 5.1. *If* $L(Z_1) = 0$ *for some* $Z_1 \in \mathfrak{B}(\mathfrak{H}_0)$, *then setting* $Z_2 = -B - Z_1^*$ *we have* $L(Z_2) = 0$. *Moreover,*

$$L(\lambda) = (\lambda I - Z_2^*)(\lambda I - Z_1) \qquad (\lambda \in \boldsymbol{C}). \tag{5.3}$$

The *proof* is simple and left to the reader. □

Lemma 5.2. *Let* $Z_0 \in \mathfrak{B}(\mathfrak{H}_0)$ *satisfy the equation*

$$Z_0^2 + BZ_0 + C = 0. \tag{5.4}$$

Suppose there exists a set σ_0 *of non-real eigenvalues of* Z_0 *with the properties*: a) $\lambda \in \sigma_0$ *implies* $\bar{\lambda} \notin \sigma_0$; b) *the span of the principal subspaces* $\mathfrak{S}_\lambda(Z_0)$ $(\lambda \in \sigma_0)$ *is dense in* $\mathfrak{H}_0$. *Then*

$$Z_0 + Z_0^* = -B, \quad Z_0^* Z_0 = C. \tag{5.5}$$

Proof. We define

$$B_1 = -Z_0 - Z_0^*, \quad C_1 = Z_0^* Z_0, \tag{5.6}$$

and verify immediately that

$$Z_0^2 + B_1 Z_0 + C_1 = 0. \tag{5.7}$$

From (5.4) we obtain

$$Z_0^2 - Z_0^{*2} = Z_0^* B - BZ_0,$$

whereas (5.7) yields

$$Z_0^2 - Z_0^{*2} = Z_0^* B_1 - B_1 Z_0 .$$

Thus for

$$D = B_1 - B$$

we have $Z_0^* D = D Z_0$, so that Z_0 is symmetric relative to the D-inner product

$$(x, y)_D = (Dx, y)_0 \qquad (x, y \in \mathfrak{H}_0) .$$

Therefore, in view of assumption a) and Theorem II.3.3, the D-inner product is identically zero on the span of the subspaces $\mathfrak{S}_\lambda(Z_0)$ $(\lambda \in \sigma_0)$. Hence, by assumption b) and the continuity of the D-inner product, $(x, y)_D = 0$ $(x, y \in \mathfrak{H}_0)$. The positive definite inner product $(. , .)_0$ being non-degenerate, D must be zero. Thus $B_1 = B$. The relation $C_1 = C$ now easily follows from (5.4) and (5.7). □

The next result connects the quadratic equation $L(Z) = 0$ with the operator $\boldsymbol{A}$ defined on the Krein space $\mathfrak{H}$ (cf. (4.5) — (4.8)).

Lemma 5.3. *Let $\mathfrak{L}$ be a subspace of the Krein space $\mathfrak{H} = \mathfrak{H}_0 \times \mathfrak{H}_0'$, where $\mathfrak{H}_0'$ stands for the anti-space of $\mathfrak{H}_0$. Suppose that $\mathfrak{L}$ has an angular operator $K = K^+(\mathfrak{L}) \in \mathfrak{B}(\mathfrak{H}^+, \mathfrak{H}^-)$ with respect to $\mathfrak{H}^+$ (see (4.6) — (4.7)). The subspace $\mathfrak{L}$ is invariant for $\boldsymbol{A}$ (cf. (4.8)) if and only if the operator $Z_0 = KC^{1/2}$ satisfies the equation $L(Z_0) = 0$.*

Proof. $\mathfrak{H}^+$ and $\mathfrak{H}^-$ being isometrically isomorphic to $\mathfrak{H}_0$ resp. its anti-space, we may regard K as an element of $\mathfrak{B}(\mathfrak{H}_0)$ and write

$$\mathfrak{L} = \{\{x^+, Kx^+\} : \quad x^+ \in \mathfrak{H}_0\} .$$

Therefore, in view of (4.8),

$$\boldsymbol{A}\mathfrak{L} = \{\{C^{1/2}Kx^+, \quad - C^{1/2}x^+ - BKx^+\} : \quad x^+ \in \mathfrak{H}_0\} .$$

Consequently, the relation $\boldsymbol{A}\mathfrak{L} \subset \mathfrak{L}$ holds if and only if

$$(5.8) \qquad - C^{1/2}x^+ - BKx^+ = KC^{1/2}Kx^+ \qquad (x^+ \in \mathfrak{H}_0) .$$

Since all operators involved are continuous and $\overline{\mathfrak{R}(C^{1/2})} = \mathfrak{N}(C^{1/2})^\perp = \mathfrak{H}_0$, (5.8) is equivalent to

$$KC^{1/2}KC^{1/2}y^+ + BKC^{1/2}y^+ + Cy^+ = 0 \qquad (y^+ \in \mathfrak{H}_0) ,$$

and this was to be proved. □

As to the interdependence between the quadratic pencil L and the roots of the operator equation $L(Z) = 0$, the simplest fact is the following.

Lemma 5.4. *If $Z_0 \in \mathfrak{B}(\mathfrak{H}_0)$ satisfies the equation $L(Z_0) = 0$, then every Jordan chain of Z_0 is a Jordan chain of the quadratic pencil L to the same eigenvalue. In particular, $\sigma_p(Z_0) \subset \sigma_p(L)$.*

Proof. Let $\{x_j\}_0^{p-1}$ be a Jordan chain of Z_0 to λ_0. Expressing $\lambda_0 x_j$ and $\lambda_0^2 x_j$ from the equations

$$Z_0 x_j = \lambda_0 x_j + x_{j-1}\,, \qquad Z_0^2 x_j = \lambda_0^2 x_j + 2\lambda_0 x_{j-1} + x_{j-2}$$
$$(j = 0, 1, \ldots, p-1; \quad x_{-1} = x_{-2} = 0)\,,$$

respectively, we obtain:

$$(\lambda_0^2 I + \lambda_0 B + C) x_j + (2\lambda_0 I + B) x_{j-1} + x_{j-2} = L(Z_0)\, x_j = 0. \quad \square$$

We are now coming to the main result of this section. It can be looked upon as a partial converse of Lemma 5.4.

Theorem 5.5. *Consider the quadratic pencil*

$$L(\lambda) = \lambda^2 I + \lambda B + C \qquad (\lambda \in \mathbf{C})$$

of operators in the Hilbert space $\mathfrak{H}_0$, *and the quadratic operator function*

$$L(Z) = Z^2 + BZ + C \qquad (Z \in \mathfrak{B}(\mathfrak{H}_0))\,.$$

Besides the conditions $B \in \mathfrak{B}(\mathfrak{H}_0)$ *selfadjoint,* $C \in \mathfrak{B}(\mathfrak{H}_0)$ *positive and invertible, imposed on* B *and* C *so far, assume that* C *is a compact operator. Let* σ_0 *be a set of non-real eigenvalues of the pencil* L *with* $\lambda \in \sigma_0$ *implying* $\bar{\lambda} \notin \sigma_0$.

Then there exists an operator $Z_0 \in \mathfrak{B}(\mathfrak{H}_0)$ *such that* a) $L(Z_0) = 0$, b) $Z_0^* Z_0 \leqq C$, *and* c) *any Jordan chain of* L *to an eigenvalue* $\lambda_0 \in \sigma_0$ *is a Jordan chain of* Z_0 *to the same* λ_0.

If the span of all Jordan chains of L *corresponding to points of* σ_0 *is dense in* $\mathfrak{H}_0$, *then requirement* c) *completely determines* Z_0; *in this case we have* $Z_0 + Z_0^* = -B$, $\quad Z_0^* Z_0 = C$.

Proof. We shall make use of the selfadjoint operator $\boldsymbol{A}$ acting in the Krein space $\mathfrak{H}$ (see the relations (4.5) — (4.8)). Owing to Corollary 4.3, σ_0 is a set of non-real eigenvalues of $\boldsymbol{A}$ such that $\lambda \in \sigma_0$ implies $\bar{\lambda} \notin \sigma_0$. Further, denoting the norm on $\mathfrak{H}_0$ by $\|\cdot\|_0$ we have: $\|C^{1/2}x\|_0^2 \leqq \|Cx\|_0 \|x\|_0$ $(x \in \mathfrak{H}_0)$. Therefore if $\{x_j\}_1^\infty \subset \mathfrak{H}_0$ is bounded and $\{Cx_j\}_1^\infty$ is convergent then $\{C^{1/2}x_j\}_1^\infty$ is convergent. Thus, along with C, also $C^{1/2}$ is compact.

Consequently, Theorem 3.2 applies to $\boldsymbol{A}$ and σ_0: there exists a maximal positive subspace $\mathfrak{L} \subset \mathfrak{H}$ that is invariant under $\boldsymbol{A}$ and contains all principal subspaces $\mathfrak{S}_\lambda(\boldsymbol{A})$ $(\lambda \in \sigma_0)$.

In view of Theorems II.11.7 and V.4.2 $\mathfrak{L}$ has an angular operator $K \in \mathfrak{B}(\mathfrak{H}^+, \mathfrak{H}^-)$ with respect to $\mathfrak{H}^+$ (cf. the definitions (4.6)), and identifying K with an element of $\mathfrak{B}(\mathfrak{H}_0)$ we have $\|K\|_0 \leqq 1$. For $Z_0 = KC^{1/2}$ Lemma 5.3 yields $L(Z_0) = 0$, whereas the inequality $Z_0^* Z_0 \leqq C$ follows from $\|Z_0 x\|_0 \leqq \|C^{1/2}x\|_0$ $(x \in \mathfrak{H}_0)$.

Let $\{x_j\}_0^{p-1}$ be a Jordan chain of the pencil L corresponding to an eigenvalue $\lambda_0 \in \sigma_0$. Then, on account of Lemma 4.2, the vectors

$$\boldsymbol{y}_j = \{C^{1/2}x_j, \lambda_0 x_j + x_{j-1}\} \in \mathfrak{H} \qquad (j = 0,1, \ldots, p-1; \quad x_{-1} = 0)$$

form a Jordan chain of $\boldsymbol{A}$ to the same λ_0. Therefore, by the construction of $\mathfrak{L}$, each $\boldsymbol{y}_j$ is contained in $\mathfrak{L}$. Recalling the definition of K we conclude that

$$KC^{1/2}x_j = \lambda_0 x_j + x_{j-1} \qquad (j = 0,1, \ldots, p-1; \quad x_{-1} = 0).$$

Hence $\{x_j\}_0^{p-1}$ is a Jordan chain of $Z_0 = KC^{1/2}$ to the eigenvalue λ_0.

If the span of all Jordan chains of L corresponding to points of σ_0 is dense in $\mathfrak{H}_0$, then condition c) completely determines the values of Z_0 on this dense subset and, by continuity, on the whole space. The rest of the statement follows from Lemma 5.2. □

6. Spectral Functions

The present-day theory of spectral functions in Krein space makes extensive use of analytic methods. Therefore we will just try to give the ideas by stating some results and neglecting the proofs.

Let $\mathfrak{H}$ be a Krein space and let $A \in \mathfrak{B}(\mathfrak{H})$ be a positive operator. There exists one and only one function E that is defined on the set of all non-zero real numbers λ and satisfies the following conditions:

1) *$E(\lambda)$ is an orthogonal projector in $\mathfrak{H}$;*
2) *$E(\lambda_1)E(\lambda_2) = E(\min\{\lambda_1, \lambda_2\})$;*
3) *$E(\lambda)\mathfrak{H}$ is uniformly negative if $\lambda < 0$,*
   $(I - E(\lambda))\mathfrak{H}$ is uniformly positive if $\lambda > 0$;
4) *$E(\lambda) = 0$ if λ is sufficiently small,*
   $E(\lambda) = I$ if λ is sufficiently large;
5) *$E(\lambda - 0) = E(\lambda)$;*
6) *$AE(\lambda) = E(\lambda)A$;*
7) *$\sigma(A|E(\lambda)\mathfrak{H}) \subset (-\infty, \lambda]$,*
   $\sigma(A|(I - E(\lambda))\mathfrak{H}) \subset [\lambda, \infty)$.

(Conditions 1), 2), 5), 6), 7) are required to hold for all admissible, i.e. non-zero real, values of λ, λ_1 and λ_2.)

E is called the *spectral function* of the positive operator A. Taking $\varkappa^-(\mathfrak{H}) = 0$ we see that this definition is consistent with Hilbert space terminology.

According to conditions 1), 6) and 2), the values of the function $E(\lambda)\mathfrak{H}$ $(\lambda \in \boldsymbol{R};\ \lambda \neq 0)$ are ortho-complemented invariant subspaces of A, and the function itself is monotone non-decreasing.

From 1)—3) it follows that the strong limits $E(\lambda-0)$, $E(\lambda+0)$ exist for every $\lambda \neq 0$. Thus condition 5) means just a norming of E.

From 1), 2), 6), 7) we easily obtain:

8) A real number λ belongs to $\sigma(A)$ if and only if $E(\lambda_2)-E(\lambda_1) \neq 0$ whenever $\lambda_1 < \lambda < \lambda_2$.

Two further properties of the spectral function:

9) $\bigcap\limits_{\lambda_1<0<\lambda_2} (E(\lambda_2) - E(\lambda_1))\mathfrak{H} = \mathfrak{S}_0(A)$

(the principal subspace of A corresponding to $\lambda = 0$);

10) $A = S + \int\limits_{-\infty}^{\infty} \lambda\, dE(\lambda)$, where S is a positive operator such that $S^2 = 0$, $S(E(\lambda_2) - E(\lambda_1)) = 0$ if $\lambda_1 < \lambda_2 < 0$ or $0 < \lambda_1 < \lambda_2$, and the integral exists in the strong topology as an improper integral with singular point $\lambda = 0$.

A few consequences of 9) and 10):

11) $A(E(\lambda_2) - E(\lambda_1)) = \int\limits_{\lambda_1}^{\lambda_2} \lambda\, dE(\lambda)$ if $\lambda_1 < \lambda_2 < 0$ or $0 < \lambda_1 < \lambda_2$;

12) $A^n = \int\limits_{-\infty}^{\infty} \lambda^n dE(\lambda) \quad (n = 2, 3, \ldots)$;

13) $A|\mathfrak{S}_0(A) = S|\mathfrak{S}_0(A)$;

14) $A^2\mathfrak{S}_0(A) = 0$.

On account of 14) the subspace $\mathfrak{S}_0(A)$ is closed, and the Jordan chains of A to the value $\lambda = 0$ consist of no more than two elements.

It can also be proved that

15) $\mathfrak{S}_0(A)$ is ortho-complemented if and only if both of the strong limits $E(-0)$, $E(+0)$ exist; in this case $\mathfrak{S}_0(A) = (E(+0) - E(-0))\mathfrak{H}$ and $S = A(E(+0) - E(-0))$.

A real λ is said to be a *critical point* of the spectral function E if the subspace $(E(\lambda_2) - E(\lambda_1))\mathfrak{H}$ is indefinite for every λ_1, λ_2 satisfying $\lambda_1 < \lambda < \lambda_2$.

In view of 3), the only possible critical point of the spectral function of a positive operator is 0. Also, by property 8), in order that $\lambda = 0$ actually be a critical point of E it is necessary that $0 \in \sigma(A)$.

Suppose 0 is not a critical point of E. Then for some $\lambda_1 < 0$ and $\lambda_2 > 0$ the subspace $(E(\lambda_2) - E(\lambda_1))\mathfrak{H}$ is semi-definite, say positive. Being ortho-complemented, it is uniformly positive. It follows that $E(-0)$ and $E(+0)$ exist. Moreover, 3) yields $E(-0) = E(\lambda_1)$. Further, owing to 15), 14) and the uniqueness of the positive square root in Hilbert space, $\mathfrak{S}_0(A)$ is ortho-complemented and $S = 0$.

If the point $\lambda = 0$ is non-critical, we agree to extend the function E setting $E(0) = E(-0)$. This convention preserves the properties 1), 2), 5), 6), 7).

Let $\mathfrak{H}$ still denote a Krein space. We say the selfadjoint operator $A \in \mathfrak{B}(\mathfrak{H})$ is *positizable*, if there exists a non-constant, real polynomial φ such that $\varphi(A)$ is a positive operator; φ will be called a *positizing polynomial* for A.

Positive operators $A \in \mathfrak{B}(\mathfrak{H})$ are positizable by the polynomial $\varphi(\lambda) = \lambda$.

If $\mathfrak{H}$ is a Hilbert space, then every selfadjoint $A \in \mathfrak{B}(\mathfrak{H})$ is positizable by the polynomial $\varphi(\lambda) = \lambda^2$.

The non-real spectrum of a positizable operator A can be proved to consist of a finite number of normal eigenvalues. It follows that the span $\mathfrak{S}$ of the principal subspaces $\mathfrak{S}_\lambda(A)$ ($\operatorname{Im} \lambda \neq 0$) is ortho-complemented and A restricted to $\mathfrak{S}^\perp$ is a positizable operator with real spectrum.

Let $A \in \mathfrak{B}(\mathfrak{H})$ be a positizable operator with real spectrum. There exists a unique function E defined in all but a finite number of points of the real line and satisfying conditions 1)—2) *and* 4)—7) *above, as well as the following ones*:

3a) *the critical points of E form a subset of the finite set $\{\lambda \in \sigma(A): \varphi(\lambda) = 0\}$, where φ is a positizing polynomial for A*;

3b) *$E(\lambda)$ is undefined if and only if λ is a critical point of E*;

3c) *if λ is a critical point of E, then for every closed interval $[\lambda_1, \lambda_2]$ lying in a sufficiently small right-hand or left-hand neighbourhood of λ the subspace $(E(\lambda_2) - E(\lambda_1))\mathfrak{H}$ is definite.*

E is called the *spectral function* of the positizable operator A.

Property 8) and the analogue of 9) remain valid in the positizable case. Instead of 10) we have:

10a) $$\varphi(A) = S + \int_{-\infty}^{\infty} \varphi(\lambda)\, dE(\lambda)\,,$$

where S is a positive operator such that $S^2 = 0$, $S(E(\lambda_2) - E(\lambda_1)) = 0$ if the closed interval $[\lambda_1, \lambda_2]$ does not contain critical points, and the integral exists in the strong topology as an improper integral with singularities at the critical points.

Notes to Chapter VIII

The first achievement in the field of fundamentally reducible operators belongs to Pesonen [1]. He proved an equivalent of Corollary 1.3 for $T = A$ in a separable Krein space. Using a simpler method, Kühne [1] extended Pesonen's theorem to selfadjoint operators T commuting with A in a possibly non-separable $\mathfrak{H}$. Hess [1] clarified the meaning of these results and obtained the corresponding fact for closed operators T. In the same context, Jalava [1] considered closed operators T_1,

T_2 that are A-adjoints of each other for some uniformly positive operator $A \in \mathfrak{B}(\mathfrak{H})$. In essence, our treatment follows Hess.

Theorem 1.4 coupled with a theorem of Sz.-Nagy [1] says that the unitary operator U in the Krein space $\mathfrak{H}$ is fundamentally reducible if and only if the sequence $\{\|U^n\|_J\}_{n=1}^{\infty}$ is bounded. This proposition, and its extension to commutative groups of unitary operators, have been proved by Phillips [3] (cf. also Kreĭn [6], [8], Daleckiĭ and Kreĭn [1]). The connection between Theorems 1.2 and 1.4 was pointed out by Hess [1].

A related type of theorem give conditions on an operator T for the existence of a reducing decomposition $\mathfrak{H} = \mathfrak{H}_1 \dotplus \mathfrak{H}_2$ where $\mathfrak{H}_1$ and $\mathfrak{H}_2$ are maximal uniformly definite (not necessarily orthogonal) subspaces with the additional property that $\sigma(T|\mathfrak{H}_1)$ is disjoint from $\sigma(T|\mathfrak{H}_2)$. These results and their applications to stability theory are largely due to Kreĭn. A detailed account and a bibliography of the subject can be found in the book of Daleckiĭ and Kreĭn [1]. Cf. also Kreĭn [6], [8] for earlier versions as well as Marksjö and Textorius [1], Langer [9], Švarcman [1] for special topics.

Theorems stating the existence of invariant maximal positive (semidefinite) subspaces for various kinds of operators in $\mathfrak{H}$ (cf. Sections 2—3) have been proved, in chronological order, by the following authors: Pontrjagin [1] ($\varkappa(\mathfrak{H}) < \infty$, selfadjoint operators, analytic method), Kreĭn (in the paper of Kreĭn and Rutman [1]; $\varkappa(\mathfrak{H}) = 1$, unitary operators, using the theory of cones in Banach spaces), Iohvidov [1] ($\varkappa(\mathfrak{H}) < \infty$, unitary operators, from Pontrjagin's result by means of the Caylcy transformations), Kreĭn [4] ($\varkappa(\mathfrak{H}) < \infty$, strict plus-operators, by means of the Brouwer fixed point principle; in a paper of Iohvidov and Kreĭn [1] reproved by the aid of the Schauder fixed point principle), Langer [1]—[3] ($\varkappa(\mathfrak{H}) = \infty$, separable $\mathfrak{H}$, selfadjoint operators A with compact P^+AP^-, by means of the Tihonov fixed point principle), Kreĭn [7] (arbitrary $\mathfrak{H}$; unitary, selfadjoint and certain plus-operators T with compact P^+TP^-), Iohvidov [9] (arbitrary $\mathfrak{H}$, arbitrary plus-operators $T \in \mathfrak{B}(\mathfrak{H})$ with compact P^+TP^-, by means of a "fixed point theorem for set-valued mappings"), Langer [5], [12] (arbitrary $\mathfrak{H}$, positizable operators, by means of the spectral function). For invariant subspaces of selfadjoint, quasi-nilpotent operators cf. Larionov [4].

Quite recently Masuda [1] has attacked the invariant subspace problem with the help of an *ad hoc* iteration procedure rather than a fixed point theorem. In this way he obtained a condition ensuring the existence of a unique invariant maximal positive subspace for a unitary operator U on $\mathfrak{H}$ in terms of the numerical ranges of $U_{11} + U_{12}K$ and $U_{22} - KU_{12}$, where K runs through $\mathfrak{K}(\mathfrak{H}^+, \mathfrak{H}^-)$.

In contrast to positizability, as pointed out by Noël [2], the compactness condition appearing in Theorems 2.1, 3.1 and 3.2 is not invariant against the choice of the fundamental decomposition. We have still decided in favour of these theorems because they automatically furnish most of the results known for the case $\varkappa(\mathfrak{H}) < \infty$ (see Chapter IX), and because their proof, as compared to that available for positizable operators, does not make use of the analytic machinery of spectral functions.

The present form of Theorem 2.1 was obtained by Langer [5] and Wittstock [4], independently of each other. In the proof we follow the unpublished lecture notes of Langer.

If U is unitary in $\mathfrak{H}$ and P^+UP^- is compact, then the perturbation theory of Hilbert space operators yields that every non-unimodular point of $\sigma(U)$ is a normal eigenvalue. Taking the set σ_0 of Theorem 3.1 to be maximal, it follows that the non-unimodular spectra of $U|\mathfrak{L}_1$ and $U|\mathfrak{L}_2$ coincide with σ_0. Moreover, it can be proved that if one merely requires the latter property (still with maximal σ_0) from the invariant maximal positive $\mathfrak{L}_1$ and maximal negative $\mathfrak{L}_2$, then they always contain the principal subspaces $\mathfrak{S}_\nu(U)$ $(\nu \in \sigma_0)$. Similar statements hold for selfadjoint operators. Theorems 3.1—3.2 have originally been proved in this form; see Langer [2], [3] and Kreĭn [7], [8]. In particular, the compactness condition appearing in Theorem 3.2 and the present proof of this theorem belong to Kreĭn [7].

For certain commuting families of unitary or selfadjoint operators the existence of a common invariant maximal positive subspace was proved by Naĭmark [1], Langer [5], [12], and Helton [1] (cf. also Larionov [1], Helton [2]). The simplest case will be treated in Chapter IX.

A strengthening of the invariant subspace theorems asserts that, for certain unitary and selfadjoint operators or commuting families of them, every alternating pair of invariant subspaces can be extended to an alternating maximal pair of invariant subspaces. Results of this kind have been obtained by Phillips [3], Langer [5], [12], and Larionov [2], [3] (cf. also Langer [11]).

For invariant subspaces of operators in Banach spaces with a J-metric see Bonsall [1], M.L. Brodskiĭ [1], Iohvidov [9], Langer [10], and Hackevič [1]. For related theorems concerning locally convex spaces cf. Fan [2]—[5].

The material of Sections 4—5, partly in a stronger form, belongs to Kreĭn and Langer [2], [3] (cf. also the lecture notes of Kreĭn [8]). From the many other facts proved by Kreĭn and Langer [3] we mention the following.

If the quadratic pencil L satisfies the conditions of Theorem 5.5 and, in addition, L is *strongly damped*, i.e.,

$$(Bx, x) > 2\sqrt{(Cx, x)(x, x)} \qquad (x \in \mathfrak{H};\ x \neq 0),$$

then a full converse of Lemma 5.4 holds; namely the equation $L(Z) = 0$ has two roots Z_1, Z_2 such that every principal vector of L is an eigenvector either of Z_1 or of Z_2.

Langer ([5] resp. [8]) extended the latter result to the case of a non-compact C and to pencils where the coefficient of λ^2 is different from I. For related investigations see Kühne [2], Larionov [5], and Eisenfeld [1].

Kühne [1] proved the completeness of the system of eigenvectors of a power-compact Pesonen operator in Krein space. A special case due to Kreĭn [1] and an extension by Harazov [1] were obtained without explicit use of the indefinite inner product. Phillips [4] has found the minimax characterization of the eigenvalues of a compact, strictly positive operator in $\mathfrak{H}$ (for the finite-dimensional case cf. also Švarcman [1]); Lax and Phillips [3] applied this result to a scattering problem. For other questions concerning compact operators see Iohvidov [7], Wendland [1].

The theory of the spectral function for selfadjoint operators in a Krein space $\mathfrak{H}$ with $\varkappa(\mathfrak{H}) < \infty$ is due to Kreĭn and Langer [1]. In Langer's dissertation this was generalized to positizable selfadjoint and unitary operators of an arbitrary Krein space, and in a paper of Langer [6] another extension to a class of continuous selfadjoint operators A with compact P^+AP^- was given. Jonas [1] proved the existence of a spectral function for unitary operators satisfying a somewhat complicated condition weaker than positizability.

A complete account of the spectral theory of positizable operators has not appeared in print. For a short treatment of continuous positive operators see Kreĭn and Šmul'jan [2]. A formulation of the main results for continuous positizable selfadjoint operators can be found in two papers of Langer [8], [12]. For an application of the theory see Langer [15].

Making use of the spectral function, Kreĭn and Šmul'jan [2] established a polar decomposition for a class of strict plus-operators. Special cases have earlier been obtained by Potapov [1] and Ginzburg [2].

In connection with the spectral theory of Krein spaces cf. also Langer [4], [13] and Naĭmark [12].

Chapter IX. Pontrjagin Spaces and Their Linear Operators

Those properties of Pontrjagin spaces (Krein spaces with a finite rank of indefiniteness) and of their linear operators are considered which are not available in arbitrary Krein spaces or can there only be established by more delicate methods. In particular, Sections 4—5 give estimates for the number of non-unimodular (resp. non-real) or non-semi-simple eigenvalues of an isometric (resp. symmetric) operator, while in Sections 7—9 the investigation of invariant maximal semi-definite subspaces is carried on. Theorems 1.4, 2.2, 2.5, 3.2, 4.3, 7.3, 8.2, 9.3, 9.4 and Corollary 8.5 are worth special mentioning.

1. The Spaces Π_k. Positive Subspaces

Krein spaces with a finite rank of indefiniteness are called *Pontrjagin spaces.*

In particular, every Pontrjagin space is a Krein space, every Hilbert space is a Pontrjagin space, and every finite-dimensional non-degenerate inner product space is also a Pontrjagin space.

A Krein space with a finite rank of positivity k ($k = 0,1,2, \ldots$) will be denoted by Π_k. Thus $\varkappa^+(\Pi_k) = k < \infty$, whereas $\varkappa^-(\Pi_k)$ is arbitrary.

Obviously, every Π_k is a Pontrjagin space and every Pontrjagin space is either a Π_k or the anti-space of a Π_k, so that in the study of Pontrjagin spaces we may restrict our attention to spaces of the type Π_k.

Leaning on the foregoing chapters it is not difficult to establish the main features of the geometry of Π_k.

From Theorem II.10.1 we obtain:

Lemma 1.1. *If* $\mathfrak{L}$ *is a positive subspace of* Π_k, *then* $\dim\mathfrak{L} \leq k$. □

Furthermore, Theorems II.10.1 and V.4.2 yield:

Lemma 1.2. *The positive subspace* $\mathfrak{L} \subset \Pi_k$ *is maximal positive if and only if* $\dim\mathfrak{L} = k$. □

Hence with the aid of Lemmas I.11.4 and I.9.8 we derive:

Lemma 1.3. *If $\mathfrak{L}$ is a k-dimensional positive definite subspace of Π_k, then there exists a fundamental decomposition*

$$\Pi_k = \Pi^+ (\dot{+}) \Pi^- ; \qquad \Pi^+ \subset \mathfrak{P}^{++} , \qquad \Pi^- \subset \mathfrak{P}^{--} \tag{1.1}$$

such that $\Pi^+ = \mathfrak{L}$. □

In Theorem VIII.3.2 among other things we have required that a certain dense subspace of $\mathfrak{H}$ would contain one component of a fundamental decomposition. In a Pontrjagin space this condition becomes superfluous.

Theorem 1.4. *Every dense subspace of Π_k contains a k-dimensional positive definite subspace.*

Proof. The case $k = 0$ is trivial. So let $k \geqq 1$, and let $\mathfrak{L}$ be a dense subspace of Π_k. Consider a fundamental decomposition (1.1) and the corresponding fundamental symmetry J. Fix an orthonormal basis $\{e_j\}_{j=1}^k$ of Π^+. Choose a positive number ε and a system $\{f_j\}_1^k \subset \mathfrak{L}$ such that

$$\|f_j - e_j\|_J < \varepsilon \qquad (j = 1, \ldots, k) .$$

Let

$$x = \sum_{j=1}^k \alpha_j e_j , \qquad y = \sum_{j=1}^k \alpha_j f_j ; \qquad \sum_{j=1}^k |\alpha_j|^2 = 1 .$$

Then

$$\|y - x\|_J \leqq \sum_{j=1}^k |\alpha_j| \, \|f_j - e_j\|_J < k\varepsilon ,$$

$$\|x\|_J = 1 , \quad \|y\|_J \leqq \sum_{j=1}^k |\alpha_j| \, \|f_j\|_J < k(1 + \varepsilon) ,$$

and therefore

$$|(y, y) - 1| = |(y, y) - (x, x)| \leqq |(y, y-x)| + |(y-x, x)| \leqq$$
$$\leqq \|y\|_J \, \|y - x\|_J + \|x\|_J \, \|y - x\|_J < (1 + k + k\varepsilon) k\varepsilon .$$

Thus for $\varepsilon < \dfrac{1}{6k^2}$ we have $(y, y) > 0$.

Next let

$$z = \sum_{j=1}^k \beta_j f_j ; \qquad \sum_{j=1}^k |\beta_j|^2 = \beta^2 > 0 .$$

Applying the previous conclusion to $y = \dfrac{1}{\beta} z$ we find that $(z, z) > 0$. Consequently, the vectors $f_1, \ldots, f_k$ are linearly independent and their span is a positive definite subspace of $\mathfrak{L}$. □

2. Closed Subspaces

The following agreeable property of Π_k has already been proved as a part of Theorem V.6.3.

Lemma 2.1. *Every closed definite subspace of Π_k is uniformly definite.* □

The next result, a consequence of Lemma 2.1, plays a decisive role in the theory of Pontrjagin spaces.

Theorem 2.2. *Every closed, non-degenerate subspace $\mathfrak{L} \subset \Pi_k$ is ortho-complemented.*

Proof. By Theorem V.3.1 and Lemma I.11.1 $\mathfrak{L}$ is the orthogonal direct sum of a closed positive definite and a closed negative definite subspace. It remains to apply Lemma 2.1 and Theorem V.5.3. □

From Theorem 2.2 with the help of Theorem V.3.4 and Lemma 1.1 we obtain:

Corollary 2.3. *Every closed, non-degenerate subspace of Π_k is a space of type $\Pi_{k'}$ for some $k' \leqq k$.* □

Example V.6.5 shows that a type $\Pi_{k'}$ subspace of Π_k, where $k' < k$, need not be closed. Nevertheless, for $k' = k$ the following converse of Corollary 2.3 holds.

Theorem 2.4. *If $\Pi_{(k)} \subset \Pi_k$ are two Pontrjagin spaces with the same finite rank of positivity k, then $\Pi_{(k)}$ is closed in Π_k.*

Proof. Let

$$\Pi_{(k)} = \Pi^{(+)} (\dot{+}) \Pi^{(-)} ; \qquad \Pi^{(+)} \subset \mathfrak{P}^{++} , \qquad \Pi^{(-)} \subset \mathfrak{P}^{--}$$

be a fundamental decomposition of $\Pi_{(k)}$. In view of Lemma 1.3 the subspace $\Pi^{(+)}$ gives rise to a fundamental decomposition of Π_k as well:

$$\Pi_k = \Pi^{(+)} (\dot{+}) \Pi^{-} .$$

Since $\Pi^{(-)}$ is an intrinsically complete subspace of the negative definite, intrinsically complete space Π^{-}, the theory of Hilbert spaces assures that $\Pi^{(-)}$ is ortho-complemented in Π^{-}. Consequently, $\Pi_{(k)}$ is ortho-complemented in Π_k. Thus by Corollary III.2.5 $\Pi_{(k)}$ is closed in Π_k. □

For possibly degenerate subspaces we have the following substitute of Theorem 2.2.

Theorem 2.5. *Let $\mathfrak{L}$ be a closed subspace of Π_k. Setting $\mathfrak{L} \cap \mathfrak{L}^{\perp} = \mathfrak{L}^0$, the space Π_k can be decomposed into a direct sum of the form*

$$\Pi_k = \mathfrak{L}_1 (\dot{+}) \mathfrak{L}_2 (\dot{+}) (\mathfrak{L}^0 \dotplus \mathfrak{L}_3) , \tag{2.1}$$

where

$$\mathfrak{L}_1 = \overline{\mathfrak{L}}_1, \qquad \mathfrak{L}^0(+)\mathfrak{L}_1 = \mathfrak{L}, \tag{2.2}$$

$$\mathfrak{L}_2 = \overline{\mathfrak{L}}_2, \qquad \mathfrak{L}^0(+)\mathfrak{L}_2 = \mathfrak{L}^\perp, \tag{2.3}$$

$$\mathfrak{L}_3 \# \mathfrak{L}^0. \tag{2.4}$$

If $\mathfrak{L}_1$ and $\mathfrak{L}_2$ are prescribed subspaces satisfying (2.2) *and* (2.3), *then apart from the conditions* (2.4) *and*

$$\mathfrak{L}_3 \subset (\mathfrak{L}_1(+)\mathfrak{L}_2)^\perp \tag{2.5}$$

the subspace $\mathfrak{L}_3$ may be chosen arbitrarily. On the other hand, for any fixed $\mathfrak{L}_3$ satisfying (2.4) *one and only one choice of $\mathfrak{L}_1$ and $\mathfrak{L}_2$ is possible*:

$$\mathfrak{L}_1 = \mathfrak{L} \cap \mathfrak{L}_3^\perp, \qquad \mathfrak{L}_2 = \mathfrak{L}^\perp \cap \mathfrak{L}_3^\perp. \tag{2.6}$$

Proof. In order to prove the existence of $\mathfrak{L}_3$ for $\mathfrak{L}_1$ and $\mathfrak{L}_2$ given, let $\mathfrak{L}_1$, $\mathfrak{L}_2$ be two subspaces of Π_k satisfying the conditions (2.2)—(2.3). (That (2.2) and (2.3) can at all be fulfilled follows from Lemma IV.8.7, Lemma III.2.4 and Corollary V.3.7.) In view of Theorem I.5.4 $\mathfrak{L}_1$ and $\mathfrak{L}_2$ are non-degenerate, hence, by Theorem 2.2, ortho-complemented. But the relations $\mathfrak{L}_1 \subset \mathfrak{L}$, $\mathfrak{L}_2 \subset \mathfrak{L}^\perp$ imply $\mathfrak{L}_1 \perp \mathfrak{L}_2$. Therefore $\mathfrak{L}_1(+)\mathfrak{L}_2$ exists and, owing to Lemma I.9.2, is ortho-complemented.

Lemma I.9.1 and Corollary I.9.5 ensure that the subspace $(\mathfrak{L}_1(+)\mathfrak{L}_2)^\perp$ is non-degenerate, and relations (2.2)—(2.3) show that it contains $\mathfrak{L}^0$. Moreover, applying Lemma 1.1 to the neutral subspace $\mathfrak{L}^0$ we obtain:

$$\dim \mathfrak{L}^0 \leqq k. \tag{2.7}$$

Thus, according to Lemma I.10.7, $\mathfrak{L}^0$ has a dual companion $\mathfrak{L}_3$ in the space $(\mathfrak{L}_1(+)\mathfrak{L}_2)^\perp$.

On account of Lemma I.10.1, relation (2.7), Lemma I.10.3 and Corollary I.11.9 the direct sum $\mathfrak{L}^0 \dotplus \mathfrak{L}_3$ exists and is ortho-complemented. By Lemma I.9.2 this implies the ortho-complementedness of the subspace

$$\mathfrak{M} = \mathfrak{L}_1 (+) \mathfrak{L}_2 (+) (\mathfrak{L}^0 \dotplus \mathfrak{L}_3).$$

Let $x \in \mathfrak{M}^\perp$. Then, in particular, $x \in (\mathfrak{L}^0(+)\mathfrak{L}_1)^\perp = \mathfrak{L}^\perp$ and $x \in (\mathfrak{L}^0(+)\mathfrak{L}_2)^\perp = \mathfrak{L}^{\perp\perp} = \mathfrak{L}$ (cf. Theorem V.3.6), so that $x \in \mathfrak{L} \cap \mathfrak{L}^\perp = \mathfrak{L}^0$. On the other hand, $x \in \mathfrak{L}_3^\perp$. Consequently, $x \in \mathfrak{L}^0 \cap \mathfrak{L}_3^\perp$. Making use of (2.4) we obtain $x = 0$. Thus (2.1) holds.

In order to prove the existence of $\mathfrak{L}_1$ and $\mathfrak{L}_2$ for fixed $\mathfrak{L}_3$, let $\mathfrak{L}_3$ denote any dual companion of $\mathfrak{L}^0$ in Π_k. (That such an $\mathfrak{L}_3$ can actually be found follows from (2.7) and Lemma I.10.7.) Set

$$\mathfrak{L}_1 = \mathfrak{L} \cap \mathfrak{L}_3^\perp.$$

Then Lemma III.2.4, the definition of $\mathfrak{L}_3$ and the relation $\mathfrak{L}^0 \perp \mathfrak{L}$, respectively, yield $\mathfrak{L}_1 = \overline{\mathfrak{L}_1}$, $\mathfrak{L}^0 \cap \mathfrak{L}_1 = 0$ and $\mathfrak{L}^0 \perp \mathfrak{L}_1$. Moreover, $\mathfrak{L}^0 \dotplus \mathfrak{L}_1 = \mathfrak{L}$, since $\mathfrak{L}^0 \dotplus \mathfrak{L}_3^\perp = \Pi_k$ (cf. Lemma I.10.8) and $\mathfrak{L}^0 \subset \mathfrak{L}$. Thus (2.2) is valid. In a similar way we obtain that the subspace $\mathfrak{L}_2 = \mathfrak{L}^\perp \cap \mathfrak{L}_3^\perp$ satisfies (2.3). Relation (2.1) is now a consequence of that part of the theorem we have already established.

Finally, in order to see the uniqueness of $\mathfrak{L}_1$ and $\mathfrak{L}_2$, let $\mathfrak{L}_3$ be fixed, $\mathfrak{L}_3 \# \mathfrak{L}^0$, and let $\mathfrak{L}_1$, $\mathfrak{L}_2$ be subspaces satisfying (2.1), (2.2) and (2.3). Then necessarily $\mathfrak{L}_1 \subset \mathfrak{L} \cap \mathfrak{L}_3^\perp$, $\mathfrak{L}_2 \subset \mathfrak{L}^\perp \cap \mathfrak{L}_3^\perp$. As, however, $\mathfrak{L} \cap \mathfrak{L}_3^\perp$ and $\mathfrak{L}^\perp \cap \mathfrak{L}_3^\perp$ themselves meet all requirements involving $\mathfrak{L}_1$ and $\mathfrak{L}_2$ (cf. the preceding paragraph), their proper subspaces cannot. This proves (2.6). □

Besides using decompositions of the form (2.1), closed degenerate subspaces of Π_k can be treated by passing to quotient spaces.

Theorem 2.6. *If $\mathfrak{L}$ is a closed subspace of Π_k, then $\mathfrak{L}/\mathfrak{L}^0$ is a space $\Pi_{k'}$ for some $k' \leqq k$.*

Proof. Lemma IV.8.7 supplies a closed, non-degenerate subspace $\mathfrak{L}^1$ such that $\mathfrak{L} = \mathfrak{L}^0 \dotplus \mathfrak{L}^1$. By Corollary 2.3 $\mathfrak{L}^1$ is of type $\Pi_{k'}$ for some $k' \leqq k$. But, according to Lemma I.5.2, $\mathfrak{L}^1$ is isometrically isomorphic to $\mathfrak{L}/\mathfrak{L}^0$. □

3. Isometric Operators: Continuity

Criteria for the continuity of isometric operators in a Krein space have been discussed in Section VI.3. The following sufficient condition is specific to Pontrjagin spaces (cf. Example VI.3.7).

Theorem 3.1. *Let U be an isometric operator in Π_k. If the closures* $\overline{\mathfrak{D}(U)}$ and $\overline{\mathfrak{R}(U)}$ *are non-degenerate, then U and U^{-1} are continuous.*

Proof. Since the inner product is continuous, along with $\overline{\mathfrak{D}(U)}$ also $\mathfrak{D}(U)$ is non-degenerate. Moreover, $\mathfrak{D}(U)$ is quasi-negative (cf. Lemma 1.1). Let $\mathfrak{D}^+$ be a maximal positive definite subspace of $\mathfrak{D}(U)$. Then, owing to Lemma I.11.4, Lemma I.9.8 and Corollary I.11.2, there exists a subspace $\mathfrak{D}^-$ such that

$$(3.1) \qquad \mathfrak{D}(U) = \mathfrak{D}^+ (\dotplus) \mathfrak{D}^-; \qquad \mathfrak{D}^+ \subset \mathfrak{P}^{++}, \quad \mathfrak{D}^- \subset \mathfrak{P}^{--}.$$

Here $\mathfrak{D}^+$ is finite-dimensional and definite, hence uniformly definite. Let us show that $\mathfrak{D}^-$, too, is uniformly definite.

By Corollary 2.3 the closure $\overline{\mathfrak{D}(U)}$ is a space of type $\Pi_{k'}$, where $k' \leqq k$. The subspace $\mathfrak{D}(U)$ is dense in this space, since $\tau_M(\Pi_{k'}) =$

$= \tau_M(\Pi_k)|\Pi_{k'}$ (cf. Theorem 2.2 and Theorem V.3.5). Thus, according to Theorem 1.4 and Lemma II.10.2, $\dim \mathfrak{D}^+ = k'$. Therefore Lemma 1.3 guarantees the existence of a subspace $\mathfrak{D}^{(-)} \subset \mathfrak{P}^{--}$ such that

$$\overline{\mathfrak{D}(U)} = \mathfrak{D}^+ (\dot{+}) \mathfrak{D}^{(-)} . \tag{3.2}$$

The subspace $\overline{\mathfrak{D}(U)}$ being ortho-complemented (Theorem 2.2), so is $\mathfrak{D}^{(-)}$ (Lemma I.9.2). Consequently, $\mathfrak{D}^{(-)}$ is uniformly definite (Theorem V.5.2). But (3.1) and (3.2) imply $\mathfrak{D}^- \subset \mathfrak{D}^{(-)}$. As a result, $\mathfrak{D}^-$ is uniformly definite.

From (3.1) and the isometric property of U we obtain:

$$\mathfrak{R}(U) = U\mathfrak{D}^+ (\dot{+}) U\mathfrak{D}^-; \qquad U\mathfrak{D}^+ \subset \mathfrak{P}^{++}, \quad U\mathfrak{D}^- \subset \mathfrak{P}^{--} .$$

Making use of the non-degeneracy of $\overline{\mathfrak{R}(U)}$ and repeating the argument just applied to $\mathfrak{D}(U)$, it follows that $U\mathfrak{D}^+$ and $U\mathfrak{D}^-$ are uniformly definite. Now Theorem VI.3.5 and Corollary II.2.2 yield the desired conclusion. □

Theorem 3.2. *Let U be an isometric operator in Π_k with $\mathfrak{D}(U) = \Pi_k$. Then U and U^{-1} are continuous.*

Proof. The image $U\Pi^+$ of a k-dimensional positive definite subspace $\Pi^+ \subset \Pi_k$ is k-dimensional (Corollary II.2.2) and positive definite. Thus $\mathfrak{R}(U)$ and, all the more, $\overline{\mathfrak{R}(U)}$ contain a maximal positive definite subspace of Π_k. It remains to apply Lemma II.10.5 and Theorem 3.1. □

4. Isometric and Symmetric Operators: Number and Length of Jordan Chains

As indicated in Chapter II (cf. Examples II.2.3—II.2.4), an isometric operator in an indefinite inner product space may have non-unimodular or non-semi-simple eigenvalues too. It turns out, however, that in a Pontrjagin space the number of these eigenvalues and the lengths of the respective Jordan chains do not exceed certain limits.

Before stating the general result, let us mention an easy special case where non-unimodular eigenvalues are only considered.

Theorem 4.1. *Let U be an isometric operator in Π_k. Let $\nu_1, \ldots, \nu_n$ be a set of non-unimodular eigenvalues of U such that $\nu_j^* \neq \nu_l$ $(j, l = 1, \ldots, n)$. Then*

$$\sum_{j=1}^{n} \dim \mathfrak{S}_{\nu_j}(U) \leq k . \tag{4.1}$$

Proof. By Theorem II.2.5 and Lemma I.3.1 the subspace $\mathfrak{S}_{\nu_1}(U) \dotplus \cdots \dotplus \mathfrak{S}_{\nu_n}(U)$ is neutral. Thus (4.1) follows from Lemma 1.1. □

Corollary 4.2. *An isometric operator in Π_k has no more than k eigenvalues inside and no more than k eigenvalues outside the unit circle.* □

The next theorem is the main result of this section. Roughly speaking, it yields a bound for the total length of all Jordan chains of an isometric operator in Π_k, excepting chains of length 1 to unimodular eigenvalues.

Theorem 4.3. *Let U be an isometric operator in Π_k.*

a) *If ε is a unimodular, non-semi-simple eigenvalue of U, then*

$$\operatorname{codim}_{\mathfrak{S}_\varepsilon(U)} \mathfrak{N}(U - \varepsilon I) < \infty . \tag{4.2}$$

In particular, the sum

$$m(\varepsilon) = \sum_{j=1}^{\infty} \operatorname{codim}_{\mathfrak{N}(U-\varepsilon I)^{2j}} \mathfrak{N}(U - \varepsilon I)^{2j-1} \tag{4.3}$$

is finite.

b) *If $\varepsilon_1, \ldots, \varepsilon_N$ are unimodular, non-semi-simple eigenvalues and $\nu_1, \ldots, \nu_n$ are non-unimodular eigenvalues of U such that $\nu_j^* \neq \nu_l$ $(j, l = 1, \ldots, n)$, then*

$$\sum_{j=1}^{N} m(\varepsilon_j) + \sum_{j=1}^{n} m_{\nu_j} \leqq k , \tag{4.4}$$

where $m_{\nu_j} = \dim \mathfrak{S}_{\nu_j}(U)$ $(j = 1, \ldots, n)$.

c) *If the operator U is closed and ε is a unimodular, non-semi-simple eigenvalue of U, then the principal subspace $\mathfrak{S}_\varepsilon(U)$ can be decomposed into an orthogonal direct sum of the form*

$$\mathfrak{S}_\varepsilon(U) = \mathfrak{S}' \,(\dotplus)\, \mathfrak{S}'' , \tag{4.5}$$

where $\mathfrak{S}'$ is non-zero, finite-dimensional, and invariant for U, while $\mathfrak{S}''$ is closed, negative definite, and contained in the eigenspace $\mathfrak{N}(U - \varepsilon I)$. Selecting a basis of $\mathfrak{S}'$ that consists of Jordan chains of U to the eigenvalue ε and discarding chains of length 1, the lengths d_j $(j = 1, \ldots, r)$ of the remaining chains depend on U and ε only; in addition,

$$\sum_{j=1}^{r} \left[\frac{d_j}{2}\right] = m(\varepsilon) , \tag{4.6}$$

where the brackets [ ] stand for "integral part".

We postpone the *proof* to Section 5. □

Besides Theorem 4.1 and Corollary 4.2 the following special cases of Theorem 4.3 should be mentioned.

Since, on account of (4.6), for a unimodular, non-semi-simple eigenvalue ε we have $m(\varepsilon) \geqq 1$, while for a non-unimodular, non-semi-simple eigenvalue ν obviously $m_\nu \geqq 2$, relation (4.4) yields:

Corollary 4.4. *The number of non-semi-simple eigenvalues of an isometric operator in Π_k does not exceed k.* □

Since in case $\mathfrak{N}(U - \varepsilon I)^{2k+1} \neq \mathfrak{N}(U - \varepsilon I)^{2k+2}$ the sum (4.3) would contain at least $k + 1$ non-zero terms, from (4.4) we deduce:

Corollary 4.5. *The length of a Jordan chain corresponding to a unimodular eigenvalue of an isometric operator in Π_k cannot be greater than $2k + 1$.* □

Below we formulate the counterparts, involving symmetric operators, of Theorem 4.1 and Corollaries 4.2, 4.4, 4.5. The analogue of Theorem 4.3 holds as well. The proofs can be accomplished either by imitating the argument applied in the isometric case, or by reduction to the isometric case via a Cayley transformation (cf. Sections II.4—5 and Theorem VI.7.3).

Theorem 4.6. *Let A be a symmetric operator in Π_k. Let $\lambda_1, \ldots, \lambda_n$ be a set of non-real eigenvalues of A such that $\overline{\lambda}_j \neq \lambda_l$ $(j,l = 1, \ldots, n)$. Then*

$$\sum_{j=1}^{n} \dim \mathfrak{S}_{\lambda_j}(A) \leqq k . \quad \square$$

Corollary 4.7. *A symmetric operator in Π_k has no more than k eigenvalues in the half-plane* $\operatorname{Im} \lambda > 0$ *and no more than k eigenvalues in the half-plane* $\operatorname{Im} \lambda < 0$. □

Theorem 4.8. *The number of non-semi-simple eigenvalues of a symmetric operator in Π_k does not exceed k.* □

Theorem 4.9. *The length of a Jordan chain corresponding to a real eigenvalue of a symmetric operator in Π_k cannot be greater than $2k + 1$.* □

Remark 4.10. All of the estimates appearing in the present section are sharp. To see this, say, for symmetric operators, recall Example II.3.2, where a symmetric operator with one pair of non-real eigenvalues λ, $\overline{\lambda}$ in Π_1 was constructed; taking the cartesian product of k copies of the space and defining in the jth copy $(j = 1, \ldots, k)$ a symmetric operator with non-real eigenvalues λ_j, $\overline{\lambda}_j$ $(\lambda_j \neq \lambda_l$ for $j \neq l)$, one obtains a symmetric operator with non-real eigenvalues $\lambda_1, \overline{\lambda}_1, \ldots, \lambda_k, \overline{\lambda}_k$ in a space Π_k. Starting from Example II.3.1, a similar process leads to a symmetric operator with k real, non-semi-simple eigenvalues in Π_k. Further, choosing two k-dimensional neutral subspaces $\mathfrak{L}_1 \# \mathfrak{L}_2$ in a $2k$-dimensional space Π_k (cf. Theorem II.11.5), defining A_1 in $\mathfrak{L}_1$ to be a linear operator with a Jordan chain of length k to a non-real eigenvalue λ, and defining a linear operator A_2 in $\mathfrak{L}_2$ so that the matrices of A_1 and A_2 relative to dual bases of $\mathfrak{L}_1$ and $\mathfrak{L}_2$ (Lemma I.10.6) would be the transposed conjugates of each other (Lemma I.10.1), we easily verify that the direct sum $A = A_1 \dotplus A_2$ is symmetric on Π_k and has a Jordan chain of length k to the non-real eigenvalue λ. Finally, the existence of a symmetric operator having a Jordan chain of length $2k + 1$ to a real eigenvalue in Π_k will be exhibited by the next example.

Example 4.11. Let Π_k $(k > 0)$ be the orthogonal direct sum of a k-dimensional positive definite subspace Π^+ and a $(k+1)$-dimensional negative definite subspace Π^-. Let $\{e_j\}_1^k$ and $\{f_j\}_1^{k+1}$ be orthonormal bases of Π^+ and Π^-, respectively. Set

$$g_j = \begin{cases} \dfrac{1}{\sqrt{2}}(f_j + e_j) & \text{if} \quad 1 \leqq j \leqq k\,, \\ f_{k+1} & \text{if} \quad j = k+1\,, \\ \dfrac{1}{\sqrt{2}}(f_{2k+2-j} - e_{2k+2-j}) & \text{if} \quad k+2 \leqq j \leqq 2k+1\,. \end{cases}$$

Then

$$(g_j, g_l) = \begin{cases} -1 & \text{if} \quad j + l = 2k+2\,, \\ 0 & \text{otherwise.} \end{cases}$$

The linear operator A defined by the equations

$$A g_j = g_{j-1} \qquad (j = 1, \ldots, 2k+1; \quad g_0 = 0)$$

turns out to be symmetric.

5. Proof of Theorem 4.3

a) The subspaces

$$\mathfrak{N}_j = \mathfrak{N}(U - \varepsilon I)^j \qquad (j = 0, 1, \ldots) \tag{5.1}$$

obviously satisfy the conditions

$$\mathfrak{N}_j \subset \mathfrak{N}_{j+1} \qquad (j = 0, 1, \ldots)\,, \tag{5.2}$$

$$\bigcup_{j=0}^{\infty} \mathfrak{N}_j = \mathfrak{S}_\varepsilon(U)\,. \tag{5.3}$$

It is also easy to see that

$$\mathfrak{N}_j = \mathfrak{N}_{j+1} \quad \text{implies} \quad \mathfrak{N}_{j+1} = \mathfrak{N}_{j+2}\,. \tag{5.4}$$

Hence part a) of our theorem holds if and only if

$$\operatorname{codim}_{\mathfrak{N}_{j+1}} \mathfrak{N}_j < \infty \qquad (j = 1, 2, \ldots)$$

and

$$\mathfrak{N}_j = \mathfrak{N}_{j+1} \qquad (j \geqq j_0)\,.$$

Let $j \geqq 1$, and let $\mathfrak{G} \subset \mathfrak{N}_{j+1}$ be a subspace linearly independent of $\mathfrak{N}_j$. Employing the identity

(5.5) $(v, (U-\varepsilon I)w) = -\bar{\varepsilon}((U-\varepsilon I)v\,,\,Uw) \quad (|\varepsilon| = 1; \quad v,w \in \mathfrak{D}(U))\,,$

which is a simple consequence of the isometric property of U, for $x \in \mathfrak{N}_1$ and $y \in \mathfrak{G}$ we find:

$$(x, (U-\varepsilon I)^j y) = -\bar{\varepsilon}((U-\varepsilon I)\,x\,,\; U(U-\varepsilon I)^{j-1}\,y) = 0\,.$$

Thus $(U-\varepsilon I)^j\mathfrak{G} \perp \mathfrak{N}_1$. On the other hand, the assumption $\mathfrak{G} \subset \mathfrak{N}_{j+1}$ yields $(U-\varepsilon I)^j\mathfrak{G} \subset \mathfrak{N}_1$. So $(U-\varepsilon I)^j\mathfrak{G}$ is orthogonal to itself and therefore, by Lemma 1.1, its dimension does not exceed the value k. Since, however, the condition $\mathfrak{G} \cap \mathfrak{N}_j = 0$ implies that $(U-\varepsilon I)^j|\mathfrak{G}$ is invertible, we have $\dim\mathfrak{G} \leqq k$. As a result,

$$\operatorname{codim}_{\mathfrak{N}_{j+1}} \mathfrak{N}_j \leqq k \qquad (j = 1,2,\ldots)\,. \tag{5.6}$$

Let $t \geqq 1$ be an index such that $\mathfrak{N}_t \neq \mathfrak{N}_{t+1}$. Then, according to (5.4), $\mathfrak{N}_j \neq \mathfrak{N}_{j+1}$ $(j = 0,1,\ldots,t)$. Thus, in view of (5.2), each $\mathfrak{N}_j$ $(j \leqq t)$ has a non-zero complementary subspace in $\mathfrak{N}_{j+1}$:

$$\mathfrak{N}_{j+1} = \mathfrak{N}_j \dotplus \mathfrak{G}_j \qquad (j = 0,1,\ldots,t)\,, \tag{5.7}$$

$$\mathfrak{G}_j \neq 0 \qquad (j = 0,1,\ldots,t)\,. \tag{5.8}$$

Put

$$\mathfrak{L}_j = (U-\varepsilon I)^j\,\mathfrak{G}_{2j-1} \qquad \left(j = 1,2,\ldots,\left[\frac{t+1}{2}\right]\right). \tag{5.9}$$

From the relations $\mathfrak{G}_{2j-1} \subset \mathfrak{N}_{2j}$, $\mathfrak{G}_{2j-1} \cap \mathfrak{N}_{2j-1} = 0$ (see (5.7)) we obtain:

$$\mathfrak{L}_j \subset \mathfrak{N}_j\,, \quad \mathfrak{L}_j \cap \mathfrak{N}_{j-1} = 0 \qquad \left(j = 1,2,\ldots,\left[\frac{t+1}{2}\right]\right). \tag{5.10}$$

Therefore, in view of (5.2), the direct sum

$$\mathfrak{M}_t = \mathfrak{L}_1 \dotplus \cdots \dotplus \mathfrak{L}_{t'} \qquad \left(t' = \left[\frac{t+1}{2}\right]\right) \tag{5.11}$$

exists. Obviously,

$$\mathfrak{M}_t \subset \mathfrak{S}_\varepsilon(U)\,. \tag{5.12}$$

Applying equation (5.5) j times and recalling (5.7), for $x \in \mathfrak{G}_{2j-1}$, $y \in \mathfrak{G}_{2q-1}$ $\left(1 \leqq j \leqq q \leqq \left[\frac{t+1}{2}\right]\right)$ we derive:

$$\begin{aligned}((U-\varepsilon I)^j x\,,\;(U-\varepsilon I)^q y) &= \\ = (-\bar{\varepsilon})^j((U-\varepsilon I)^{2j} x\,,\; U^j(U-\varepsilon I)^{q-j} y) &= 0\,.\end{aligned}$$

Hence $\mathfrak{L}_j \perp \mathfrak{L}_q$ $\left(j, q = 1,2, \ldots, \left[\frac{t+1}{2}\right]\right)$ and, consequently, $\mathfrak{M}_t$ is a neutral subspace:

$$(5.13) \qquad \mathfrak{M}_t \perp \mathfrak{M}_t \,.$$

In particular, by Lemma 1.1,

$$(5.14) \qquad \dim \mathfrak{M}_t \leqq k \,.$$

On the other hand, since $\mathfrak{G}_{2j-1} \cap \mathfrak{N}_j \subset \mathfrak{G}_{2j-1} \cap \mathfrak{N}_{2j-1} = 0$ (cf. (5.7)), the restriction of $(U - \varepsilon I)^j$ to $\mathfrak{G}_{2j-1}$ is invertible. Therefore, owing to (5.9),

$$(5.15) \qquad \dim \mathfrak{L}_j = \dim \mathfrak{G}_{2j-1} \qquad \left(j = 1,2, \ldots, \left[\frac{t+1}{2}\right]\right),$$

so that (5.8) and (5.11) yield

$$(5.16) \qquad \dim \mathfrak{M}_t \geqq \left[\frac{t+1}{2}\right].$$

From (5.14) and (5.16) we obtain the inequality $t \leqq 2k$. Thus, on account of the definition of t,

$$(5.17) \qquad \mathfrak{N}_j = \mathfrak{N}_{j+1} \qquad (j = 2k+1,\ 2k+2, \ldots)\,.$$

Relations (5.6) and (5.17) imply (4.2).

b) Pursuing the investigation of a single unimodular, non-semi-simple eigenvalue ε for a while, and adhering to the notation introduced above, we set (cf. (5.17), (5.11))

$$(5.18) \qquad t_0 = \max\{t\colon\ \mathfrak{N}_t \neq \mathfrak{N}_{t+1}\}\,,$$

$$(5.19) \qquad \mathfrak{M}(\varepsilon) = \mathfrak{M}_{t_0}\,.$$

Then, in accordance with (5.12) and (5.13),

$$(5.20) \qquad \mathfrak{M}(\varepsilon) \subset \mathfrak{S}_\varepsilon(U), \qquad \mathfrak{M}(\varepsilon) \perp \mathfrak{M}(\varepsilon)\,.$$

Moreover, taking into account relations (5.19), (5.11), (5.15) and (5.7) as well as the special choice (5.18) of t_0 we find:

$$\dim \mathfrak{M}(\varepsilon) = \sum_{j=1}^{t_0'} \dim \mathfrak{L}_j = \sum_{j=1}^{t_0'} \dim \mathfrak{G}_{2j-1} =$$
$$= \sum_{j=1}^{t_0'} \operatorname{codim}_{\mathfrak{N}_{2j}} \mathfrak{N}_{2j-1} = \sum_{j=1}^{\infty} \operatorname{codim}_{\mathfrak{N}_{2j}} \mathfrak{N}_{2j-1}\,.$$

Consequently (see (4.3)),

$$(5.21) \qquad \dim \mathfrak{M}(\varepsilon) = m(\varepsilon)\,.$$

Passing to the situation described in part b) of the statement, we consider the direct sum

$$(5.22) \quad \mathfrak{M} = \mathfrak{M}(\varepsilon_1) \dotplus \cdots \dotplus \mathfrak{M}(\varepsilon_N) \dotplus \mathfrak{S}_{\nu_1}(U) \dotplus \cdots \dotplus \mathfrak{S}_{\nu_n}(U)\,,$$

whose existence follows from the fact that the principal subspaces of an operator are linearly independent. By Theorem II.2.5 the components of $\mathfrak{M}$ are pairwise orthogonal. The same theorem assures that each $\mathfrak{S}_{\nu_j}(U)$ is orthogonal to itself. Finally, (5.20) yields $\mathfrak{M}(\varepsilon_j) \perp \mathfrak{M}(\varepsilon_j)$ $(j = 1, \ldots, N)$. As a result, $\mathfrak{M}$ is a neutral subspace. Hence, in view of Lemma 1.1,

$$\dim \mathfrak{M} \leqq k\,. \tag{5.23}$$

Relations (5.21)—(5.23) imply (4.4).

c) For a closed U the eigenspace $\mathfrak{N}_1 = \mathfrak{N}(U - \varepsilon I)$ is closed. Therefore, owing to Theorem V.3.1, $\mathfrak{N}_1$ has a fundamental decomposition of the form

$$\mathfrak{N}_1 = \mathfrak{N}_1^0 \,(\dot{+})\, \mathfrak{N}_1^+ \,(\dot{+})\, \mathfrak{N}_1^-; \tag{5.24}$$
$$\mathfrak{N}_1^0 \subset \mathfrak{P}^0\,, \qquad \mathfrak{N}_1^+ \subset \mathfrak{P}^{++}\,, \qquad \mathfrak{N}_1^- \subset \mathfrak{P}^{--}\,,$$

where $\mathfrak{N}_1^-$ is closed.

Choose

$$\mathfrak{S}'' = \mathfrak{N}_1^-\,. \tag{5.25}$$

Then $\mathfrak{S}'' \subset \mathfrak{S}_\varepsilon(U)$ and, by Theorem 2.2, $\mathfrak{S}''$ is ortho-complemented in Π_k. Consequently, there exists a subspace $\mathfrak{S}'$ satisfying

$$\mathfrak{S}_\varepsilon(U) = \mathfrak{S}' \,(\dot{+})\, \mathfrak{S}''\,.$$

As ε is a non-semi-simple eigenvalue, $\mathfrak{S}'$ cannot be zero. Moreover, ε being non-zero and $\mathfrak{S}''$ being contained in the eigenspace $\mathfrak{N}_1 = \mathfrak{N}(U - \varepsilon I)$, we have $U\mathfrak{S}'' = \mathfrak{S}''$, so that Lemma II.2.9 and the invariance of $\mathfrak{S}_\varepsilon(U)$ yield $U\mathfrak{S}' \subset \mathfrak{S}'$.

Set

$$U' = U|\mathfrak{S}'\,, \qquad \mathfrak{N}_j' = \mathfrak{N}(U' - \varepsilon I)^j \qquad (j = 0,1, \ldots)\,. \tag{5.26}$$

Applying assertion a), already established, to the operator U' we find:

$$\operatorname{codim}_{\mathfrak{S}'} \mathfrak{N}_1' < \infty\,. \tag{5.27}$$

On the other hand, the relations $\mathfrak{N}_1' \subset \mathfrak{N}_1$, $\mathfrak{N}_1' \cap \mathfrak{N}_1^- \subset \mathfrak{S}' \cap \mathfrak{S}'' = 0$ imply $\dim \mathfrak{N}_1' \leqq \operatorname{codim}_{\mathfrak{N}_1} \mathfrak{N}_1^-$; hence, in view of (5.24), Lemma I.3.1 and Lemma 1.1,

$$\dim \mathfrak{N}_1' \leqq k\,. \tag{5.28}$$

From (5.27) and (5.28) it follows that $\mathfrak{S}'$ is finite-dimensional.

Since $\mathfrak{S}''$ is ortho-complemented whereas $\mathfrak{S}'' \subset \mathfrak{N}_1 \subset \mathfrak{N}_j$ $(j = 1,2, \ldots)$ and, according to (5.26), $\mathfrak{N}_j' = \mathfrak{N}_j \cap \mathfrak{S}'$ $(j = 0,1, \ldots)$, we have

$$\mathfrak{N}_j = \mathfrak{N}_j' \,(\dot{+})\, \mathfrak{S}'' \qquad (j = 1,2, \ldots)\,. \tag{5.29}$$

In particular, the definitions (5.18) and

$$t_0 = \max\{t\colon\ \mathfrak{N}'_t \neq \mathfrak{N}'_{t+1}\} \tag{5.30}$$

are equivalent.

Let $\mathfrak{G}'_j$ $(j \leqq t_0)$ denote a complementary subspace to $\mathfrak{N}'_j$ in $\mathfrak{N}'_{j+1}$:

$$\mathfrak{N}'_{j+1} = \mathfrak{N}'_j \dotplus \mathfrak{G}'_j \qquad (j = 0, 1, \ldots, t_0)\,. \tag{5.31}$$

Selecting first $\mathfrak{G}'_{t_0}$ and then proceeding to smaller and smaller indices it can be achieved that

$$(U - \varepsilon I)\mathfrak{G}'_j \subset \mathfrak{G}'_{j-1} \qquad (j = 1, \ldots, t_0)\,. \tag{5.32}$$

Really, from $\mathfrak{G}'_j \subset \mathfrak{N}'_{j+1}$ and $\mathfrak{G}'_j \cap \mathfrak{N}'_j = 0$ we deduce $(U - \varepsilon I)\mathfrak{G}'_j \subset \mathfrak{N}'_j$ and $(U - \varepsilon I)\mathfrak{G}'_j \cap \mathfrak{N}'_{j-1} = 0$, respectively.

Formulas (4.5), (5.26), (5.30) and (5.31) yield:

$$\mathfrak{S}' = \mathfrak{S}_\varepsilon(U') = \mathfrak{N}'_{t_0+1} = \mathfrak{G}'_0 \dotplus \mathfrak{G}'_1 \dotplus \cdots \dotplus \mathfrak{G}'_{t_0}\,.$$

Therefore we can build up a basis of $\mathfrak{S}'$ from bases of the $\mathfrak{G}'_j$.

Let $e_{j,1}, \ldots, e_{j,g_j}$ be a basis of $\mathfrak{G}'_j$, where $1 \leqq j \leqq t_0$. Then, owing to (5.32), the vectors

$$e_{j-1,1} = (U - \varepsilon I)e_{j,1}\,, \quad \ldots, \quad e_{j-1,g_j} = (U - \varepsilon I)e_{j,g_j} \tag{5.33}$$

belong to $\mathfrak{G}'_{j-1}$. Moreover, they are linearly independent, since by (5.31) the restriction of $U - \varepsilon I$ to $\mathfrak{G}'_j$ is invertible:

$$\mathfrak{N}(U - \varepsilon I) \cap \mathfrak{G}'_j = \mathfrak{N}'_1 \cap \mathfrak{G}'_j \subset \mathfrak{N}'_j \cap \mathfrak{G}'_j = 0\,.$$

Consequently, the system (5.33) can be extended to a basis of $\mathfrak{G}'_{j-1}$.

Starting from a basis of $\mathfrak{G}'_{t_0}$ and applying the above procedure successively t_0 times, we obtain a basis of $\mathfrak{S}'$ which consists of Jordan chains of U to the eigenvalue ε. Simultaneously it turns out that the number p_j of Jordan chains of a given length j can be expressed as

$$p_j = \dim \mathfrak{G}'_{j-1} - \dim \mathfrak{G}'_j \qquad (j = 1, \ldots, t_0 + 1;\quad \mathfrak{G}'_{t_0+1} = 0)\,. \tag{5.34}$$

But, in view of (5.31) and (5.29),

$$\dim \mathfrak{G}'_j = \operatorname{codim}_{\mathfrak{N}_{j+1}} \mathfrak{N}_j \qquad (j = 1, \ldots, t_0)\,. \tag{5.35}$$

Therefore the values $p_2, p_3, \ldots, p_{t_0+1}$ depend neither on the choice of the decomposition (4.5) nor on the selection of the Jordan basis of $\mathfrak{S}'$.

The proof of equation (4.6) is now within reach. In fact, (4.3) and (5.35) along with (5.18) imply that

$$m(\varepsilon) = \sum_{j=1}^{t'_0} \dim \mathfrak{G}'_{2j-1} \qquad \left(t'_0 = \left[\frac{t_0 + 1}{2}\right]\right). \tag{5.36}$$

Further, from (5.34) we derive

$$\dim \mathfrak{G}'_s = \sum_{j=s+1}^{t_0+1}{}' p_j \qquad (s = 0,1,\ldots,t_0)\,.$$

In particular,

$$\begin{aligned}
\dim \mathfrak{G}'_1 &= p_2 + p_3 + p_4 + p_5 + p_6 + p_7 + \cdots + p_{t_0+1}\,,\\
\dim \mathfrak{G}'_3 &= \qquad\qquad p_4 + p_5 + p_6 + p_7 + \cdots + p_{t_0+1}\,,\\
\dim \mathfrak{G}'_5 &= \qquad\qquad\qquad\qquad p_6 + p_7 + \cdots + p_{t_0+1}\,,\\
&\ldots\ldots\ldots\ldots\ldots\ldots\ldots\ldots
\end{aligned}$$

Putting these expressions into (5.36) we finally obtain the relation

$$m(\varepsilon) = p_2 + p_3 + 2p_4 + 2p_5 + 3p_6 + 3p_7 + \cdots + \left[\frac{t_0+1}{2}\right]p_{t_0+1} = \\ = \sum_{j=2}^{t_0+1}\left[\frac{j}{2}\right]p_j\,,$$

which coincides with (4.6). □

6. Regular Symmetric Extensions

Let A be a symmetric operator with dense domain in the Pontrjagin space Π_k. Let Π'_k be a Pontrjagin space containing Π_k and having the same rank of positivity $\varkappa^+(\Pi'_k) = \varkappa^+(\Pi_k) = k < \infty$. Let A_1 be a symmetric operator in Π'_k satisfying the conditions $\mathfrak{D}(A) \subset \mathfrak{D}(A_1)$, $A_1 x = Ax$ $(x \in \mathfrak{D}(A))$. Then A_1 is called a *regular symmetric extension* of A.

In other words, a regular symmetric extension is a symmetric extension in a possibly larger space of type Π_k, where k is unaltered.

Let, again, A be symmetric and densely defined in Π_k. Owing to Theorem 1.4 and Lemma 1.3 there exists a fundamental decomposition

$$\text{(6.1)} \qquad \Pi_k = \Pi^+ (\dot{+}) \Pi^-; \qquad \Pi^+ \subset \mathfrak{P}^{++}\,, \quad \Pi^- \subset \mathfrak{P}^{--}$$

such that

$$\text{(6.2)} \qquad \Pi^+ \subset \mathfrak{D}(A)\,.$$

Denote the corresponding fundamental projectors by P^+, P^-. With respect to (6.1) the operator A can be represented by a matrix

$$\text{(6.3)} \qquad A = \begin{pmatrix} A_{11} & A_{12} \\ A_{21} & A_{22} \end{pmatrix},$$

where

$$(6.4)\qquad \begin{aligned} A_{11} &= P^+A|\Pi^+ , & A_{12} &= P^+A|\Pi^- , \\ A_{21} &= P^-A|\Pi^+ , & A_{22} &= P^-A|\Pi^- . \end{aligned}$$

A few properties of A_{jl} $(j,l = 1,2)$ have been enumerated in Lemma VI.6.7.

Consider a space Π_0 (i.e., a negative definite, intrinsically complete space) containing Π^-, and a symmetric extension A_{22}' of A_{22} in Π_0. Set

$$(6.5)\qquad A_{12}' = \overline{A}_{12} P_{\Pi^-} | \mathfrak{D}(A_{22}') ,$$

where P_{Π^-} is the orthogonal projector from Π_0 onto Π^-. It is easy to see that the matrix operator

$$(6.6)\qquad A' = \begin{pmatrix} A_{11} & A_{12}' \\ A_{21} & A_{22}' \end{pmatrix},$$

acting in the space $\Pi_k' = \Pi^+ \times \Pi_0$, is a regular symmetric extension of A. Conversely, every regular symmetric extension can be obtained by selecting Π_0 and A_{22}' in a suitable way. Moreover, according to Lemma VI.6.7, A' is selfadjoint in Π_k' if and only if A_{22}' is selfadjoint in Π_0.

Thus the problem of finding regular symmetric extensions of operators in Π_k is equivalent to that of finding symmetric extensions (within or beyond the space) of Hilbert space operators.

Next we examine non-real eigenvalues of regular symmetric extensions.

Theorem 6.1. *Consider a densely defined symmetric operator A in Π_k and a non-real number λ. The following conditions are equivalent:*

a) $\mathfrak{R}(A - \overline{\lambda} I)^\perp \cap \mathfrak{P}^0 \neq 0$;

b) *A has a symmetric extension A_0 in Π_k such that λ is an eigenvalue of A_0;*

c) *A has a regular symmetric extension A_1 such that λ is an eigenvalue of A_1 and one of the corresponding eigenvectors belongs to Π_k.*

Proof. Let $x_0 \perp \mathfrak{R}(A - \overline{\lambda} I)$, $(x_0, x_0) = 0$, $x_0 \neq 0$. If $x_0 \in \mathfrak{D}(A)$, then λ is an eigenvalue of A itself, since

$$(6.7)\qquad (Ax_0 - \lambda x_0, y) = \big(x_0, (A - \overline{\lambda} I)y\big) = 0 \qquad (y \in \mathfrak{D}(A)) .$$

If $x_0 \notin \mathfrak{D}(A)$, we define A_0 as the linear extension of A to $\langle \mathfrak{D}(A), x_0 \rangle$ satisfying $A_0 x_0 = \lambda x_0$, and establish that A_0 is symmetric:

$$(A_0 x_0, x_0) - (x_0, A_0 x_0) = \lambda (x_0, x_0) - \overline{\lambda} (x_0, x_0) = 0 ,$$

$$(A_0 x_0, y) - (x_0, A_0 y) = -\big(x_0, (A - \overline{\lambda} I)y\big) = 0 \qquad (y \in \mathfrak{D}(A)) .$$

Thus a) implies b). Further, b) obviously implies c) with $A_1 = A_0$.

Finally, suppose that c) holds. Then for some non-zero element $x_1 \in \Pi_k$ we have $A_1 x_1 = \lambda x_1$. Corollary II.3.4 yields $(x_1, x_1) = 0$. On the other hand,

$$(x_1, (A - \bar{\lambda} I) y) = (A_1 x_1 - \lambda x_1, y) = 0 \qquad (y \in \mathfrak{D}(A)) .$$

Hence a) is valid. □

Theorem 6.2. *Consider a densely defined symmetric operator A in Π_k and a non-real number λ. The following conditions are equivalent:*

a) $\mathfrak{R}(A - \bar{\lambda} I)^{\perp} \cap \mathfrak{P}^{++} \neq 0$;

b) *A has a regular symmetric extension A_2 such that λ is an eigenvalue of A_2 and one of the corresponding eigenvectors does not belong to Π_k.*

Proof. Let $x_0 \perp \mathfrak{R}(A - \bar{\lambda} I)$, $(x_0, x_0) > 0$. Then $x_0 \notin \mathfrak{D}(A)$, since otherwise the equation $A x_0 = \lambda x_0$ (see (6.7)) coupled with Corollary II.3.4 would imply $(x_0, x_0) = 0$. Set $\Pi_k' = \langle \Pi_k, u_0 \rangle$, where

$$(u_0, x) = 0 \quad (x \in \Pi_k), \qquad (u_0, u_0) = -(x_0, x_0) ,$$

and define A_2 as the linear extension of A to $\langle \mathfrak{D}(A), x_0 + u_0 \rangle$ satisfying $A_2(x_0 + u_0) = \lambda(x_0 + u_0)$. On account of the relations

$$\begin{aligned} &(A_2(x_0 + u_0), x_0 + u_0) - (x_0 + u_0, A_2(x_0 + u_0)) = \\ &\qquad = (\lambda - \bar{\lambda})(x_0 + u_0, x_0 + u_0) = 0 , \\ &(A_2(x_0 + u_0), y) - (x_0 + u_0, A_2 y) = -(x_0 + u_0, (A - \bar{\lambda} I) y) = \\ &\qquad = -(x_0, (A - \lambda I) y) = 0 \qquad (y \in \mathfrak{D}(A)) \end{aligned}$$

the operator A_2 is symmetric.

Conversely, let b) be satisfied. Then there exists an $x_2 \notin \Pi_k$ with $A_2 x_2 = \lambda x_2$. According to Theorems 2.4 and 2.2 x_2 has a projection x_0 on Π_k. Since, by Corollary II.3.4, x_2 is neutral and, in view of Lemma 1.1 and Remark I.3.2, $x_2 - x_0$ is negative, x_0 must be positive. Moreover,

$$(x_0, (A - \bar{\lambda} I) y) = (x_2, (A - \bar{\lambda} I) y) = (A_2 x_2 - \lambda x_2, y) = 0$$
$$(y \in \mathfrak{D}(A)) . \quad □$$

Theorems 6.1—6.2 yield:

Corollary 6.3. *Let A be a densely defined symmetric operator in Π_k, and let λ be a non-real number. Unless the orthogonal companion of $\mathfrak{R}(A - \bar{\lambda} I)$ is negative definite, λ is an eigenvalue of some regular symmetric extension of A.* □

7. Invariant Positive Subspaces: Existence

Since every continuous operator of finite rank is compact, from Theorem VIII.3.1 and Lemma 1.2 we obtain:

Theorem 7.1. *Let U be a unitary operator in Π_k. Then U has a k-dimensional invariant positive subspace $\mathfrak{L}_1$ which contains the principal subspaces $\mathfrak{S}_\nu(U)$ ($|\nu| > 1$). If ν is a normal eigenvalue with $|\nu| > 1$, then $\mathfrak{S}_{\nu^*}(U) \cap \mathfrak{L}_1 = 0$.*

Similar statements hold for $|\nu| < 1$. □

The analogue, involving selfadjoint operators, of this result reads as follows.

Theorem 7.2. *Let A be a selfadjoint operator in Π_k. Then $\varrho(A)$ contains at least one non-real point. Moreover, A has a k-dimensional invariant positive subspace $\mathfrak{L}_1$ such that $\mathfrak{L}_1 \subset \mathfrak{D}(A)$ and $\mathfrak{S}_\lambda(A) \subset \mathfrak{L}_1$ ($\operatorname{Im} \lambda > 0$). If λ is a normal eigenvalue with $\operatorname{Im} \lambda > 0$, then $\mathfrak{S}_{\bar\lambda}(A) \cap \mathfrak{L}_1 = 0$.*

Similar statements hold for $\operatorname{Im} \lambda < 0$.

Proof. Making use of Theorem 1.4 and Lemma 1.3 we can find a fundamental decomposition (6.1) with the property $\Pi^+ \subset \mathfrak{D}(A)$. Introducing the matrix elements (6.4), from Lemma VI.6.7 it follows that A_{12} is continuous on $\mathfrak{D}(A_{12}) = \mathfrak{D}(A_{22})$, while $\varrho(A_{22})$ is non-empty. Thus for $\zeta_0 \in \varrho(A_{22})$ the operator $A_{12}(A_{22} - \zeta_0 I)^{-1}$ is continuous; having a finite rank, it is compact. Now we apply Theorem VIII.3.2 and Lemma 1.2. □

Since positizability plays a basic role in the spectral theory of Krein spaces (cf. Section VIII.6), the next result is particularly important.

Theorem 7.3. *Every continuous selfadjoint operator in Π_k is positizable.*

Proof. Let $A \in \mathfrak{B}(\Pi_k)$, $A = A^*$. According to Theorem 7.2 A has a k-dimensional invariant positive subspace $\mathfrak{L}_1$. As it is well known from linear algebra, there exists a non-constant polynomial φ (e. g. the characteristic polynomial or the minimal polynomial of $A|\mathfrak{L}_1$ in case $k > 0$, and $\varphi(\lambda) = \lambda$ if $k = 0$) such that $\varphi(A|\mathfrak{L}_1) = 0$. Let $\bar\varphi$ be the polynomial whose coefficients are complex conjugate to those of φ. Then

$$(\bar\varphi(A)x, y) = (x, \varphi(A)y) = 0 \qquad (x \in \Pi_k, \quad y \in \mathfrak{L}_1),$$

i. e., $\mathfrak{R}(\bar{\varphi}(A)) \perp \mathfrak{L}_1$. Hence, by Lemma 1.2 and Lemma I.6.3,

$$(\bar{\varphi}(A)x, \bar{\varphi}(A)x) \leqq 0 \qquad (x \in \Pi_k).$$

Therefore A is positizable by the real, non-constant polynomial $-\varphi\bar{\varphi}$. □

The following consequence of Theorem 7.1 will be needed when proving the existence of common invariant subspaces for commuting systems of unitary operators in Π_k.

Lemma 7.4. *Let U be a unitary operator in Π_k. Let $\mathfrak{M}$ be a closed subspace of Π_k such that $U\mathfrak{M} = \mathfrak{M}$. Unless $\mathfrak{M}$ is negative definite, $\mathfrak{M}$ contains a non-zero positive subspace which is invariant for U.*

Proof. If $\mathfrak{M}$ is degenerate, then its isotropic part $\mathfrak{M}^0 = \mathfrak{M} \cap \mathfrak{M}^\perp$ is non-zero, neutral and, by Corollary VI.4.6, invariant for U. If $\mathfrak{M}$ is non-degenerate and not negative definite then, in view of Corollary 2.3, $\mathfrak{M}$ is a space of type $\Pi_{k'}$, where $0 < k' \leqq k$; thus we may apply Theorem 7.1 to the restriction $U|\mathfrak{M}$. □

8. Invariant Positive Subspaces: Uniqueness

We are going to examine the question whether the k-dimensional invariant positive subspace of a selfadjoint operator in Π_k is unique or not.

Beginning the study with the case of a Pesonen operator, we first need the following lemma, interesting on its own right.

Lemma 8.1. *Let $A \in \mathfrak{B}(\Pi_k)$ be a Pesonen operator such that $(Ax, x) > 0$ $(x \in \mathfrak{P}^{00})$. Then the value*

$$\mu_A = \inf_{(x,x)>0} \frac{(Ax, x)}{(x, x)} \tag{8.1}$$

is finite and it is an eigenvalue of A; at the corresponding eigenvectors x, and only there, the infimum (8.1) is attained. Moreover, the eigenspace $\mathfrak{N}(A - \mu_A I)$ is contained in every k-dimensional invariant positive definite subspace of A.

Proof. Owing to Theorem 7.2 and Corollary II.9.6 A has a k-dimensional invariant positive definite subspace $\mathfrak{L}_1$. By Lemma 1.3 there exists a fundamental decomposition

$$\Pi_k = \Pi^+ (\dotplus) \Pi^-; \quad \Pi^+ \subset \mathfrak{P}^{++}, \quad \Pi^- \subset \mathfrak{P}^{--} \tag{8.2}$$

such that $\Pi^+ = \mathfrak{L}_1$. Since a Pesonen operator is always understood to act in an indefinite space, Π^+ and Π^- cannot be zero.

Let $(x, x) > 0$, $x \notin \Pi^+$. Then for the decomposition $x = x^+ + x^-$ $(x^\pm \in \Pi^\pm)$ we have $x^+ \neq 0$, $x^- \neq 0$. Since Lemma II.6.4 applied to the inner product $(.\,,.)_1 = (.\,,.)_A$ yields

$$\frac{(Ax^-, x^-)}{(x^-, x^-)} < \frac{(Ax^+, x^+)}{(x^+, x^+)},$$

making use of Lemma II.3.9 we obtain:

$$(Ax, x) = (Ax^+, x^+) + (Ax^-, x^-) > \frac{(Ax^+, x^+)}{(x^+, x^+)}[(x^+, x^+) + (x^-, x^-)].$$

Consequently,

$$\frac{(Ax, x)}{(x, x)} > \frac{(Ax^+, x^+)}{(x^+, x^+)} \qquad (x \in \mathfrak{P}^{++},\ x \notin \Pi^+). \tag{8.3}$$

Thus the infimum (8.1) can only be attained on the finite-dimensional, invariant, positive definite subspace Π^+. There, however, it is really attained, namely at the eigenvectors of $A|\Pi^+$ corresponding to the eigenvalue μ_A, as stated by a well-known theorem of linear algebra and easily verified with the help of the eigenvector expansion of $A|\Pi^+$.

Let x_0 be any eigenvector of A to the eigenvalue μ_A. We have just seen that A has at least one positive eigenvector to μ_A; therefore, on account of Corollary II.9.8, x_0 is positive. Moreover,

$$\frac{(Ax_0, x_0)}{(x_0, x_0)} = \frac{(\mu_A x_0, x_0)}{(x_0, x_0)} = \mu_A. \tag{8.4}$$

Relations (8.4), (8.1) and (8.3) imply $x_0 \in \Pi^+$. □

Theorem 8.2. *Let A be a Pesonen operator in Π_k. Then the k-dimensional invariant positive subspace of A is unique.*

Proof. Since A and $-A$ have the same invariant subspaces, by Theorem II.9.1 we may assume that $(Ax, x) > 0$ for every $x \in \mathfrak{P}^{00}$.

Let $\mathfrak{L}$ be a k-dimensional invariant positive subspace of A (cf. Theorem 7.2). According to Corollary II.9.6 and Lemma 8.1 $\mathfrak{L}$ is definite and contains the non-zero subspace $\mathfrak{N}(A - \mu_A I)$.

If $\mathfrak{N}(A - \mu_A I) \neq \mathfrak{L}$, then the orthogonal companion of $\mathfrak{N}(A - \mu_A I)$ is a space of type $\Pi_{k'}$ with $0 < k' < k$. Furthermore,

$$\mathfrak{L} = \mathfrak{N}(A - \mu_A I)\,(\dot{+})\,\mathfrak{L}',$$

where $\mathfrak{L}' = \mathfrak{L} \cap \Pi_{k'}$, $\dim \mathfrak{L}' = k'$. Also, by Theorem II.3.7, $A\Pi_{k'} \subset \Pi_{k'}$ and $A\mathfrak{L}' \subset \mathfrak{L}'$. Setting $A' = A|\Pi_{k'}$, introducing $\mu_{A'}$ in conformity with (8.1) and applying Lemma 8.1 again, we obtain that $\mu_{A'}$ is an eigenvalue of A' and $\mathfrak{N}(A' - \mu_{A'} I) \subset \mathfrak{L}'$.

Next, supposing that $\mathfrak{N}(A' - \mu_{A'}I) \neq \mathfrak{L}'$, we consider the space

$$\Pi_{k''} = (\mathfrak{N}(A - \mu_A I) \dot{+} \mathfrak{N}(A' - \mu_{A'}I))^{\perp},$$

the operator $A'' = A|\Pi_{k''}$, and the invariant subspace $\mathfrak{L}'' = \mathfrak{L} \cap \Pi_{k''}$; etc. After no more than k steps the process breaks off. Thus $\mathfrak{L}$ is the direct sum of a certain number of subspaces uniquely defined by A. □

Owing to Theorem 7.2 and Theorem VI.7.5, the number of k-dimensional invariant positive subspaces for a selfadjoint operator having a non-real normal eigenvalue in Π_k is not less than two. Below we give a non-trivial example of a selfadjoint operator with real spectrum and several k-dimensional invariant positive subspaces in Π_k.

Example 8.3. Let Π_k $(k > 0)$ be the orthogonal direct sum of a k-dimensional positive definite subspace Π^+ and a k-dimensional negative definite subspace Π^-. Let $\{e_j\}_1^k$ and $\{f_j\}_1^k$ be orthonormal bases of Π^+ and Π^-, respectively. Set

$$g_j = \begin{cases} \frac{1}{\sqrt{2}}(e_j + f_j) & \text{if} \quad 1 \leqq j \leqq k, \\ \frac{1}{\sqrt{2}}(e_{2k+1-j} - f_{2k+1-j}) & \text{if} \quad k+1 \leqq j \leqq 2k. \end{cases}$$

Then

$$(g_j, g_l) = \begin{cases} 1 & \text{if} \quad j + l = 2k + 1, \\ 0 & \text{otherwise}. \end{cases}$$

The linear operator A defined by the relations

$$Ag_j = g_{j-1} \quad (j \neq 1, k+1); \qquad Ag_1 = Ag_{k+1} = 0$$

is selfadjoint, has the only eigenvalue 0, and each of the subspaces

$$\langle g_1, g_2, \ldots, g_k \rangle, \langle g_1, g_2, \ldots, g_{k-1}, g_{k+1} \rangle,$$
$$\langle g_1, g_2, \ldots, g_{k-2}, g_{k+1}, g_{k+2} \rangle, \ldots, \langle g_{k+1}, g_{k+2}, \ldots, g_{2k} \rangle$$

is neutral and invariant under A.

Although the k-dimensional invariant positive subspace need not be unique, certain parameters connected with it are uniquely determined.

Lemma 8.4. *Let A be a selfadjoint operator in Π_k. Let $\mathfrak{L}_1 \subset \mathfrak{D}(A)$ and $\mathfrak{L} \subset \mathfrak{D}(A)$ be a k-dimensional and an arbitrary invariant positive subspace of A, respectively. Then for every $\lambda \in \mathbf{R}$ we have*

$$\dim \mathfrak{S}_\lambda(A|\mathfrak{L}) \leqq \dim \mathfrak{S}_\lambda(A|\mathfrak{L}_1).$$

In particular, every real eigenvalue of $A|\mathfrak{L}$ is an eigenvalue of $A|\mathfrak{L}_1$ too.

Proof. The subspace $\mathfrak{L}_1$, of dimension k, is the direct sum of the principal subspaces of $A|\mathfrak{L}_1$. Replacing the component $\mathfrak{S}_{\lambda_0}(A|\mathfrak{L}_1)$, where λ_0 is real, by $\mathfrak{S}_{\lambda_0}(A|\mathfrak{L})$, the direct sum remains positive because of Theorem II.3.3 and Lemma I.3.1; thus, on account of Lemma 1.1, its dimension cannot increase. (If $\lambda_0 \notin \sigma_p(A|\mathfrak{L}_1)$, then the "component" $\mathfrak{S}_{\lambda_0}(A|\mathfrak{L}_1)$ is understood to be zero; cf. Section II.1.) □

Corollary 8.5. *Let A be a selfadjoint operator in Π_k. Then the real eigenvalues $\lambda_1 < \lambda_2 < \cdots < \lambda_r$ of A in the k-dimensional invariant positive subspace $\mathfrak{L}_1 \subset \mathfrak{D}(A)$ and the algebraic multiplicities $m_{\lambda_1}(A|\mathfrak{L}_1), \ldots, m_{\lambda_r}(A|\mathfrak{L}_1)$ do not depend on the special choice of $\mathfrak{L}_1$.* □

Remark 8.6. The conclusions of the items 8.3—8.5 carry over to unitary operators in a natural way. Really, the proof of Lemma 8.4 can be repeated almost unaltered (substituting "unimodular" for "real"), while in the case of Example 8.3 it is easier to apply a Cayley transformation and refer to Lemma II.5.4.

9. Common Invariant Positive Subspaces for Commuting Operators

Lemma 9.1. *Let $U_1, \ldots, U_n$ be commuting unitary operators in Π_k, where $k > 0$. Then Π_k has a non-zero positive subspace $\mathfrak{L}$ such that $U_j\mathfrak{L} \subset \mathfrak{L}$ $(j = 1, \ldots, n)$.*

Proof. We proceed by induction. Owing to Theorem 7.1 the result holds for $n = 1$. Assume that it holds for n.

Let $U_1, \ldots, U_{n+1}$ be commuting unitary operators in Π_k $(k > 0)$. On account of the induction hypothesis there exists a non-zero positive subspace $\mathfrak{L} \subset \Pi_k$ such that $U_j\mathfrak{L} \subset \mathfrak{L}$ $(j = 1, \ldots, n)$. Since, by Lemma 1.1, $\dim \mathfrak{L} \leq k$, Lemma II.1.2 applies to the operators $U_1|\mathfrak{L}, \ldots, U_n|\mathfrak{L}$. Consequently, for a non-zero $x_0 \in \mathfrak{L}$ and suitable numbers λ_j we have

$$U_jx_0 = \lambda_jx_0 \qquad (j = 1, \ldots, n)\,. \tag{9.1}$$

Set

$$\mathfrak{M} = \bigcap_{j=1}^{n} \mathfrak{N}(U_j - \lambda_jI)\,. \tag{9.2}$$

Lemma II.1.1 yields $U_{n+1}\mathfrak{M} \subset \mathfrak{M}$. On the other hand, U_{n+1}^{-1} also commutes with U_j. (Really, if A and B are commuting, completely invertible operators, then $AB^{-1} = B^{-1}BAB^{-1} = B^{-1}ABB^{-1} = B^{-1}A$.) Hence $U_{n+1}^{-1}\mathfrak{M} \subset \mathfrak{M}$. In addition, $\mathfrak{M}$ is closed and contains the non-negative vector $x_0 \neq 0$. Therefore, according to Lemma 7.4, $\mathfrak{M}$ has a non-zero positive subspace $\mathfrak{L}'$ which is invariant under U_{n+1}. In view of (9.2) the subspace $\mathfrak{L}' \subset \mathfrak{M}$ is invariant for $U_1, \ldots, U_n$ as well. □

Lemma 9.2. *Let $U_1, \ldots, U_n$ be commuting unitary operators in Π_k. Then Π_k has a k-dimensional positive subspace $\mathfrak{L}_1$ such that $U_j\mathfrak{L}_1 \subset \mathfrak{L}_1$ $(j = 1, \ldots, n)$.*

Proof. The case $k = 0$ is trivial, so let $k > 0$. Then Lemma 9.1 provides a non-zero positive subspace $\mathfrak{L}$ satisfying the relations

$$U_j\mathfrak{L} \subset \mathfrak{L} \qquad (j = 1, \ldots, n)\,. \tag{9.3}$$

Suppose that

$$\dim \mathfrak{L} < k\,. \tag{9.4}$$

Since U_j is invertible (Theorem VI.4.2) and $\dim\mathfrak{L} < \infty$, in (9.3) inclusion can be replaced by equality. Therefore, on account of Corollary VI.4.6,

$$U_j\mathfrak{L}^\perp = \mathfrak{L}^\perp\,, \quad U_j\mathfrak{L}^0 = \mathfrak{L}^0 \qquad (j = 1, \ldots, n)\,, \tag{9.5}$$

where $\mathfrak{L}^0 = \mathfrak{L} \cap \mathfrak{L}^\perp$.

From Theorem 2.6 with the help of Lemma III.2.4, relation (9.4) and Corollary V.3.7 it follows that for some $k' \leq k$ the quotient space $\mathfrak{L}^\perp/\mathfrak{L}^0$ is a space of type $\Pi_{k'}$. Moreover, owing to (9.4), Lemma 1.2 and Lemma V.4.5, k' cannot be zero.

In view of (9.5) the operators U_j induce commuting unitary operators $\hat{U}_j$ on $\mathfrak{L}^\perp/\mathfrak{L}^0$. Consequently, by Lemma 9.1, there exists a non-zero positive subspace $\hat{\mathfrak{M}} \subset \mathfrak{L}^\perp/\mathfrak{L}^0$ with the property

$$\hat{U}_j\hat{\mathfrak{M}} \subset \hat{\mathfrak{M}} \qquad (j = 1, \ldots, n)\,. \tag{9.6}$$

Lemma 1.1 assures that $\dim\hat{\mathfrak{M}} \leq k'$. Let $\hat{x}_1, \ldots, \hat{x}_s$ be a basis of $\hat{\mathfrak{M}}$, and let $x_j \in \mathfrak{L}^\perp$ be a representative of the equivalence class $\hat{x}_j$ $(j = 1, \ldots, s)$. Then for the subspace

$$\mathfrak{M} = \langle x_1, \ldots, x_s \rangle \subset \mathfrak{L}^\perp$$

we have

$$U_j(\mathfrak{L} + \mathfrak{M}) \subset \mathfrak{L} + \mathfrak{M} \qquad (j = 1, \ldots, n)\,. \tag{9.7}$$

Further, since $\mathfrak{M}$ is positive, non-zero and linearly independent of $\mathfrak{L}^0$, the subspace $\mathfrak{L} + \mathfrak{M}$ is positive (cf. Lemma I.3.1) and different from $\mathfrak{L}$.

The invariant positive subspace $\mathfrak{L}$ satisfying (9.4) could thus be extended to an invariant positive subspace $\mathfrak{L}+\mathfrak{M}$ with $\dim(\mathfrak{L}+\mathfrak{M}) > \dim \mathfrak{L}$. Iteration, at most $k - \dim\mathfrak{L}$ times, of the process yields a k-dimensional invariant positive subspace $\mathfrak{L}_1$ as required. □

Theorem 9.3. *Let $\mathfrak{U}$ be a family of commuting unitary operators in Π_k. Then Π_k has a k-dimensional positive subspace $\mathfrak{L}_1$ such that $U\mathfrak{L}_1 \subset \mathfrak{L}_1$ $(U \in \mathfrak{U})$.*

Proof. Consider a fundamental decomposition

$$\Pi_k = \Pi^+ (\dot{+}) \Pi^-; \quad \Pi^+ \subset \mathfrak{P}^{++}, \quad \Pi^- \subset \mathfrak{P}^{--}, \tag{9.8}$$

and denote the corresponding fundamental symmetry by J. Let $(U_{jl})_{j,l=1,2}$ be the matrix of a $U \in \mathfrak{U}$ with respect to (9.8). Consider the space $\mathfrak{B}_1 = \mathfrak{B}(\Pi^+, \Pi^-)$ and its weak operator topology τ_{wo}. Set

$$\mathfrak{K} = \{X \in \mathfrak{B}_1: \ \|X\|_J \leqq 1\}. \tag{9.9}$$

The assertions a) — c) below can be obtained by specializing the proof of Theorem VIII.2.1 to the case $\mathfrak{L} = 0$, $\mathfrak{H} = \Pi_k$, $T = U$.

a) The formula

$$\mathfrak{L}_1 = \{x^+ + K_1 x^+: \ x^+ \in \Pi^+\} \tag{9.10}$$

defines a one-to-one correspondence between the class of all k-dimensional invariant positive subspaces $\mathfrak{L}_1$ of U and the set $\mathfrak{K}_U$ of all operators K_1 satisfying the conditions

$$K_1 \in \mathfrak{K}, \tag{9.11}$$

$$U_{21} + U_{22} K_1 - K_1 (U_{11} + U_{12} K_1) = 0. \tag{9.12}$$

b) $\mathfrak{K}$ is τ_{wo}-compact.

c) The left-hand side of (9.12) is a τ_{wo}-continuous function of the variable $K_1 \in \mathfrak{K}$.

According to a), the conclusion of our theorem holds if and only if

$$\bigcap_{U \in \mathfrak{U}} \mathfrak{K}_U \neq \emptyset. \tag{9.13}$$

But, on account of a) and c), every $\mathfrak{K}_U$ is contained in $\mathfrak{K}$ and closed with respect to the induced topology $\tau_{\text{wo}} | \mathfrak{K}$. Therefore, by assertion b), instead of (9.13) it is sufficient to show that the intersection of any finite subfamily of the family $\{\mathfrak{K}_U\}_{U \in \mathfrak{U}}$ is non-void. In other words (see a)), it is sufficient to prove the theorem for finite families $\mathfrak{U}$. This, however, has been previously done in Lemma 9.2. □

Theorem 9.4. *Let $\mathfrak{A}$ be a family of commuting continuous selfadjoint operators in Π_k. Then Π_k has a k-dimensional positive subspace $\mathfrak{L}_1$ such that $A\mathfrak{L}_1 \subset \mathfrak{L}_1$ $(A \in \mathfrak{A})$.*

Proof. Since the invariant subspaces of A coincide with those of λA $(\lambda \in \boldsymbol{C};\ \lambda \neq 0)$, we may assume that

$$\|A\|_J \leqq 1 \quad (A \in \mathfrak{A}), \tag{9.14}$$

where J is a fixed fundamental symmetry.

Let $\zeta \in \boldsymbol{C}$, $|\zeta| > 1$, $\zeta \neq \bar{\zeta}$. On account of (9.14) we have $\zeta \in \varrho(A)$ for every $A \in \mathfrak{A}$. Thus, according to Theorem VI.7.1, the Cayley transforms

$$(9.15) \qquad U_A = (A - \bar{\zeta}I)(A - \zeta I)^{-1} = I + (\zeta - \bar{\zeta})(A - \zeta I)^{-1} \qquad (A \in \mathfrak{A})$$

exist and are unitary. In addition, the operators U_A $(A \in \mathfrak{A})$ are easily seen to commute. Therefore Theorem 9.3 guarantees the existence of a k-dimensional positive subspace $\mathfrak{L}_1 \subset \Pi_k$ with the property

$$(9.16) \qquad U_A \mathfrak{L}_1 \subset \mathfrak{L}_1 \qquad (A \in \mathfrak{A}) .$$

Let us prove that $\mathfrak{L}_1$ is invariant under the operators A as well. Solving (9.15) for A we find (cf. Lemma II.4.1):

$$(9.17) \qquad A = (\zeta U_A - \bar{\zeta}I)(U_A - I)^{-1} \qquad (A \in \mathfrak{A}) .$$

Furthermore, the relations $(U_A - I)\mathfrak{L}_1 \subset \mathfrak{L}_1$ (see (9.16)), $\dim \mathfrak{L}_1 < \infty$, $\mathfrak{N}(U_A - I) = 0$ yield

$$(9.18) \qquad (U_A - I)\mathfrak{L}_1 = \mathfrak{L}_1 .$$

From (9.16) — (9.18) we obtain the desired inclusions $A\mathfrak{L}_1 \subset \mathfrak{L}_1$ $(A \in \mathfrak{A})$. □

Notes to Chapter IX

Sequence spaces of type Π_k were introduced by Pontrjagin [1]. Theorem 1.4, Theorem 2.2 and the "symmetric version" of Theorem 4.3 belong to this author.

The axiomatic definition of Π_k is due to Iohvidov and Kreĭn [1], [2], who established Theorems 2.5, 3.1, 4.3 and Example 4.11. Another treatment of Theorem 2.5 was given by Bognár [1]. For Theorem 3.2 and further discussion of the continuity of isometric operators cf. Iohvidov [11].

The results of Section 6 were obtained by Kreĭn and Langer [4] in the context of a detailed study of defect subspaces and generalized resolvents. For generalized resolvents in Π_k cf. also Kreĭn and Langer [5], Langer [14], Sorjonen [1].

Making use of analytic methods, Theorem 7.1 can be significantly improved (cf. also the Notes to Chapter VIII). Namely, if U is unitary in Π_k, then from Hilbert space perturbation theory it follows that the non-unimodular points of $\sigma(U)$ are normal eigenvalues. Therefore 1) the non-unimodular points of $\sigma(U)$ form at most k pairs, the members of each pair lying symmetrically with respect to the unit circle; 2) if the k-dimensional invariant positive subspace $\mathfrak{L}_1$ contains all principa subspaces $\mathfrak{S}_\nu(U)$ with $|\nu| > 1$, then for $|\nu| < 1$ necessarily $\mathfrak{L}_1 \cap \mathfrak{S}_\nu(U) = 0$ These facts have further implications.

Another way of obtaining the above improvements of Theorem 7.' is to prove first, again by perturbation theoretic methods, the existence of two k-dimensional invariant positive subspaces $\mathfrak{L}_1$ and $\mathfrak{L}_1'$ such tha all non-unimodular eigenvalues of $U|\mathfrak{L}_1$ (resp. $U|\mathfrak{L}_1'$) are locate outside (inside) the unit circle; everything else follows then by relatively simple geometric arguments. Cf. especially Iohvidov and Kreĭn [1].

Similar comments can be made in connection with Theorem 7.2.

For historical remarks on invariant subspace theorems see the Notes to Chapter VIII.

Azizov and Iohvidov [2] showed that Theorem 7.2 does not extend to non-degenerate quasi-negative or quasi-positive spaces ("non complete Pontrjagin spaces"), not even in case they admit a Hilbert majorant.

Iohvidov and Kreĭn [2] (cf. also Iohvidov [2]) proved an analogue of Theorem 7.3 for unitary operators and gave a characterization of the „positizing functions" involved. Their results were carried over to (possibly discontinuous) selfadjoint operators by Lo [1]. For an application to differential operators see Langer [15].

In Lemma 7.4 even the existence of a maximal positive subspace of $\mathfrak{M}$ invariant under U could be stated (see Iohvidov and Kreĭn [2]).

Lemma 8.1 and Theorem 8.2 belong to Kühne [1]. In the rest of Section 8 we follow Iohvidov and Kreĭn [2].

For further properties of spectra and invariant subspaces cf Iohvidov [2], Iohvidov and Kreĭn [1], [2].

The results of Section 9 are due to Naĭmark [1] (cf. also Helton [2]) Extensions to Krein spaces have been given by Langer [5], [12] and Helton [1].

Much of the theory of indefinite inner product spaces, especially of Pontrjagin spaces, was modelled in a finite-dimensional non-degenerate space and presented as a series of problems by Glazman and Ljubič [1; Chapter X].

Bognár [1] investigated square roots of selfadjoint operators in Π_k His results were carried over to more general functions by Peĭsahovič [1]. Operators of the form T^*T in Π_k have been considered by Holevo [1] as well as by Bognár and Krámli [1]. For other special questions concerning selfadjoint operators in Π_k cf. Bognár [3]—[6], Lo [2].

Shah [1] and independently, Naĭmark [12] generalized Stone's theorem to one-parameter unitary groups in Π_k.

Representations by unitary groups and symmetric algebras in Π_k as well as reductions and canonical models of the latter have been studied by Naĭmark [2]—[11], Ismagilov [1]—[4], Langer [7], Liberzon and Šul'man [1], Loginov [1]—[3], Šul'man [1]. Cf. also the survey articles Naĭmark [13], Naĭmark and Ismagilov [1].

Bognár [8] gave a characterization of the mapping $T \to T^*$ ($T \in \mathfrak{B}(\Pi_k)$) regarded as an operation in the algebra of all continuous linear operators on a Banach space.

Iohvidov [2], Azizov and Iohvidov [1], Azizov [2] found conditions in order that the principal vectors of a compact selfadjoint or compact "dissipative" operator in Π_k form a complete system.

The theory of Pontrjagin spaces has been applied to obtain integral representations, extensions and other properties of sequences, matrices, functions and kernels with a finite rank of indefiniteness in the following papers: Kreĭn [3], [5], Iohvidov [3], [13], Iohvidov and Kreĭn [2], Daleckiĭ [1], Shah [1], V. I. Gorbačuk [1]—[7], V. I. Gorbačuk and M. L. Gorbačuk [1].

Applications of Pontrjagin spaces to concrete problems of mechanics have been given by S. G. Kreĭn and Moiseev [1], Sobolev [1], M. L. Gorbačuk, Slepcova and Temčenko [1], to orthogonal polynomials by M. G. Kreĭn [9], and to numerical methods by Gerisch and Gerisch [1].

Bibliography

Arons, M. E., Han, M. Y., Sudarshan, E. C. G.: [1] Finite quantum electrodynamics: a field theory using an indefinite metric. Phys. Rev. (2) **137**, B 1085—B 1104 (1965).

Aronszajn, N.: [1] Quadratic forms on vector spaces. In: Proc. Internat. Sympos. Linear Spaces, pp. 29—87. Jerusalem and Oxford: Jerusalem Academic Press and Pergamon 1961.

Azizov, T. Ja.: [1] The spectra of certain operator classes in Hilbert space. Mat. Zametki **9**, 303—310 (1971) [Russian].

— [2] Invariant subspaces and criteria of completeness for the system of root vectors of J-dissipative operators in the Pontrjagin space $\Pi_\varkappa$. Dokl. Akad. Nauk SSSR **200**, 1015—1017 (1971) [Russian].

Azizov, T. Ja., Iohvidov, I. S.: [1] A criterion, in order to form a complete system or a basis, for the root vectors of a completely continuous J-selfadjoint operator in the Pontrjagin space $\Pi_\varkappa$. Mat. Issled. **6**, no. 1, 158—161 (1971) [Russian].

— [2] Linear operators in Hilbert spaces with a G-metric. Uspehi Mat. Nauk **26**, no. 4, 43—92 (1971) [Russian].

Berezin, F. A.: [1] On the Lee model. Mat. Sb. **60**, 425—446 (1963) [Russian].

Berge, C.: [1] Espaces topologiques: fonctions multivoques, Paris: Dunod 1959.

Bleuler, K.: [1] Eine neue Methode zur Behandlung der longitudinalen und skalaren Photonen. Helvetica Phys. Acta **23**, 567—586 (1950).

Bognár, J. (= Bognar, Ja.): [1] On the existence of square roots of an operator which is self-adjoint with respect to an indefinite metric. Magyar Tud. Akad. Mat. Kutató Int. Közl. **6**, 351—363 (1961) [Russian].

— [2] On a discontinuity property of the inner product in spaces with indefinite metric. Uspehi Mat. Nauk **17**, no. 1, 157—159 (1962) [Russian].

— [3] Non-negativity properties of operators in spaces with indefinite metric. Ann. Acad. Sci. Fenn. Ser. A I, no. 336/10 (1963).

— [4] Certain relations among the non-negativity properties of operators in spaces with an indefinite metric. Magyar Tud. Akad. Mat. Kutató Int. Közl. **8**, 201—212 (1963) [Russian].

— [5] Certain relations among the non-negativity properties of operators in spaces with an indefinite metric. II. Studia Sci. Math. Hungar. **1**, 97—102 (1966) [Russian].

— [6] Certain relations among the non-negativity properties of operators in spaces with an indefinite metric. III. Studia Sci. Math. Hungar. **1**, 419—426 (1966) [Russian].

— [7] On decomposition majorants of an indefinite metric. Math. Z. **101**, 65—67 (1967).

Bognár, J.: [8] Involution as operator conjugation. In: Colloquia Math. Soc. János Bolyai, Vol. 5, Hilbert space operators and operator algebras, pp. 53—64. Amsterdam/London: North-Holland 1972.

— [9] A remark on doubly strict plus-operators. Mat. Issled. (to appear) [Russian].

Bognár, J., Krámli, A.: [1] Operators of the form C^*C in indefinite inner product spaces. Acta Sci. Math. (Szeged) **29**, 19—29 (1968).

Bogoljubov, N. N., Medvedev, B. V., Polivanov, M. K.: [1] On the question of an indefinite metric in quantum field theory. Naučnye Doklady Vysšeĭ Školy, Fiz.-Mat. Nauki (1958), no. 2, 137—142 (1958) [Russian].

Bonsall, F. F.: [1] Indefinitely isometric linear operators in a reflexive Banach space. Quart. J. Math. Oxford Ser. (2) **6**, 179—187 (1955).

Bourbaki, N.: [1] Éléments de mathématique. XV, XVIII, XIX. Espaces vectoriels topologiques. Actualités Sci. Ind., nos. 1189, 1229, 1230. Paris: Hermann 1953 and 1955.

Brodskiĭ, M. L.: [1] On properties of operators mapping the non-negative part of a space with indefinite metric into itself. Uspehi Mat. Nauk **14**, no. 1, 147—152 (1959) [Russian].

Brodskiĭ, V. M.: [1] Operator colligations and their characteristic functions. Dokl. Akad. Nauk SSSR **198**, 16—19 (1971) [Russian].

Browder, F. E.: [1] A remark on the Dirichlet problem for non-elliptic self-adjoint partial differential operators. Rend. Circ. Mat. Palermo (2) **6**, 249—253 (1957).

— [2] On the Dirichlet problem for linear non-elliptic partial differential equations. II. Rend. Circ. Mat. Palermo (2) **7**, 303—308 (1958).

Cordes, H. O.: [1] On maximal first order partial differential operators. Amer. J. Math. **82**, 63—91 (1960).

Crandall, M. G., Phillips, R. S.: [1] On the extension problem for dissipative operators. J. Functional Analysis **2**, 147—176 (1968).

Daleckiĭ, J. u. L.: [1] Differentiation of non-hermitian matrix functions depending on a parameter. Izv. Vysš. Učebn. Zaved. Matematika 2, 52—64 (1962) [Russian].

Daleckiĭ, Ju. L., Fadeeva, E. A.: [1] Hyperbolic equations with operator coefficients, and ultra-parabolic systems. Ukrain. Mat. Ž. **24**, 92—95 (1972) [Russian].

Daleckiĭ, Ju. L., Kreĭn, M. G.: [1] The stability of the solutions of differential equations in a Banach space, Moscow: Nauka 1970 [Russian].

Davis, Ch.: [1] J-unitary dilation of a general operator. Acta Sci. Math. (Szeged) **31**, 75—86 (1970).

— [2] Dilation of uniformly continuous semi-groups. Rev. Roumaine Math. Pures Appl. **15**, 975—983 (1970).

Davis, Ch., Foiaş, C.: [1] Operators with bounded characteristic function and their J-unitary dilation. Acta Sci. Math. (Szeged) **32**, 127—139 (1971).

Dirac, P. A. M.: [1] The physical interpretation of quantum mechanics. Proc. Roy. Soc. London Ser. A **180**, 1—40 (1942).

Dolph, C. L.: [1] Recent developments in some non-self-adjoint problems of mathematical physics. Bull. Amer. Math. Soc. **67**, 1—69 (1961).

Dunford, N., Schwartz, J.T.: [1] Linear operators. I. General theory, New York/London: Interscience 1958.

Eisenfeld, J.: [1] On symmetrization and roots of quadratic eigenvalue problems. J. Functional Analysis **9**, 410—422 (1972).

Fan, K.: [1] Fixed-point and minimax theorems in locally convex topological linear spaces. Proc. Nat. Acad. Sci. U.S.A. **38**, 121—126 (1952).
— [2] Invariant subspaces of certain linear operators. Bull. Amer. Math. Soc. **69**, 773—777 (1963).
— [3] Invariant cross-sections and invariant linear subspaces. Israel J. Math. **2**, 19—26 (1964).
— [4] Invariant subspaces for a semigroup of linear operators. Indag. Math. **27**, 447—451 (1965).
— [5] Applications of a theorem concerning sets with convex sections. Math. Ann. **163**, 189—203 (1966).

Fischer, H.R., Gross, H.: [1] Quadratic forms and linear topologies. I. Math. Ann. **157**, 296—325 (1964).

Gerisch, A. (= Geriš, A.), Gerisch, W. (= Geriš, V.): [1] Pontrjagin's space and convergence of the Bubnov-Galerkin method. Dokl. Akad. Nauk SSSR **193**, 1218—1221 (1970) [Russian].

Ginzburg, Ju. P.: [1] On *J*-contractive operator functions. Dokl. Akad. Nauk SSSR **117**, 171—173 (1957) [Russian].
— [2] On *J*-contractive operators in Hilbert space. Odess. Gos. Ped. Inst. Naučn. Zap. Fiz.-Mat. Fak. **22**, no. 1, 13—20 (1958) [Russian].
— [3] Subspaces of a Hilbert space with indefinite metric. Odess. Ped. Inst. Naučn. Zap. Kaf. Mat. Fiz. Estestv. **25**, no. 2, 3—9 (1961) [Russian].
— [4] Projections in a Hilbert space with bilinear metric. Dokl. Akad. Nauk SSSR **139**, 775—778 (1961) [Russian].

Ginzburg, Ju. P., Iohvidov, I. S.: [1] A study of the geometry of infinite-dimensional spaces with bilinear metric. Uspehi Mat. Nauk **17**, no.4, 3—56 (1962) [Russian].

Glazman, I. M., Ljubič, Ju. I.: [1] Finite-dimensional linear analysis, Moscow: Nauka 1969 [Russian].

Glicksberg, I. L.: [1] A further generalization of the Kakutani fixed point theorem, with application to Nash equilibrium points. Proc. Amer. Math. Soc. **3**, 170—174 (1952).

Gorbačuk, M. L., Slepcova, G. P., Temčenko, M. E.: [1] Stability of motion of a rigid body suspended on a string and filled with fluid. Ukrain. Mat. Ž. **20**, 586—602 (1968) [Russian].

Gorbačuk, V. I. (= Pljuščeva, V. I.): [1] The integral representation of hermitian-indefinite matrices with $\varkappa$ negative squares. Ukrain. Mat. Ž. **14**, 30—39 (1962) [Russian].
— [2] The integral representation of continuous hermitian-indefinite kernels. Dokl. Akad. Nauk SSSR **145**, 534—537 (1962) [Russian].
— [3] The integral representation of hermitian-indefinite kernels (the case of several variables). Ukrain. Mat. Ž. **16**, 232—236 (1964) [Russian].
— [4] The integral representation of hermitian-indefinite kernels. Ukrain. Mat. Ž. **17**, no. 3, 43—58 (1965) [Russian].
— [5] On the uniqueness of the representation of hermitian-indefinite functions and sequences. Ukrain. Mat. Ž. **18**, no. 2, 107—113 (1966) [Russian].
— [6] Extensions of a real hermitian-indefinite function with one negative square. Ukrain. Mat. Ž. **19**, no. 4, 119—125 (1967) [Russian].
— [7] Self-adjoint extensions of some Hermitian operators in a space with indefinite metric. In: Colloquia Math. Soc. János Bolyai, Vol. 5, Hilbert space operators and operator algebras, pp. 265—269. Amsterdam/London: North-Holland 1972.

Gorbačuk, V. I., Gorbačuk, M. L.: [1] Representation of the vacuum-mean of field operators in a space with an indefinite metric. Ukrain. Mat. Ž. **18**, no. 6, 108—111 (1966) [Russian].

Greub, W. H.: [1] Linear algebra, 2nd edition, New York and Berlin/Göttingen/Heidelberg: Academic Press and Springer 1963.

Gupta, S. N.: [1] Theory of longitudinal photons in quantum electrodynamics. Proc. Phys. Soc. Sect. A **63**, 681—691 (1950).

Hackevič, V. A.: [1] Invariant subspaces for certain classes of linear operators in normed spaces with an indefinite metric. Mat. Issled. **6**, no. 3, 133—147 (1971) [Russian].

Harazov, D. F.: [1] Symmetrizable operators that do not satisfy the conditions of positive-definiteness, and their applications. Studia Math. **34**, 241—252 (1970) [Russian].

Heisenberg, W.: [1] Erweiterungen des Hilbert-Raums in der Quantentheorie der Wellenfelder. Z. Physik **144**, 1—8 (1956).

— [2] Hilbert space II and the "ghost" states of Pauli and Källén. Nuovo Cimento (10) **4**, supplemento, 743—747 (1956).

— [3] Lee model and quantisation of non linear field equations. Nuclear Phys. **4**, 532—563 (1957).

— [4] Introduction to the unified field theory of elementary particles, London/New York/Sydney: Interscience 1966.

Helton, J. W.: [1] Unitary operators on a space with an indefinite inner product. J. Functional Analysis **6**, 412—440 (1970).

— [2] Operators unitary in an indefinite metric and linear fractional transformations. Acta Sci. Math. (Szeged) **32**, 261—266 (1971).

Hess, P.: [1] Zur Theorie der linearen Operatoren eines J-Raumes. Operatoren die von kanonischen Zerlegungen reduziert werden. Math. Z. **106**, 88—96 (1968).

— [2] Über Polynome J-symmetrischer Operatoren in J-Räumen. Math. Z. **114**, 271—277 (1970).

Hestenes, M. R.: [1] Applications of the theory of quadratic forms in Hilbert space to the calculus of variations. Pacific J. Math. **1**, 525—581 (1951).

Hildebrandt, S.: [1] Rand- und Eigenwertaufgaben bei stark elliptischen Systemen linearer Differentialgleichungen. Math. Ann. **148**, 411—429 (1962).

Holevo, A. S.: [1] Generalization to spaces with an indefinite metric of a theorem of von Neumann on the operator T^*T. Azerbaĭdžan. Gos. Univ. Učen. Zap. Ser. Fiz.-Mat. Nauk (1965), no. 2, 45—48 (1965) [Russian].

Iohvidov, I. S.: [1] Unitary operators in a space with an indefinite metric. Zap. Mat. Otd. Fiz.-Mat. Fak. i Har'kov. Mat. Obšč. (4) **21**, 79—86 (1949) [Russian].

— [2] On the spectra of hermitian and unitary operators in a space with indefinite metric. Dokl. Akad. Nauk SSSR **71**, 225—228 (1950) [Russian].

— [3] On the theory of indefinite Toeplitz forms. Dokl. Akad. Nauk SSSR **101**, 213—216 (1955) [Russian].

— [4] Boundedness of J-isometric operators. Uspehi Mat. Nauk **16**, no. 4, 167—170 (1961) [Russian].

— [5] Regular and projection-complete linear manifolds in spaces with a general hermitian bilinear metric. Dokl. Akad. Nauk SSSR **139**, 791—794 (1961) [Russian].

— [6] Singular linear manifolds in spaces with an arbitrary hermitian bilinear metric. Uspehi Mat. Nauk **17**, no. 4, 127—133 (1962) [Russian].

Iohvidov, I. S.: [7] Operators with completely continuous iterations. Dokl. Akad. Nauk SSSR **153**, 258–261 (1963) [Russian].
— [8] Singular linear manifolds in the space $\Pi_\varkappa$. Ukrain. Mat. Ž. **16**, 300–308 (1964) [Russian].
— [9] On a lemma of Ky Fan generalizing the fixed-point principle of A. N. Tihonov. Dokl. Akad. Nauk SSSR **159**, 501–504 (1964) [Russian].
— [10] On maximal definite linear manifolds in a Hilbert space with a G-metric. Ukrain. Mat. Ž. **17**, no. 4, 22–28 (1965) [Russian].
— [11] G-isometric and J-semiunitary operators in Hilbert space. Uspehi Mat. Nauk **20**, no. 3, 175–181 (1965) [Russian].
— [12] Linear fractional transformations of J-contractive operators. Akad. Nauk Armjan. SSR Dokl. **42**, 3–8 (1966) [Russian].
— [13] Unitary extensions of isometric operators in the Pontrjagin space Π_1 and continuations in the $\mathfrak{P}_1$ class of finite sequences of the class $\mathfrak{P}_{1;n}$. Dokl. Akad. Nauk SSSR **173**, 758–761 (1967) [Russian].
— [14] Banach spaces with a J-metric and certain classes of linear operators in these spaces. Bul. Akad. Štiince RSS Moldoven. 1, 60–80 (1968) [Russian].
— [15] On a class of linear fractional operator transformations. Voronež. Gos. Univ. Trudy Sem. Funkcional. Anal. 18–44 (1970) [Russian].

Iohvidov, I. S., Kreĭn, M. G.: [1] Spectral theory of operators in spaces with an indefinite metric. I. Trudy Moskov. Mat. Obšč. **5**, 367–432 (1956) and **6**, 486 (1957) [Russian].
— [2] Spectral theory of operators in spaces with an indefinite metric. II. Trudy Moskov. Mat. Obšč. **8**, 413–496 (1959) and **15**, 452–454 (1966) [Russian].

Iohvidov, I. S., Senderov, V. A.: [1] The boundedness of J-semiunitary operators in Banach spaces with a J-metric. Mat. Issled. **5**, no. 4, 166–170 (1970) [Russian].

Ismagilov, R. S.: [1] Description of the unitary representations of the Lorentz group in a space with indefinite metric. Dokl. Akad. Nauk SSSR **158**, 268–270 (1964) [Russian].
— [2] Unitary representations of the Lorentz group in a space with indefinite metric. Izv. Akad. Nauk SSSR Ser. Mat. **30**, 497–522 (1966) [Russian].
— [3] Irreducible representations of the discrete group $SL(2, P)$ that are unitary with respect to an indefinite metric. Izv. Akad. Nauk SSSR Ser. Mat. **30**, 923–950 (1966) [Russian].
— [4] Rings of operators in a space with an indefinite metric. Dokl. Akad. Nauk SSSR **171**, 269–271 (1966) [Russian].

Jalava, V.: [1] On spectral decompositions of operators in J-space. Ann. Acad. Sci. Fenn. Ser. A I, no. 446 (1969).
— [2] On operators in a linear space with a non-degenerate sesquilinear form. Univ. Jyväskylä Dept. Math., Report 5 (1969).
— [3] On the square root of a self-adjoint operator in J-space. Ann. Acad. Sci. Fenn. Ser. A I, no. 468 (1970).

Jarchow, H.: [1] Stetigkeit hermitescher Formen. Ann. Acad. Sci. Fenn. Ser. A I, no. 441 (1969).
— [2] Topologisch stetige hermitesche Formen. Math. Z. **113**, 326–334 (1970).

Jelínek, J., Virsik, J.: [1] Pseudo-unitary spaces. Časopis Pěst. Mat. **91**, 18–33 (1966).

Jonas, P.: [1] Eine Bedingung für die Existenz einer Eigenspektralfunktion für gewisse Automorphismen lokalkonvexer Räume. Math. Nachr. **45**, 143—160 (1970).

Källén, G., Pauli, W.: [1] On the mathematical structure of T. D. Lee's model of a renormalizable field theory. Danske Vid. Selsk. Mat.-Fys. Medd. **30**, no. 7 (1955).

Karrer, G.: [1] Spektraltheorie der Automorphismen Hermite'scher Formen. Ann. Acad. Sci. Fenn. Ser. A I, no. 237 (1957).

Köthe, G.: [1] Topologische lineare Räume, Vol. 1, Berlin/Göttingen/Heidelberg: Springer 1960.

Kraljević, H.: [1] Simultaneous diagonalisation of two symmetric bilinear functionals. Glasnik Mat. Ser. III **1**, 57—63 (1966).

Kreĭn, M. G.: [1] On weighted integral equations the distribution functions of which are not monotonic. In: Memorial volume dedicated to D. A. Grave, pp. 88—103. Moscow/Leningrad: Gostehizdat 1940 [Russian].

— [2] Completely continuous linear operators in function spaces with two norms. Akad. Nauk Ukrain. RSR. Zbirnik Prac' Inst. Mat. no. 9, 104—129 (1947) [Ukrainian].

— [3] Helices in the infinite-dimensional Lobačevskiĭ space. Uspehi Mat. Nauk **3**, no. 3, 158—160 (1948) [Russian].

— [4] An application of the fixed-point principle in the theory of linear transformations of spaces with an indefinite metric. Uspehi Mat. Nauk **5**, no. 2, 180—190 (1950) [Russian].

— [5] Integral representation of a continuous hermitian-indefinite function with a finite number of negative squares. Dokl. Akad. Nauk SSSR **125**, 31—34 (1959) [Russian].

— [6] Lectures on the theory of the stability of solutions of differential equations in a Banach space. Kiev: Akad. Nauk Ukrain. SSR Inst. Mat. 1964 [Russian].

— [7] A new application of the fixed-point principle in the theory of operators in a space with indefinite metric. Dokl. Akad. Nauk SSSR **154**, 1023—1026 (1964) [Russian].

— [8] Introduction to the geometry of indefinite J-spaces and to the theory of operators in those spaces. In: Second mathematical summer school, Part I, pp. 15—92. Kiev: Naukova Dumka 1965 [Russian].

— [9] Distribution of roots of polynomials orthogonal on the unit circle with respect to a sign-alternating weight. Teor. Funkciĭ Funkcional. Anal. i Priložen. no. 2, 131—137 (1966) [Russian].

Kreĭn, M. G., Langer, H. (= Langer, G. K.): [1] On the spectral function of a selfadjoint operator in a space with indefinite metric. Dokl. Akad. Nauk SSSR **152**, 39—42 (1963) [Russian].

— [2] On the theory of quadratic pencils of self-adjoint operators. Dokl. Akad. Nauk SSSR **154**, 1258—1261 (1964) [Russian].

— [3] Certain mathematical principles of the linear theory of damped vibrations of continua. In: Applications of the theory of functions in continuum mechanics, Vol. II, pp. 283—322. Moscow: Nauka 1965 [Russian].

— [4] The defect subspaces and generalized resolvents of a hermitian operator in the space $\Pi_\varkappa$. Funkcional. Anal. i Priložen. **5**, no. 2, 59—71 and no. 3, 54—69 (1971) [Russian].

— [5] Über die verallgemeinerten Resolventen und die characteristische Funktion eines isometrischen Operators im Raume $\Pi_\varkappa$. In: Colloquia Math. Soc. János Bolyai, Vol. 5, Hilbert space operators and operator algebras, pp. 353—399. Amsterdam/London: North-Holland 1972.

Kreĭn, M. G., Rutman, M. A.: [1] Linear operators leaving invariant a cone in a Banach space. Uspehi Mat. Nauk **3**, no. 1, 3—95 (1948) [Russian].

Kreĭn, M. G., Šmul'jan, Ju. L.: [1] Plus-operators in a space with an indefinite metric. Mat. Issled. **1**, no. 1, 131—161 (1966) [Russian].
— [2] *J*-polar representations of plus-operators. Mat. Issled. **1**, no. 2, 172—210 (1966) [Russian].
— [3] Linear fractional transformations with operator coefficients. Mat. Issled. **2**, no. 3, 64—96 (1967) [Russian].

Kreĭn, S. G., Moiseev, N. N.: [1] On oscillations of a vessel containing a liquid with a free surface. Prikl. Mat. Meh. **21**, 169—174 (1957) [Russian].

Kühne, R.: [1] Über eine Klasse *J*-selbstadjungierter Operatoren. Math. Ann. **154**, 56—69 (1964).
— [2] Minimaxprinzipe für stark gedämpfte Scharen. Acta Sci. Math. (Szeged) **29**, 39—68 (1968).

Kužel', A. V.: [1] The spectral analysis of quasi-unitary operators in a space with indefinite metric. Teor. Funkciĭ Funkcional. Anal. i Priložen. no. 4, 3—27 (1967) [Russian].

Langer, H. (= Langer, G. K.): [1] On *J*-hermitian operators. Dokl. Akad. Nauk SSSR **134**, 263—266 (1960) [Russian].
— [2] Zur Spektraltheorie *J*-selbstadjungierter Operatoren. Math. Ann. **146**, 60—85 (1962).
— [3] Eine Verallgemeinerung eines Satzes von L. S. Pontrjagin. Math. Ann. **152**, 434—436 (1963).
— [4] Eine Erweiterung der Spurformel der Störungstheorie. Math. Nachr. **30**, 123—135 (1965).
— [5] Invariant subspaces of linear operators acting in a space with indefinite metric. Dokl. Akad. Nauk SSSR **169**, 12—15 (1966) [Russian].
— [6] Spektralfunktionen einer Klasse *J*-selbstadjungierter Operatoren. Math. Nachr. **33**, 107—120 (1967).
— [7] Über einen Satz von M. A. Neumark. Math. Ann. **175**, 303—314 (1968).
— [8] Über stark gedämpfte Scharen im Hilbertraum. J. Math. Mech. **17**, 685—705 (1968).
— [9] Über die schwache Stabilität linearer Differentialgleichungen mit periodischen Koeffizienten. Math. Scand. **22**, 203—208 (1968).
— [10] A remark on invariant subspaces of linear operators in Banach spaces with an indefinite metric. Mat. Issled. **4**, no. 1, 27—34 (1969) [Russian].
— [11] Maximal dual pairs of invariant subspaces of *J*-selfadjoint operators. Mat. Zametki **7**, 443—447 (1970) [Russian].
— [12] Invariante Teilräume definisierbarer *J*-selbstadjungierter Operatoren. Ann. Acad. Sci. Fenn. Ser. A I, no. 475 (1971).
— [13] Verallgemeinerte Resolventen eines *J*-nichtnegativen Operators mit endlichem Defekt. J. Functional Analysis **8**, 287—320 (1971).
— [14] Generalized co-resolvents of a π-isometric operator with unequal defect numbers. Funkcional. Anal. i Priložen. **5**, no. 4, 73—75 (1971) [Russian].
— [15] Zur Spektraltheorie verallgemeinerter gewöhnlicher Differentialoperatoren zweiter Ordnung mit einer nichtmonotonen Gewichtsfunktion. Univ. Jyväskylä Dept. Math., Report 14 (1972).

Larionov, E. A.: [1] A commutative family of operators in a space with indefinite metric. Mat. Zametki **1**, 589—594 (1967) [Russian].

Larionov, E. A.: [2] The extension of dual subspaces. Dokl. Akad. Nauk SSSR **176**, 515–517 (1967) [Russian].
— [3] The extension of dual subspaces invariant under an algebra. Mat. Zametki 3, 253–260 (1968) [Russian].
— [4] Nilpotent J-selfadjoint operators. Dokl. Akad. Nauk SSSR **183**, 768–771 (1968) [Russian].
— [5] Selfadjoint quadratic pencils. Izv. Akad. Nauk SSSR Ser. Mat. **33**, 138–154 (1969) [Russian].

Lax, P. D., Phillips, R. S.: [1] The acoustic equation with an indefinite energy form and the Schrödinger equation. J. Functional Analysis **1**, 37–83 (1967).
— [2] Scattering theory, New York/London: Academic Press 1967.
— [3] Decaying modes for the wave equation in the exterior of an obstacle. Comm. Pure Appl. Math. **22**, 737–787 (1969).

Lee, T. D., Wick, G. C.: [1] Negative metric and the unitarity of the S-matrix. Nuclear Phys. B **9**, 209–243 (1969).

Liberzon, V. I., Šul'man, V. S.: [1] Operator-irreducible symmetric operator algebras in the Pontrjagin space Π^1. Izv. Akad. Nauk SSSR Ser. Mat. **35**, 1159–1170 (1971) [Russian].

Littman, W.: [1] Remarks on the Dirichlet problem for general linear partial differential equations. Comm. Pure Appl. Math. **11**, 145–151 (1958).

Lo, C.-Y.: [1] A class of polynomials in self-adjoint operators in spaces with an indefinite metric. Canad. J. Math. **20**, 673–678 (1968).
— [2] On polynomials in self-adjoint operators in spaces with an indefinite metric. Trans. Amer. Math. Soc. **134**, 297–304 (1968).

Loginov, A. I.: [1] Semidegenerate algebras in a Pontrjagin space. Mat. Zametki **6**, 73–80 (1969) [Russian].
— [2] Commutative symmetric operator algebras in a Pontrjagin space. Izv. Akad. Nauk SSSR Ser. Mat. **33**, 549–569 (1969) [Russian].
— [3] Complete commutative symmetric operator algebras in the Pontrjagin space Π_1. Mat. Sb. **84**, 575–582 (1971) [Russian].

Louhivaara, I. S.: [1] Über das erste Randwertproblem für die Differentialgleichung $u_{xx} + u_{yy} + q\,u + f = 0$. Ann. Acad. Sci. Fenn. Ser. A I, no. 183 (1955).
— [2] Über das zweite und dritte Randwertproblem für die Differentialgleichung $u_{xx} + u_{yy} + q\,u + f = 0$. Ann. Acad. Sci. Fenn. Ser. A I, no. 203 (1955).
— [3] Über das Dirichletsche Problem für die selbstadjungierten linearen partiellen Differentialgleichungen zweiter Ordnung. Rend. Circ. Mat. Palermo (2) **5**, 260–274 (1956).
— [4] Bemerkung zur Theorie der Nevanlinnaschen Räume. Ann. Acad. Sci. Fenn. Ser. A I, no. 232 (1956).
— [5] Zur Theorie der Unterräume in linearen Räumen mit indefiniter Metrik. Ann. Acad. Sci. Fenn. Ser. A I, no. 252 (1958).
— [6] Über verschiedene Metriken in linearen Räumen. Ann. Acad. Sci. Fenn. Ser. A I, no. 282 (1960).
— [7] Über die neuere Entwicklung der Theorie der linearen Räume mit indefiniten Bilinearformen. In: Festband 70. Geburtstag R. Nevanlinna, pp. 66–81. Berlin/Heidelberg/New York: Springer 1966.

Mal'cev, A. I.: [1] Foundations of linear algebra, San Francisco/London: Freeman 1963.

Marksjö, B., Textorius, B.: [1] On the stability of linear differential equations in spaces with an indefinite metric. Math. Scand. **20**, 177–192 (1967).

Masuda, K.: [1] On the existence of invariant subspaces in spaces with indefinite metric. Proc. Amer. Math. Soc. **32**, 440—444 (1972).

Murray, F. J.: [1] On complementary manifolds and projections in spaces L_p and l_p. Trans. Amer. Math. Soc. **41**, 138—152 (1937).

Nagy, K. L.: [1] Indefinite metric in quantum field theory. Nuovo Cimento (10) **17**, supplemento, 92—131 (1960).

— [2] State vector spaces with indefinite metric in quantum field theory, Groningen and Budapest: Noordhoff and Akadémiai Kiadó 1966.

— [3] Complex poles, cuts, indefinite metric and unitarity. Acta Phys. Acad. Sci. Hungar. **29**, 251—265 (1970).

Naĭmark, M. A.: [1] On commuting unitary operators in spaces with indefinite metric. Acta Sci. Math. (Szeged) **24**, 177—189 (1963).

— [2] Unitary representations of solvable groups in spaces with indefinite metric. Izv. Akad. Nauk SSSR Ser. Mat. **27**, 1181—1185 (1963) [Russian].

— [3] Commutative algebras of operators in the space Π_1. Rev. Roumaine Math. Pures Appl. **9**, 499—528 (1964) [Russian].

— [4] Unitary representations of the Lorentz group in spaces with indefinite metric. Mat. Sb. **65**, 198—211 (1964) [Russian].

— [5] On unitary group representations in spaces with indefinite metric. Acta Sci. Math. (Szeged) **26**, 201—209 (1965).

— [6] Kommutative symmetrische Operatorenalgebren in Pontryaginschen Räumen Π_k. Math. Ann. **162**, 147—171 (1965) and **170**, 166 (1967).

— [7] On the structure of the unitary representations of locally compact groups in the space Π_1. Izv. Akad. Nauk SSSR Ser. Mat. **29**, 689—700 (1965) [Russian].

— [8] Conditions for the unitary equivalence of commutative symmetric algebras in the Pontrjagin space Π_k. Trudy Moskov. Mat. Obšč. **15**, 383—399 (1966) [Russian].

— [9] Structure of unitary representations of locally compact groups and symmetric representations of algebras in the Pontrjagin space Π_k. Izv. Akad. Nauk SSSR Ser. Mat. **30**, 1111—1132 (1966) [Russian].

— [10] Degenerate operator algebras in the Pontrjagin space Π_k. Izv. Akad. Nauk SSSR Ser. Mat. **30**, 1229—1256 (1966) [Russian].

— [11] Representations of commutative symmetric Banach algebras and commutative topological groups in the space Π_k. Dokl. Akad. Nauk SSSR **170**, 271—274 (1966) [Russian].

— [12] Analog of Stone's theorem for a space with an indefinite metric. Dokl. Akad. Nauk SSSR **170**, 1259—1261 (1966) [Russian].

— [13] On unitary group representations and symmetric algebra representations in spaces with indefinite metric. In: Proceedings of the symposium in analysis, pp. 145—156. Kingston, Ontario: Queen's University 1967.

Naĭmark, M. A., Ismagilov, R. S.: [1] Representations of groups and algebras in a space with indefinite metric. In: Itogi Nauki, Mathematical analysis 1968, pp. 73—105. Moscow: Akad. Nauk SSSR Inst. Naučn. Informacii 1969 [Russian].

Nevanlinna, R.: [1] Erweiterung der Theorie des Hilbertschen Raumes. Comm. Sém. Math. Univ. Lund, Tome Supplémentaire, 160—168 (1952).

— [2] Über metrische lineare Räume. II. Bilinearformen und Stetigkeit. Ann. Acad. Sci. Fenn. Ser. A I, no. 113 (1952).

— [3] Über metrische lineare Räume. III. Theorie der Orthogonalsysteme. Ann. Acad. Sci. Fenn. Ser. A I, no. 115 (1952).

— [4] Über metrische lineare Räume. IV. Zur Theorie der Unterräume. Ann. Acad. Sci. Fenn. Ser. A I, no. 163 (1954).

Noël, G.: [1] Topologies sur un vectoriel hermitien non dégénéré. C. R. Acad. Sci. Paris **257**, 2785—2787 (1963).
— [2] Opérateurs fortement (J)-normaux dans un espace de type (J). Acad. Roy. Belg. Bull. Cl. Sci. (5) **51**, 570—585 (1965).

Olubummo, A., Phillips, R. S.: [1] Dissipative ordinary differential operators. J. Math. Mech. **14**, 929—949 (1965).

Ovčinnikov, V. I.: [1] The decomposability of spaces with indefinite metric. Mat. Issled. **3**, no. 4, 175—177 (1968) [Russian].

Pauli, W.: [1] On Dirac's new method of field quantization. Rev. Modern Phys. **15**, 175—207 (1943).

Peĭsahovič, È. È.: [1] Sufficient conditions for the existence of a solution of the equation $A = f(B)$ for selfadjoint operators in a space with indefinite metric. Vestnik Moskov. Univ. Ser. I Mat. Meh. **21**, no. 4, 47—53 (1966) [Russian].

Pesonen, E.: [1] Über die Spektraldarstellung quadratischer Formen in linearen Räumen mit indefiniter Metrik. Ann. Acad. Sci. Fenn. Ser. A I, no. 227 (1956).

Phillips, R. S.: [1] Dissipative operators and parabolic partial differential equations. Comm. Pure Appl. Math. **12**, 249—276 (1959).
— [2] Dissipative operators and hyperbolic systems of partial differential equations. Trans. Amer. Math. Soc. **90**, 193—254 (1959).
— [3] The extension of dual subspaces invariant under an algebra. In: Proc. Internat. Sympos. Linear Spaces, pp. 366—398. Jerusalem and Oxford: Jerusalem Academic Press and Pergamon 1961.
— [4] A minimax characterization for the eigenvalues of a positive symmetric operator in a space with an indefinite metric. J. Fac. Sci. Univ. Tokyo Sect. I A Math. **17**, 51—59 (1970).

Phillips, R. S., Sarason, L.: [1] Singular symmetric positive first order differential operators. J. Math. Mech. **15**, 235—271 (1966).

Pontrjagin, L. S.: [1] Hermitian operators in spaces with indefinite metric. Izv. Akad. Nauk SSSR Ser. Mat. **8**, 243—280 (1944) [Russian].

Potapov, V. P.: [1] The multiplicative structure of J-contractive matrix functions. Trudy Moskov. Mat. Obšč. **4**, 125—236 (1955) [Russian].

Reid, W. T.: [1] Symmetrizable completely continuous linear transformations in Hilbert space. Duke Math. J. **18**, 41—56 (1951).

Riesz, F., Sz.-Nagy, B.: [1] Leçons d'analyse fonctionnelle, 4e édition, Paris and Budapest: Gauthier-Villars and Akadémiai Kiadó 1965.

Robertson, A. P., Robertson, W.: [1] Topological vector spaces, New York: Cambridge University Press 1964.

Savage, L. J.: [1] The application of vectorial methods to metric geometry. Duke Math. J. **13**, 521—528 (1946).

Schaefer, H. H.: [1] Topological vector spaces, New York and London: Macmillan and Collier-Macmillan 1966.

Scheibe, E.: [1] Über hermitische Formen in topologischen Vektorräumen. I. Ann. Acad. Sci. Fenn. Ser. A I, no. 294 (1960).

Senderov, V. A.: [1] Operators that are absolute indefinitely bounded from below in spaces with indefinite metric. Mat. Zametki **10**, 301—305 (1971) [Russian].

Shah Tao-shing: [1] On conditionally positive-definite generalized functions. Sci. Sinica **11**, 1147—1168 (1962).

Šmul'jan, Ju. L.: [1] Contractive operators in a finite-dimensional space with indefinite metric. Uspehi Mat. Nauk **18**, no. 6, 225—230 (1963) [Russian].
— [2] Division in the class of J-expansive operators. Mat. Sb. **74**, 516—525 (1967) [Russian].
— [3] J-expansive operators in J-spaces. Ukrain. Mat. Ž. **20**, 352—362 (1968) [Russian].
— [4] J-majorizing and modular operators in J-spaces. Mat. Issled. **3**, no. 1, 198—214 (1968) [Russian].
— [5] Linear fractional transformations with operator coefficients, and operator balls. Mat. Sb. **77**, 335—353 (1968) [Russian].
— [6] Linear fractional transformations of the upper half plane of operators. Izv. Vysš. Učebn. Zaved. Matematika (1969), no. 1, 97—105 (1969) [Russian].
— [7] Linear fractional transformations in a space with involution. Izv. Vysš. Učebn. Zaved. Matematika (1969), no. 2, 117—126 (1969) [Russian].
— [8] A certain class of holomorphic operator-valued functions. Mat. Zametki **5**, 351—359 (1969) [Russian].

Sobolev, S. L.: [1] The motion of a symmetric top containing a cavity filled with a liquid. Ž. Prikl. Meh. i Tehn. Fiz. (1960), no. 3, 20—55 (1960) [Russian].

Sorjonen, P.: [1] Verallgemeinerte Resolventen eines symmetrischen Operators im Pontrjaginraum. Univ. Jyväskylä Dept. Math., Report 15 (1972).

Šul'man, V. S.: [1] Operator algebras in the space Π_1 with indefinite metric. Dokl. Akad. Nauk SSSR **201**, 44—47 (1971) [Russian].

Švarcman, P. A.: [1] Inequalities for the eigenvalues of J-hermitian and J-unitary operators. I. Mat. Issled. **4**, no. 4, 33—45 (1969) [Russian].

Sz.-Nagy, B.: [1] On uniformly bounded linear transformations in Hilbert space. Acta Sci. Math. (Szeged) **11**, 152—157 (1947).

Sz.-Nagy, B., Foiaş, C.: [1] Harmonic analysis of operators on Hilbert space, Amsterdam/London and New York and Budapest: North-Holland and American Elsevier and Akadémiai Kiadó 1970.

Wendland, W.: [1] Die Fredholmsche Alternative für Operatoren, die bezüglich eines bilinearen Funktionals adjungiert sind. Math. Z. **101**, 61—64 (1967).

Wittstock, G.: [1] Über koerzive indefinite Metriken. Ann. Acad. Sci. Fenn. Ser. A I, no. 347 (1964).
— [2] Über Zerlegungsmajoranten indefiniter Metriken. Math. Z. **91**, 421—430 (1966).
— [3] Über Majoranten indefiniter Bilinearformen. Ann. Acad. Sci. Fenn. Ser. A I, no. 381 (1966).
— [4] Über invariante Teilräume zu positiven Transformationen in Räumen mit indefiniter Metrik. Math. Ann. **172**, 167—175 (1967).
— [5] Über indefinit symmetrisierbare lineare Abbildungen. Math. Z. **111**, 131—144 (1969).

Wonenburger, M. J.: [1] Simultaneous diagonalization of symmetric bilinear forms. J. Math. Mech. **15**, 617—622 (1966).

Index of Terms

Index of Symbols

721/17/73 V/12/6

Ergebnisse der Mathematik und ihrer Grenzgebiete

1. Bachmann: Transfinite Zahlen. DM 48,—
2. Miranda: Partial Differential Equations of Elliptic Type. DM 58,—
4. Samuel: Méthodes d'Algèbre Abstraite en Géométrie Algébrique. DM 34,—
5. Dieudonné: La Géométrie des Groupes Classiques. DM 42,—
7. Ostmann: Additive Zahlentheorie. 1. Teil: Allgemeine Untersuchungen. DM 42,—
8. Wittich: Neuere Untersuchungen über eindeutige analytische Funktionen. DM 36,—
11. Ostmann: Additive Zahlentheorie. 2. Teil: Spezielle Zahlenmengen. DM 34,—
13. Segre: Some Properties of Differentiable Varieties and Transformations. DM 46,—
14. Coxeter/Moser: Generators and Relations for Discrete Groups 3rd edition. DM 42,—
15. Zeller/Beckmann: Theorie der Limitierungsverfahren. DM 64,—
16. Cesari: Asymptotic Behavior and Stability Problems in Ordinary Differential Equations. DM 54,—
17. Severi: Il teorema di Riemann-Roch per curve, superficic e varietà questioni collegate. DM 30,—
18. Jenkins: Univalent Functions and Conformal Mapping. DM 37,—
19. Boas/Buck: Polynomial Expansions of Analytic Functions. DM 24,—
20. Bruck: A Survey of Binary Systems. DM 46,—
21. Day: Normed Linear Spaces. 3rd edition. DM 42,—
23. Bergmann: Integral Operators in the Theory of Linear Partial Differential Equations. DM 40,—
25. Sikorski: Boolean Algebras. DM 42,—
26. Künzi: Quasikonforme Abbildungen. DM 43,—
27. Schatten: Norm Ideals of Completely Continuous Operators. DM 30,—
28. Beckenbach/Bellmann: Inequalities. DM 38,—
29. Wolfowitz: Coding Theorems of Information Theory. DM 30,—
30. Constantinescu/Cornea: Ideale Ränder Riemannscher Flächen. DM 75,—
31. Conner/Floyd: Differentiable Periodic Maps. DM 34,—
32. Mumford: Geometric Invariant Theory. DM 24,—
33. Gabriel/Zisman: Calculus of Fractions and Homotopy Theory. DM 42,—
34. Putnam: Commutation Properties of Hilbert Space Operators and Related Topics. DM 31,—
35. Neumann: Varieties of Groups. DM 51,—
36. Boas: Integrability Theorems for Trigonometric Transforms. DM 20,—
37. Sz.-Nagy: Spektraldarstellung linearer Transformationen des Hilbertschen Raumes. DM 24,—
38. Seligman: Modular Lie Algebras. DM 43,—
39. Deuring: Algebren. DM 30,—
40. Schütte: Vollständige Systeme modaler und intuitionistischer Logik. DM 30,—

41. Smullyan: First-Order Logic. DM 36,–
42. Dembowski: Finite Geometries. DM 68,–
43. Linnik: Ergodic Properties of Algebraic Fields. DM 44,–
44. Krull: Idealtheorie. DM 34,–
45. Nachbin: Topology on Spaces of Holomorphic Mappings. DM 18,–
46. A. Ionescu Tulcea/C. Ionescu Tulcea: Topics in the Theory of Lifting. DM 36,–
47. Hayes/Pauc: Derivation and Martingales. DM 48,–
48. Kahane: Séries de Fourier absolument convergentes. DM 44,–
49. Behnke/Thullen: Theorie der Funktionen mehrerer komplexer Veränderlichen. DM 48,–
50. Wilf: Finite Sections of Some Classical Inequalities. DM 28,–
51. Ramis: Sous-ensembles analytiques d'une variété banachique complexe. DM 36,–
52. Busemann: Recent Synthetic Differential Geometry. DM 32,–
53. Walter: Differential and Integral Inequalities. DM 74,–
54. Monna: Analyse non-archimédienne. DM 38,–
55. Alfsen: Compact Convex Sets and Boundary Integrals. DM 46,–
56. Greco/Salmon: Topics in m-Adic Topologies. DM 24,–
57. Lopez de Medrano: Involutions on Manifolds. DM 36,–
58. Sakai: C*-Algebras and W*-Algebras. DM 68,–
59. Zariski: Algebraic Surfaces. DM 54,–
60. Robinson: Finiteness Conditions and Generalized Soluble Groups, Part 1. DM 48,–
61. Robinson: Finiteness Conditions and Generalized Soluble Groups, Part 2. DM 64,–
62. Hakim: Topos annelés et schémas relatifs. DM 48,–
63. Browder: Surgery on Simply-Connected Manifolds. DM 42,–
64. Pietsch: Nuclear Locally Convex Spaces. DM 48,–
65. Dellacherie: Capacités et processus stochastiques. DM 44,–
66. Raghunathan: Discrete Subgroups of Lie Groups. DM 56,–
67. Rourke/Sanderson: Introduction to Piecewise-Linear Topology. DM 42,–
68. Kobayashi: Transformation Groups in Differential Geometry. DM 52,–
69. Tougeron: Idéaux de fonctions différentiables. DM 69,–
70. Gihman/Skorohod: Stochastic Differential Equations. DM 88,–
71. Milnor/Husemoller: Symmetric Bilinear Forms. DM 42,–
72. Fossum: The Divisor Class Group of a Krull Domain. DM 44,–
73. Springer: Jordan Algebras and Algebraic Groups. DM 48,–
74. Wehrfritz: Infinite Linear Groups. DM 59,–
75. Radjavi/Rosenthal: Invariant Subspaces. DM 50,–
76. Bognár: Indefinite Inner Product Spaces. DM 48,–
77. Skorohod: Integration in Hilbert Space. In preparation
78. Bonsall/Duncan: Complete Normed Algebras. DM 68,–